Benchmark Papers
in Animal Behavior

Series Editor: Martin W. Schein
West Virginia University

Published Volumes and Volumes in Preparation

HORMONES AND SEXUAL BEHAVIOR
  Carol Sue Carter
TERRITORY
  Allen W. Stokes
SOCIAL HIERARCHY AND DOMINANCE
  Martin W. Schein
CRITICAL PERIODS
  J. P. Scott
MIMICRY
  Joseph A. Marshall
IMPRINTING
  E. H. Hess
VERTEBRATE SOCIAL ORGANIZATION
  Edwin M. Banks
PLAY
  Dietland Müller-Schwarze
HUMAN SEXUALITY
  Milton Diamond
ABORTION
  Milton Diamond

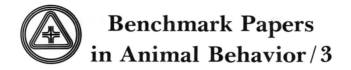

Benchmark Papers
in Animal Behavior / 3

A BENCHMARK® Books Series

SOCIAL HIERARCHY
AND DOMINANCE

Edited by
MARTIN W. SCHEIN
West Virginia University

Dowden, Hutchinson
& Ross, Inc.
Stroudsburg, Pennsylvania

Distributed by
HALSTED PRESS *A Division of John Wiley & Sons, Inc.*

LIBRARY OF CONGRESS CATALOGING IN PUBLICATION DATA

Exclusive Distributor: **Halsted Press**
A Division of John Wiley & Sons, Inc.

Acknowledgments
and Permissions

ACKNOWLEDGMENTS

AGRICULTURAL EXPERIMENT STATION, KANSAS STATE UNIVERSITY—*Social Behavior of the
Domestic Fowl*

AMERICAN ORNITHOLOGISTS' UNION—*Auk*
  The Social Order in Flocks of the Common Chicken and Pigeon
  Territory as a Result of Despotism and Social Organization in Geese

THE POULTRY SCIENCE ASSOCIATION—*Poultry Science*
  Productivity of Pullets Influenced by Genetic Selection for Social Dominance Ability and by Stability of
    Flock Membership
  Selective Breeding for Aggressiveness in Chickens

PERMISSIONS

The following papers have been reprinted with the permission of the authors and copyright holders.

AMERICAN PSYCHOLOGICAL ASSOCIATION—*Journal of Abnormal and Social Psychology*
  Problems in the Biopsychology of Social Organization

BALLIÈRE TINDALL, LONDON—*British Journal of Animal Behaviour*
  Social Dominance Relationships in a Herd of Dairy Cattle

JOHANN AMBROSIUS BARTH, LEIPZIG—*Zeitschrift für Psychologie*
  Beiträge zur Sozialpsychologie des Haushuhns

THE JOURNAL PRESS
  Journal of General Psychology
    The Experimental Measurement of a Social Hierarchy in *Gallus domesticus*: I. The Direct Identification
      and Direct Measurement of Social Reflex No. 1 and Social Reflex No. 2
  Journal of Genetic Psychology
    A Study of a Social Hierarchy in the Lizard, *Anolis carolinensis*
  Journal of Social Psychology
    Dominance-Quality and Social Behavior in Infra-human Primates
    The Experimental Measurement of a Social Hierarchy in *Gallus domesticus*: V. The Postmortem Mea-
      surement of Anatomical Features

NEW YORK ZOOLOGICAL SOCIETY—*Zoologica*
  Social Behavior of the American Buffalo (*Bison bison bison*)

UNIVERSITY OF CHICAGO PRESS
  American Naturalist
    Statistical Analysis of Factors Which Make for Success in Initial Encounters Between Hens

Physiological Zoology

Series Editor's Preface

Not too many years ago, virtually all research publications dealing with animal behavior could be housed within the covers of a very few hardbound volumes easily accessible to the few workers in the field. Times have changed! Today's student of animal behavior has all he can do to keep abreast of developments within his area of special interest, let alone the field as a whole.

Even fewer years ago, those who taught animal behavior courses could easily choose a suitable textbook from among the very few that were available; all "covered" the field, according to the bias of the author. Students working on a special project used *the* text and *the* journal as reference sources, and for the most part successfully covered their assigned topics. Times have changed! Today's teacher of animal behavior is confronted with a bewildering array of books among which to choose, some purporting to be all-encompassing, others confessing to limited coverage, and still others simply collections of recent and profound writings.

In response to the problem of the steadily increasing and overwhelming volume of information in the area, the Benchmark Papers on Animal Behavior was launched as a series of single-topic volumes designed to be some things to some people. Each volume contains a collection of what an expert considers to be *the* significant research papers in a given topic area. Each volume serves several purposes. To the teacher, a Benchmark volume serves as a supplement to other written materials assigned to students; it permits in-depth consideration of a particular topic while confronting the student (often for the first time) with original research papers of outstanding quality. To the researcher, a Benchmark volume serves to save countless hours digging through the various journals to find *the* basic articles in his area of interest; often the journals are not easily available. To the student, a Benchmark volume provides a readily accessible set of original papers on the topic, a set that forms the core of the more extensive bibliography that he is likely to compile; it also permits him to see at first hand what an "expert" thinks is important in the area, and to react accordingly. To the librarian, a Benchmark volume represents a collection of important papers from many diverse sources, thus making readily available materials that might otherwise not be economically possible to obtain or physically possible to keep in stock.

The ordering of topics to be covered in the series was no easy matter. Eventually, of course, as many topics as possible will be included in what must be an open-ended series. However, at the outset, it was decided to aim primarily at the more fundamental topic areas, topics that erupted into print some years ago and have now subsided into a semidormant but still rumbling state. Social hierarchy and dominance is clearly a "classic" topic in animal behavior. Along with its closely related companion topic, territory (see *Territory,* edited by A. W. Stokes, Volume 2 of this series), it forms one of the basic pillars of many contemporary descriptions of animal societies.

Martin W. Schein

Preface

Dominance and dominance hierarchies received wide attention among ecologically oriented behaviorists (or behaviorally oriented ecologists) from the 1930s to the 1960s, with perhaps a peak around the late 1940s and early 1950s. The concept was useful as a shorthand term that, when applied to a specific group of animals, implied a social structure of known and defined characteristics; it does have valuable descriptive properties and was widely used as such. The concept was also useful as an explanation of behavior in various social interactions. For example, if an individual avoided one but did not avoid another member of the group (that is, was subordinate to the first but not to the second), the outcome of subsequent interactions in other situations could be predicted. The concept was especially useful because it allowed predictions to be made with respect to third parties.

The use of dominance hierarchies to describe, explain, and/or predict behavior tended to decline in the 1960s—or if not decline to at least be applied more sparingly. There were too many variations of the classic hierarchial system to permit a single term, such as "rank order," to be applied too broadly. Further, more and more cases turned up in which dominance was specific to a particular situation or context and not applicable "across the board" in all interactions between a pair of individuals. Finally, it is becoming increasingly apparent that, at least in some instances, *subordination* rather than *dominance* might be the more important concept: dominance may accrue by default rather than positive action. The lower-ranking individual may have to signal subordination to the higher-ranking individual, who may need to do nothing in return.

The collection of papers assembled here traces the development of the concept of dominance hierarchies from its firm beginnings in 1922 with the work of Thorleif Schjelderup-Ebbe, through the peak of activities in the 1940s that centered around the work of W. C. Allee and his students. In the brief review that serves as an introduction, no attempt is made to be exhaustive. Instead, I have tried to cite only representative examples of studies published, in general, earlier than the mid-1960s. Selection of specific papers for reproduction in this collection was much more difficult: page limitations imposed by practicality forced the choice to only 20 or so papers, the most representative of the representative. In all instances, choice favored the generally less accessible and older articles. Thus, except for two articles at the end, selection of articles was restricted to papers published prior to 1960. The two exceptions were published in *Poultry Science,* a journal that is not likely to be on the shelves of many researchers. It is hoped that this collection of papers will rekindle interest in an area that once was very fruitful and perhaps now needs reexamination.

This book is dedicated to A. M. Guhl, a friend of long standing, a colleague of world repute, and an inspiration over the years.

Martin W. Schein

Contents

III. ANALYSIS

IV. APPLICATIONS

Contents by Author

Introduction

Observations on the social world of infrahuman species have probably been made since the earliest times, as long as somebody was around to do the observing, but serious propositions concerning animal societies did not come to the fore until the late 1800s. With the advent of Darwinism and the concept of survival of the fittest, interactions among group members was explained by placing emphasis on intense competition that translated into a continual and unrelenting bitter struggle for survival.

Yet observant naturalists of the day saw many instances of animal groupings and pair interactions that were *not* based on a struggle to the death. Espinas (1878) described a number of examples of cooperation, even altruism, among members of social groups, and based his ideas of social systems more on the benefits accruing to individuals than on the elimination of the less successful. Espinas's ideas were largely ignored at the time of publication, and it was not until after the turn of the century that nonhuman societies began to again receive serious attention (e.g., Alverdes, 1927). The *cooperative* basis of animal societies was emphasized by W. C. Allee in his search for explanations underlying the evolution of sociality and social systems (Allee, 1938).

Some sort of reconciliation between the apparent extremes of cooperation versus competition was necessary. Obviously individuals did not spend all their waking hours fighting each other to the death; just as obviously, they did not protect and care for each other to the exclusion of their own safety and survival. Social systems that stressed a balance between cooperation and competition were a natural development. One of these was the concept of simple dominance wherein one individual has preferential rights over another: the dominant animal has first choice of desired limited resources, and this arrangement is accepted by the subordinate. It is not necessary for the dominant animal to physically destroy the subordinate, as long as the latter does not challenge the rights of the former. If the sought-after resource is plentiful, both individuals can exist relatively harmoniously; if the resource is limited, the subordinate does without.

It remained to Thorleif Schjelderup-Ebbe to discover that within a group, dominance and subordination may be interlocked and arranged hierarchially so that an individual subordinate to one may dominate another. His classic paper (Schjelderup-Ebbe, Paper 1) on dominance hierarchies in chickens led to a number of studies elaborating on the hierarchial arrangement in chickens (Schjelderup-Ebbe, 1923; Fischel, 1927; Masure and Allee, Paper 2) or describing similar social arrangements in other species. During the period from 1920 into the 1960s, the existence of dominance hierarchies or a related system was reported in pigeons (Masure and Allee, Paper 2), parakeets (Masure and Allee, 1934b), lizards (Evans, Paper 3), red deer (Darling, 1937), white mice (Uhrich, 1938), night herons (Noble, Wurm, and Schmidt, 1938), dogs (James, 1939), canaries (Shoemaker, 1939), ring doves (Bennett, 1939), various primates (Maslow, Paper 4; Carpenter, 1942a and 1942b), geese (Jenkins, Paper 7), sheep (Scott, 1945), hermit crabs (Allee and Douglis, 1945), lobsters (Douglis, 1946), goats (Scott, 1946), wasps (Pardi, 1948), horses (Grzimek, 1949), elk (Altmann, 1952), crayfish (Bovbjerg, Paper 8), dairy cows (Schein and Fohrman, Paper 10), feral cattle (Schloeth, 1958), jungle fowl (Banks, 1956), and buffalo (McHugh, Paper 11). This listing is far from complete; further, new species have been added to the list during the 1960s and are still being added. Indeed, the concept of a dominance hierarchy became so widely recognized that "peck order" is now a commonplace household phrase in our language. (Other terms used interchangeably with "dominance hierarchy" include "rank order," "dominance order," "hook order" and "bunt order" (in cattle), "social dominance," "social rank," "social hierarchy," and "dominance rank.") Dominance and its role in social organization was excellently reviewed in general by Allee (Paper 5), in birds by Noble (1939), in vertebrates by Collias (1944), and in chickens by Guhl (Paper 9). But even at this early date, Schneirla (Paper 6) introduced a word of caution concerning the universality and general usefulness of the concept of dominance in explaining social systems.

It was only natural that attention should be drawn early to the analysis of dominance hierarchy, particularly with respect to measurable and identifiable characteristics that could be correlated with and causally related to social status. Murchison, in a now-all-but-forgotten series of papers, attempted to analyze social hierarchies in chickens by means of covariant techniques applied to behavioral and physical quanta such as social approach, fighting, mating, social discrimination, and body size in time and space categories (Murchison, Paper 12, 1935a, 1935b, 1935c; Murchison, Pomerat, and Zarrow, Paper 13). By calculating the regression equations for each variable with respect to social rank and by combining equations, Murchison arrived at a single general equation that accounted for the contributions of each of the variables in a ranking system (Murchison, 1935d, 1936). The rather elaborate and perhaps laborious approach suggested by Murchison did not survive the test of time; few if any workers picked up where he left off, and his concept of Social Reflexes and Social Law fell into disuse.

Other researchers applied more direct techniques to examining the relationship between rank orders and various observable characteristics. Collias (Paper 16) identified male hormone output and thyroxin secretion as important factors relating to the outcome of initial encounters between hens; other such obvious influences as sex and territorial familiarity were controlled. Carpenter (1942b) noted that among the primates, males dominated females. Bovbjerg (Paper 8, 1956) listed size, sex, and age as being of primary importance in dominance among crayfish. Schein and Fohrman (Paper 10) found that age and weight, but more significantly, seniority in the herd, were highly correlated with rank position in dairy cows. Banks and Allee (Paper 17) reported a correlation between flock size and levels of agonistic behavior in domestic hens. Allee et al. (1949), in summarizing the many observations of correlations between position in the rank order and various factors, list (with qualifications) strength and experience, relative maturity, size, territorial familiarity, hormones, social facilitation, sex, consort relationships and heredity as among the most important.

The relationship between hormone output and social rank was an early and major concern of Allee and his students. They systematically examined the effects of injected epinephrine (Allee and Collias, 1938), androgens (Allee, Collias, and Lutherman, Paper 14; Allee et al., 1955), estradial (Allee and Collias, 1940), and thyroxin (Allee, Collias, and Beeman, 1940) on position in the hierarchy. The male sex hormones had by far the most profound effect: androgens and rank position were highly correlated. Translated into actual matings, high-status (high-androgen) female chickens tended to be courted and mated less frequently than low-status (low-androgen) females, and high-status males mated more frequently and produced more offspring than did low-status males; indeed, low-status males that performed normally when alone with females did not mate at all, i.e., were "psychologically castrated," when more dominant males were present along with the females (Guhl, Collias, and Allee, Paper 18; Guhl and Warren, 1946; Guhl, 1949, 1950, 1964). That the correlation between rank position and male sex hormones holds in a number of other species besides *Gallus* has been reported many times since (e.g., Thomas, Robinson, and Marburger, 1965, in white tail deer). However, that other hormones are also in some way related to rank position in a number of species is equally clear (e.g., estrogens in chimpanzees, Birch and Clark, 1950; epinephrine in starlings, Davis, 1957; estrogens in chickens, Guhl, 1961, 1968).

The role of experience in determining the outcome of interactions between individuals was also studied by Allee and a number of his students (e.g., Guhl, 1942; Ginsburg and Allee, Paper 15; Guhl and Allee, 1944; Douglis, 1948) as well as others (e.g., Radlow, Hale, and Smith, 1958; Maroney, Warren, and Sinha, 1959). Guhl (1958) traced the ontogeny of social organization in domestic chicks, with particular attention to the interaction between experience and sex hormones; he also found that social stability (based on experience) could on occasion mask the effects of hormonal treatments (Guhl, 1968).

Many investigators have examined the role of heredity in dominance interactions (e.g., Fennell, 1945; Guhl and Eaton, 1948; Potter, 1949; Craig and Baruth, 1965). Even casual observations reveal that within a single species, some strains are invariably dominant over other strains. Indeed, the development of "winner" versus "loser" strains of chickens has been accomplished in as little as four or five generations of selection (Guhl, Craig, and Mueller, Paper 19; Craig, Ortman, and Guhl,

1965). The differences between "winner" and "loser" strains are based on complex changes and physiological responsiveness to social stimuli rather than merely to changes in endogenous androgen production (Ortman and Craig, 1968).

Since the domestic chicken was and still is so extensively used in studies of dominance hierarchies, it is surprising that attention was not drawn at the outset to the economic impact of this behavioral phenomenon. Schein, Hyde, and Fohrman (1954) suggested a 5 percent decline in milk production as a consequence of continuous disruptions to an established social hierarchy in dairy cows, but the definitive experiment on cows remains to be done. In chickens, the relationship between social order and egg and meat production has been more extensively studied by Craig and his associates (e.g., Tindall and Craig, 1959; Craig, 1968; Craig, Biswas, and Guhl, 1969; Craig and Guhl, 1969; Craig and Tóth, Paper 20; Mueller, 1970; Biswas and Craig, 1971). The direct relationship between behavioral studies and poultry management is emphasized by McBride (1958, 1960, 1962); in fact, even adult hens housed alone in cages were reported to establish social relationships with neighbors in adjoining cages, and the relationship affected the onset of egg laying (James and Foenander, 1961).

This review, abbreviated as it is, would be misleading if it conveyed the impression that the phenomenon of dominance begins and ends with pure dominance–subordination and that a paired relationship holds constant over all possible contexts. Indeed, Rowell (1966) working with captive baboons, reverses the usual emphasis in hierarchies: the initiative in a dominant–subordinate relationship rests more with the subordinate than with the dominant, and so the hierarchy is maintained by the *lower*-ranking animals. Further, an increasing number of observations are leading to the conclusion that dominance is more context-related than was previously supposed: an individual dominant in one particular situation might not be dominant in others. This observation led Drews (1973) to rightly suggest caution in the use of dominance to describe or explain behavior.

I
In the Beginning

Editor's Comments on Papers 1 Through 6

1 **Schjelderup-Ebbe:** *Beiträge zur Sozialpsychologie des Haushuhns*
   English Translation: *Contributions to the Social Psychology of the Domestic Chicken*

2 **Masure and Allee:** *The Social Order in Flocks of the Common Chicken and the Pigeon*

3 **Evans:** *A Study of a Social Hierarchy in the Lizard,* Anolis carolinensis

4 **Maslow:** *Dominance-Quality and Social Behavior in Infra-human Primates*

5 **Allee:** *Social Dominance and Subordination Among Vertebrates*

6 **Schneirla:** Excerpts from *Problems in the Biopsychology of Social Organization*

Rank orders burst upon the scene with the 1922 paper of Thorleif Schjelderup Ebbe (Paper 1). An earlier paper by Schjelderup-Ebbe is mentioned in Allee (Paper 5), but the 1922 paper remains the definitive "starting point" for dominance hierarchy studies. In view of its importance as a primary citation in many subsequent studies, I have chosen to reprint the paper here in its original "Norwegian German" so that readers competent in the language can savor the style of that great student of animal behavior. The original "Norwegian German" was translated into "Austrian English" by Dr. and Mrs. Wolfgang Schleidt (University of Maryland), and finally into "American English" by myself. Both the Schleidts and I have tried to retain the charming style of the original writing in the translation while adhering as closely as possible to the text; any "misses" in the translation are purely my own fault.

In the two decades following Schjelderup-Ebbe's 1922 paper, a number of excellent descriptive "rank order" papers were published, three of which are reprinted here. Masure and Allee's 1934 study of chickens and pigeons (Paper 2) is included in this collection because it was one of the first of a number of careful studies to come out of the laboratory of W. C. Allee at the University of Chicago. As such, it served as a model for subsequent workers in that laboratory (several of whom are represented in this volume) as well as elsewhere. Evans' 1936 study of lizards (Paper 3) was among the first to break away from birds and apply the concept of dominance hierarchies to poikilothermic forms. Maslow's 1940 study (Paper 4) extended the concept to include nonhuman primates, but not without difficulties (see footnote 3 at the beginning of his paper). Allee, then, in his classic 1942 review (Paper 5) brought up to date some 20 years of rank order studies, and it served thereafter as a renewed starting point for subsequent and more detailed analytic studies of the phenomenon. But the too liberal application of the concept began to draw fire from some biopsychologists, most notably Schneirla; that portion of his 1946 article dealing with dominance concepts is reprinted here (Paper 6).

Reprinted from Z. Psychol., **88**, 225–252 (1922)

Beiträge zur Sozialpsychologie des Haushuhns.

Von

THORLEIF SCHJELDERUP-EBBE.[1]

Anhang: Tierpsychologie und Soziologie des Menschen.

Von

DAVID KATZ.

1. Kapitel.

Die soziale Stellung jedes Individuums.

Orientierende Bemerkungen. Wenn man in einen Hühnerhof geht und seine Einwohner beobachtet, wird man die Hühner entweder ruhend oder in Bewegung sehen. Einige Tiere stehen still oder liegen faul ausgestreckt da, während andere in der Erde nach Nahrung scharren, sich im Sande baden, am Gitter hin- und herlaufen, ruhig umhergehen, sich gegenseitig vom Trinkgefäfs verdrängen usw.

Beobachtet man Hühner, die sich frei bewegen (also nicht in einem umzäunten Hof), so wird man sehen, dafs sich die Tiere ebenso wie oben beschrieben benehmen. Doch ist die Übersichtlichkeit über alle Individuen auf einmal nicht so grofs, als wenn man die Hühner in einem Hühnerhof von nicht zu grofsem Flächeninhalt beobachtet. Es empfiehlt sich daher, Hühner in einem derartigen Hühnerhof zu beobachten, weil sie auf demselben ja keine Gelegenheit haben, sich so weit voneinander zu entfernen wie auf einem freien Platz. Das Verhältnis der Tiere zueinander wird leichter in einem ein-

---

[1] Herrn Prof. KATZ in Rostock sage ich meinen herzlichen Dank dafür, dafs er meiner als Greifswalder Dissertation erschienenen Arbeit „Gallus domesticus in seinem täglichen Leben" die geeignete Form für die Aufnahme in *dieser Zeitschrift* gegeben hat.

gezäunten Hühnerhof studiert werden können. Bei meinen Untersuchungen beobachtete ich darum die Tiere hauptsächlich im Hühnerhof, aber auch auf freiem Gebiet. Ich studierte die Tiere sowohl in kleineren Scharen (2—25 Stück) als auch in gröfseren (25—150 Stück).

Wir halten im folgenden, um Mifsverständnissen zu entgehen, die Ausdrücke Hennen und Hähne auseinander, im Gegensatz zu Hühner, worunter wir Hennen und Hähne verstehen.

Die Verteidigungs- und Angriffswaffen der Henne. Wenn man am Tage den Hennen Futter hinstreut, das sie gern haben, kommen sie von allen Seiten herbeigestürzt, um sich über die Nahrung zu werfen und schleunigst das gröfstmögliche Quantum zu verzehren. Sogar die phlegmatischsten, die sich vielleicht liegend ausgeruht haben, erheben sich in der Regel und kommen angelaufen, um ihr Teil und am liebsten noch mehr zu erlangen. Diese Nahrungssucht ist für alle Hennen ganz natürlich und hat ihre Wurzel im Selbsterhaltungstriebe. Aber geht nun die Mahlzeit ganz ruhig vonstatten? Nein, es herrscht ein beständiger Kampf um das Futter; jede Henne legt deutlich zutage, dafs sie sich selbst am nächsten ist.

Um möglichst viel Futter während einer solchen Mahlzeit zu bekommen, beobachtet die Henne folgendes Verhalten: 1. Sie mufs das Futter, das sie schon erschnappt hat und vielleicht noch im Schnabel hält, verteidigen, 2. sie mufs anderen Hennen das wegschnappen, was sie noch im Schnabel tragen und aus Mangel an Zeit noch nicht ganz hinunterschlucken konnten. Beides kommt nicht in Frage, wenn man blofs Körner oder so kleine Futterstücke gibt, dafs die Hennen sie nicht mit dem Schnabel zu zerteilen brauchen. 3. Die Henne mufs soweit als möglich die anderen vom Futterplatze fortjagen und selbst so schnell wie möglich fressen.

Diese Mafsregeln verursachen beständig Zank und Streit. Auf welche Weise versucht nun die Henne die erwähnten Verhaltungsmafsregeln durchzuführen? Wie jagt sie z. B. ihre Mitschwestern weg? Die Antwort lautet: indem sie ihnen einen körperlichen Schmerz zufügt, der so stark ist, dafs sie, um ihm zu entgehen, einen Teil oder sogar das ganze Futter aufgeben, das ihnen zukommen sollte — oder durch Androhung eines Schmerzes. Das Verteidigungs- und Angriffsmittel der Henne ist das Hacken mit dem Schnabel. Viele Hacke werden bei einer solchen Mahlzeit ausgeteilt. Die Hennen, die gehackt werden, stoſsen oft kleine Schreie oder Jammertöne aus. (Ein Mensch fühlt das Hacken wie einen kräftigen Kniff oder eine schmerzende Schramme.) Eine Feder kann beim Hacken herausgerissen werden und trifft der Hack das Gesicht oder den Kamm des Angegriffenen, so kann der betreffende Körperteil zerkratzt werden oder sogar zu bluten anfangen. Ehe eine Henne hackt, stöſst sie oft einen eigenartigen „drohenden" Laut aus, und diese Drohlaute zusammen mit anderen kleinen Schrei- und Jammertönen lassen eine solche Mahlzeit nichts weniger als lautlos von statten gehen. Auch wenn die Hennen nicht fressen, sondern nur zusammen im Hühnerhof sind oder sich im oder beim Neste befinden oder am Abend auf die Sitzstangen fliegen, kann man sehr oft Hacke beobachten.

Die Hackliste und ihre Eigentümlichkeiten. Wir fragen nun: Sind alle Hennen gleichgroſsen Angriffen ausgesetzt? Sind alle Individuen gleichwertig im Kampf um das Dasein? Man meint gewöhnlich entweder, daſs sich die Hennen gegenseitig hacken (damit meint man, daſs Henne A einmal Henne B hackt, ein anderes Mal von Henne B gehackt wird und so fort betreffs aller übrigen Hennen, so daſs das Hacken ganz regellos sein sollte) oder, daſs die Stärksten die Schwächsten hacken und sie verdrängen.

Was die erste Ansicht betrifft, so muſs ich bestimmt hervorheben, daſs sie grundfalsch ist. Ich habe nämlich in über 1900 Fällen beobachtet, daſs, wenn von 2 Hennen A und B, die einige Zeit zusammen gelebt haben, Henne A Henne B hackt, man nicht finden wird, daſs B die A hackt, sondern B weicht A aus und hat Angst vor ihr. Allerdings zeigt sich, daſs eine Henne B, die von einer anderen A schlecht behandelt worden ist, nach einiger Zeit gegen A opponieren kann und sie zu hacken versucht. Das führt dann augenblicklich zu einer Schlägerei, weil A absolut nicht ihre alte Herrschaft über B aufgeben will. Eine solche Schlägerei wird meistens damit enden, daſs A gewinnt und der alte Zustand wieder eintritt: A jagt B beständig weg. Gewinnt B, so wird sie Despot werden,

15*

vielleicht für immer, auf alle Fälle vorläufig, bis es A bei einem
eventuellen späteren Kampf gelingt, den Sieg davon zu tragen.
Meistens tritt jedoch eine solche Opposition wie die eben beschriebene nicht ein, und einzelne Hennen machen jahrelang,
ja ihr ganzes Leben lang keinen Aufruhrversuch.

Was die andere oben mitgeteilte Auffassung betrifft, daſs
die stärksten Individuen die schwächsten hacken und verdrängen, so sei hervorgehoben, daſs sie — so wahrscheinlich
die Auffassung auch von vornherein klingt — zu einseitig ist
und nicht immer mit der Wirklichkeit übereinstimmt.   Ich
werde später Beweise gegen die Berechtigung der letzteren
Auffassung bringen.

Es sollen einige „Hacklisten" angeführt werden, die sich
auf längere Untersuchungen gründen.[1]   Die Buchstaben sind
Abkürzungen für die Namen der betreffenden Hühner.

Hackliste Nr. 1. 7 Hennen (Hvl, Pl, Oml, Byl, Sml, Afl, FrSl).

Hvl   hackt alle anderen.
Pl   „   Oml, Byl, Sml, Afl.
Oml   „   Byl, Sml, Afl, FrSl.
FrSl   „   Pl, Byl, Sml, Afl.
Byl   „   Sml, Afl.
Sml   „   Afl.
Afl   „   keine.

Betreffs dieser 7 Hennen ersieht man schon aus der Zu-
sammenstellung, daſs von ihnen unbedingt die Henne Hvl die
am günstigsten gestellte ist, denn sie „hackt ja alle anderen",
d. h. alle anderen gingen ihr aus dem Wege, hatten Angst
vor ihren Hacken, die sie zahlreich austeilte; keine Henne
wagte ihr das Futter fortzunehmen, und sie übte Justiz gegen
alle aus.

Am schlechtesten ist Henne Afl gestellt, die niemanden zu

---

[1] Derartige Hacklisten lassen sich nicht so schnell aufstellen, weil
sich ihr Inhalt auf stundenlanges Zusammensein mit den Tieren und die
Beobachtung derselben mehrere Tage lang hintereinander gründet (für
15 Hennen mindestens 14 Tage — den ganzen Tag lang).   Es ist sehr
empfehlenswert, die Hennen mit besonderer Aufmerksamkeit zu stu-
dieren, wenn sie gefüttert werden und wenn sie abends auf die Sitzstangen hüpfen, weil man bei diesen Gelegenheiten das Hacken am
häufigsten beobachten kann.

hacken wagt. Die Zwischenstellungen in der Hackliste nehmen die anderen 5 ein, von denen doch Pl, Oml und FrSl bedeutend besser gestellt sind als Byl, der es wiederum besser ergeht als Sml.

Es liegt nahe anzunehmen, daſs die Hennen je nach ihren höheren oder niedrigeren Plätzen auf der Hackliste ein mehr oder weniger sorgenfreies Dasein führen, da es doch wahrscheinlich ist, daſs eine Henne, die von vielen gehackt wird, ein weniger behagliches Leben hat als eine, die von wenigen gehackt wird und selbst viele hackt. Das zeigte sich auch wirklich. Ich beobachtete z. B., daſs in der genannten Hennenschar Afl am wenigsten Futter bekam und oft nervös wurde wegen der vielen Hacke. Ich hatte den Eindruck, daſs ihre Aufmerksamkeit, um schmerzenden Hacken zu entgehen und bei der Fütterung genügend viel Futter zu bekommen, in so hohem Grade in Anspruch genommen wurde, daſs es die Henne etwas anstrengte. Ähnliches habe ich auch sonst bei Hennen beobachtet, die niedrig auf der Hackliste stehen. Die Despotin Hvl dagegen, die stets ihren Willen durchsetzte, die anderen vom Futter und vom Nest wegjagte und selbst ungestört sein konnte, wo sie sich auch befand, fühlte sich scheinbar äuſserst wohl.

Die Hackliste Nr. 1 ist nicht — wenn man so sagen darf — kontinuierlich. Wäre sie das, so würde die mächtigste von den 7 Hennen Despot über 6 Stück gewesen sein, die zweitmächtigste über 5, die nächste über 4, die darauf folgende über 3, die nächste über 2, die vorletzte über 1 und die letzte über keins. Aber die wirkliche Sachlage ist anders; die Tiere Pl, Oml und FrSl bewirken Abweichungen: Pl ist Despot über Oml und Oml über FrSl; es müſste dann zu erwarten sein, daſs Pl auch Despot über FrSl wäre, aber stattdessen ist FrSl Despot über Pl!

Man kann hierin einen Beweis dafür sehen, daſs die Hennen ihre Hackliste nicht ausschlieſslich nach der individuellen Stärke ordnen.

Es würde demnach übereilt sein zu behaupten, daſs der Despotismus der Hvl über all die anderen 6 Hennen darauf beruhe, daſs Hvl so viel stärker sein müsse als die anderen 6. Es kann Stärke sein, aber auch andere Gründe können die

stark hervortretende soziale Stellung der Hvl erklären. Darauf
werden wir später zurückkommen und jetzt eine andere Hackliste besehen.

Hackliste Nr. 2. 10 Hennen: Om2, Attl, Spl, Gaml, P2,
Lvkl, Sm2, Bll, Gr, Hv2.

Om2	hackt	alle aufser Spl.
Attl	„	alle aufser Om2.
Spl	„	alle aufser Attl.
Gaml	„	P2, Lvkl, Sm2, Bll, Gr, Hv2.
P2	„	Lvkl, Sm2, Bll, Gr, Hv2.
Lvkl	„	Sm2, Bll, Gr, Hv2.
Sm2	„	Bll, Gr, Hv2.
Bll	„	Gr, Hv2.
Gr	„	Hv2.
Hv2	„	keine.

Betrachtet man Hackliste 2, so sieht man, dafs hier
k e i n e Henne vorkommt, die Despot über alle anderen Hennen
ist. Das ist ein Gegensatz zu dem, was wir in Hackliste 1
fanden. 3 Hennen in Hackliste 2, Om2, Attl und Spl, sind
gleich günstig gestellt, indem sie alle 3 jede 9 Hennen hacken.
Zu einander verhalten sich die 3 Hennen wie die Hennen Pl,
Oml und FrSl in Hackliste 1, insofern Om2 die Attl hackt,
welche ihrerseits Spl hackt, während Spl Om2 [1] hackt. Diese
3 Hennen hacken sich gegenseitig im Dreieck, was Fig. 1 zu
veranschaulichen sucht, bei der die Richtung der Pfeilspitzen
die Richtung bezeichnet, in welcher das Hacken verläuft.

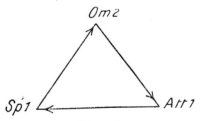

Figur 1.

Es sei hier erwähnt, dafs in anderen Scharen viele andere
Hackkombinationen vorkommen, wodurch die Hackliste Abweichungen von dem kontinuierlichen Verlauf erfährt.

---

[1] Wieder ein Beispiel dafür, dafs Hennen nicht immer der Stärke
nach hacken.

Wenn man von den 3 Hennen absieht, die das Dreieck bilden, ist das Hacken für die übrigen 7 Hennen in Hackliste 2 völlig kontinuierlich und könnte am leichtesten durch eine gerade Linie abgebildet werden wie in Fig. 2.

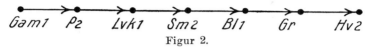

Figur 2.

Die Pfeilspitzen bezeichnen die Hackrichtung. Längs der geraden Linie von links nach rechts stehen die Namenbezeichnungen für die Hennen; jede Henne ist Despot über all die anderen Hennen, deren Namen rechts von ihrem eigenen Namen stehen. Diese gerade Linie kann Grundstrich genannt werden. Das einfachste (das völlig kontinuierliche) Hackschema ist das, wo man blofs einen derartigen Grundstrich aufzuführen hat. So einfach ist aber die Hackordnung selten bei einer Hennenschar. Es kommen sehr häufig Dreiecke, Karrees usw. beim Hacken vor, und auf diese mufs man bei der Konstruktion Rücksicht nehmen.

Bei der Kompliziertheit mancher Hacklisten kann man sich nicht wundern, dafs man bei der oberflächlichen Beobachtung des Hackens in einer Hennenschar den Eindruck erhält, dafs die Hennen sich gegenseitig hacken, trotzdem das völlig unrichtig ist.

Zusammenfassende Betrachtung über das Hacken in der Hennenschar.

Bei einer Übersicht über das Hacken in der Hennenschar ergibt sich auf Grund der zitierten Hacklisten und noch vieler anderer:

1. Es kommt vor, dafs in einer Hennenschar eine Henne ist, die eine besondere Stellung einnimmt, indem sie von keiner ihrer Mithennen gehackt wird, sondern Despot über alle ist.

2. Aber es gibt auch Scharen, wo keine Henne Despot über alle anderen Hennen im Hühnerhof ist.

3. In einzelnen Hennenscharen kommt eine Henne vor, die sozusagen Aschenbrödel ist, da sie von allen anderen Hennen gehackt wird.

4. Aber in anderen Hennenscharen gibt es keine Henne, die von allen anderen gehackt wird. Dieses Verhältnis kommt häufig vor, und doch habe ich folgendes Extrem beobachtet: Von 21 Hennen hackte die Henne Z 19; diese 19 hackten die Henne Y; aber Y hackte Z, und Z hatte schreckliche Angst vor ihr!

5. Selten ist eine Hackliste einer Hennenschar von über 10 Individuen kontinuierlich.

6. Es k ö n n e n mit der Zeit Veränderungen in der Hackordnung eintreten. Wie das geschieht, versteht man am leichtesten im Zusammenhang mit dem, was in Kapitel 2 ausgeführt werden wird, und wir werden deshalb dort auf diesen Punkt zurückkommen. Es sei erwähnt, daſs sich die Hackordnung unter jungen Tieren auf ganz dieselbe Weise entwickelt, gleichgültig ob sie isoliert oder nicht isoliert ausgebrütet und erzogen wurden — was zahlreiche Experimente zeigen.

Sozialpsychologisch verdient noch folgendes Beachtung: Gibt es in einer Hennenschar eine Henne, die alle anderen hackt (wir wollen sie eine α-Henne nennen), so hat das Einfluſs auf das Benehmen dieser Henne, indem sie nun gewöhnlich nicht mehr besonders gehässig gegen die anderen Hennen ist. Unterdessen kann eine derartige α-Henne doch eine Art Führerstellung im Hühnerhof einnehmen, besonders wenn sie etwas älter ist und kein Hahn da ist. Die α-Henne kann dann sogar anfangen, eine von den anderen Hennen zu kurtisieren. Es soll nicht verneint werden, daſs das manchmal auf einer Veränderung der Generationsorgane beruht. Es muſs aber doch hervorgehoben werden, daſs stark legende α-Hennen sich sehr wohl auf die beschriebene Weise benehmen können.

Im allgemeinen ist es nicht schwer, dem Benehmen einer Henne anzusehen, daſs sie fühlt, sie hat eine sichere Stellung; und kaum je ist eine Schlägerei erbitterter, als wenn zwei α-Hennen von zwei verschiedenen Scharen aneinander geraten.

Es zeigt sich, daſs eine Henne, die tief auf der Hackliste steht, durchweg grausamer gegen die wenigen ist, die sie hacken kann, als eine Henne, die hoch auf der Hackliste steht, gegenüber den vielen ist, über die sie Despot ist. Offenbar hängt damit folgendes zusammen: Eine Henne, die in einer

Schar eine niedrige Stelle in der Hackliste einnimmt, wird, wenn sie in eine andere Schar versetzt wird, wo sie eine höhere Stelle einnimmt, fast immer viel gemäßigter in ihrem Benehmen den vielen anderen gegenüber, die sie jetzt hacken kann, als gegenüber den wenigen, die sie früher hacken konnte. — Ein typisches Beispiel ist Henne Gr in Hackliste Nr. 2, wo sie, wie man sieht, blofs 1 Henne (Hv2) hackt. Das Jahr darauf wurde Gr in eine andere Schar, die 11 Individuen zählte, versetzt. Von diesen 11 hackte sie 9 Stück. Tatsache war, dafs Gr früher sehr unangenehm und garstig Hv2 gegenüber war, während Gr sehr mäßig gegen alle die Hennen war, die sie in der neuen Schar Gelegenheit zu hacken hatte.

Das Hacken in der Hähneschar.

Für das Hacken in der Hähneschar gilt ganz Entsprechendes wie für das Hacken in der Hennenschar, nur zeigen die Hähne gröfsere Wildheit.

2. Kapitel.

Die Kämpfe der Hühner.

Rauferei und Hackgesetz. Wenn man glaubt, dafs die Einwohner eines Hühnerhofes gedankenlose, frohe Wesen seien, für die das tägliche Leben eine ungemischte Freude ist und die in Frieden miteinander und unbesorgt um die ganze Welt krähen, Eier legen und fressen, dann ist man auf dem Holzwege. Ein tiefer Ernst liegt über dem Hühnerhof, und die Hennen haben viel Kummer, viel Ärger und Angst auszustehen. Sogar die Lebhaftigkeit, die die Hennen während der Mahlzeiten beweisen, ist eigentlich keine frohe Lebhaftigkeit, da sich, wie schon angedeutet, das Leben der ganzen Szene auf dem Eifer der einzelnen Henne gründet, soviel wie möglich selbst zu konsumieren und die anderen soviel wie möglich zu verjagen.

Eine der ernstlichsten, bedeutungsvollsten, im Gesellschaftsleben wichtigsten Begebenheiten für die Hennen ist die Rauferei, dieser plötzliche, oft furchtbare Kampf zwischen 2 Geschöpfen. Meist, wenn auch nicht immer entscheidet es sich beim Kampf, welche von 2 Hennen Despot über eine andere sein soll.

Wir wollen nun den Kampf der Hennen näher betrachten. Eine Bemerkung schicke ich voraus: Aus meinen Beobachtungen geht hervor, d a f s n i c h t 2 H e n n e n A u n d B e x i s t i e r e n , d i e , w e n n s i e z u s a m m e n l e b e n , d i e F r a g e , o b A D e s p o t ü b e r B , o d e r B D e s p o t ü b e r A s e i n s o l l , n i c h t u n t e r s i c h a b g e m a c h t h a b e n . Zu welcher Zeit wird die Sache abgemacht? Es geschieht auf einem frühen Bekanntschaftsstadium.

Treffen sich 2 Hennen, die sich nicht von früher her kennen, so entscheidet es sich sehr schnell, wer von beiden Despot sein wird. Die 2 Tiere fassen sich gegenseitig ins Auge und folgendes geschieht.

1. E n t w e d e r erschrickt die eine Henne beim Anblick der anderen so sehr, dafs die andere darauf aufmerksam wird und sich drohend nähert; dann ist die Sache abgemacht — und zwar ohne Kampf. Die Furchtsame wird danach von der anderen gehunzt, entweder für immer oder auf alle Fälle vorläufig. Das Geschilderte ist ein häufig vorkommendes Ereignis, wenn sich 2 Hennen treffen; es sieht so aus, als habe die furchtsame Henne Angst, den Kampf, wenn sie sich darauf einliefse, so gründlich zu verlieren, dafs es besser ist, sich gleich zu ergeben.

2. O d e r beide Hennen haben sofort Angst. Diejenige, w e l c h e d a n n z u e r s t i h r e A n g s t ü b e r w i n d e n k a n n , w i r d D e s p o t .

3. O d e r — sehr allgemein — keine von beiden fürchtet sich, sondern sie zielen aufeinander zum Kampf. Dann kommt bald eine regelrechte Rauferei zustande und die Siegerin wird entweder für immer oder auf alle Fälle vorläufig Tyrann über die andere, und die erste kann nun die andere jagen oder hacken, ohne dafs diese es erwidert. — Oft ist also der augenblickliche Mut oder das Glück im Kampf das Entscheidende für eine zukünftige Gewaltherrschaft.

Die 3 genannten Regeln bilden den Grund für d a s H a c k g e s e t z d e r H e n n e n . [1] Wir wollen hier folgen-

---

[1] Das Vorhandensein eines ähnlichen Gesetzes habe ich auch bei verschiedenen wilden Vögeln beobachtet, z. B. beim Sperling (Passer domesticus) und beim Eisvogel (Alcedo ispida).

des bemerken. Dafs eine Henne F wirklich stärker ist als eine andere Henne H, kann indirekt so bewiesen werden: F wird mit einer fremden Henne, die vorher H im Kampf besiegt hat, zusammen geführt. F und die Fremde fangen an zu kämpfen, und F besiegt leicht die Fremde. Nach den oben genannten 3 Regeln kann man auch verstehen, wie z. B. ein Dreieck im Hacken entsteht, selbst wenn wirklich A stärker als B und B stärker als C ist und man demzufolge eigentlich Kontinuität erwarten sollte. Erhält man z. B. das nebenstehende Dreieck, so kann die Unregelmäfsigkeit — dafs C

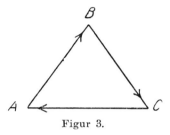

Figur 3.

die A hackt (obgleich C schwächer als A ist) — so entstanden sein, dafs A furchtsam war, als sich A und C zum erstenmal trafen, und dafs C die Gelegenheit benutzte, sich zum Herrscher über A zu machen, obgleich ihre eigenen Körperkräfte faktisch geringer waren als die der A. — Ähnliches habe ich viele Male beobachtet, und ohne Zweifel entstehen Dreiecke, Karrees usw. aus den genannten oder ähnlichen Ursachen.

Diejenige Henne, welche so glücklich ist, „über" zu werden, darf monatelang, vielleicht für immer ein Herrscherregiment über die andere führen, die sich darein finden mufs, auf alle Weise gehunzt und sowohl vom Futtertrog wie vom Nest weggejagt zu werden. Häufig sieht man, dafs eine grofse oder kräftige Henne F sich ohne zu mucken jeden Tag von einer schwächlicher gebauten oder kleinen Henne H hacken läfst. Griffe F zur Wehr gegen H, so würde F sicher den Kampf gewinnen und dadurch „über" die andere kommen; aber F denkt nicht daran, das Hacken zu erwidern![1]

---

[1] Analogie: Das alte arabische Sprichwort: „Wehe dem Reiter, wenn das Pferd seine Stärke fühlte."

Die Gewohnheit macht sicher viel; F ist gewohnt, vor H zu weichen, und da fällt es ihr nicht ein, Aufruhr zu machen. Nun wird man fragen: Wie ist es aber von Anfang an gekommen, daſs sich die starke F von der schwachen H kujonieren läſst? Es zeigt sich, daſs das auf verschiedenen Ursachen beruhen kann:

a) Es ist ja stets so, daſs die erwachsenen Hennen ein Schreckensregiment über die Küchel führen (mit Ausnahme der Henne, die sie versorgt). Die Furcht, welche die heranwachsenden Küchel vor den erwachsenen Hennen hegen, bleibt bei den Kücheln meistens bestehen, auch wenn sie dann selbst erwachsen sind. Auf diese Weise wurden die ältesten Hennen ohne weiteres Despoten — gleichgültig ob die ältesten oder die jüngsten in Wirklichkeit am kräftigsten waren. Das wiederholte sich bei jeder neuen Küchelschar. Die jüngsten Tiere lieſsen sich gewöhnlich, auch wenn sie herangewachsen waren, nicht aus ihrem gewohnten Zustand vor den älteren auszuweichen herausreiſsen.

b) Wenn Hennen neu in den Hühnerhof kamen, weit gereist waren und verängstigt, müde oder hungrig (oder alles zusammen) bei dem ersten Zusammentreffen mit anderen Hennen waren, wurden diese letzteren leicht Despoten über sie und behielten später in der Regel die Oberhand. Das war sehr auffallend. — Wurden Hennen krank und in ihrem Schwächezustand von gesunden Hennen überfallen, so setzte sich oft die Gewaltherrschaft der Gesunden fort, nachdem die Kranken wieder ganz gesund geworden waren.

Es gibt also verschiedene Umstände, welche die Hennen undisponiert zum Kampf machen. Das nutzen dann die anderen Hennen aus; in einem solchen Augenblick ist es ihnen leicht, die Übermacht zu gewinnen, die sie dann behalten.

c) Eine Henne konnte von mehreren, weniger kräftigen (oder einigen kräftigen und einigen schwachen) Hennen auf einmal angegriffen werden; ihre vereinten Kräfte trugen den Sieg davon. Nachher hatte die Henne vor jeder einzelnen von diesen Hennen Angst.

Die 3 Fälle waren häufig vertreten, besonders die ersten 2.

Wir haben also nun festgestellt, daſs das Hacken der Hennen nicht zufällig ist, wie es beim ersten Blick erscheint;

es ist von bestimmten Gesetzen abhängig, denen sich alle Hennen unterordnen. Es gibt nicht 2 in derselben Gesellschaft lebende Hennen, die nicht genau wissen, wer von ihnen „über" und wer „unter" ist. Und diese Rangordnung hängt nicht so sehr von der Stärke der Hennen ab, als von der Gewohnheit oder von gewissen Umständen bei dem ersten Zusammentreffen der Hennen. (Für die Hähneschar gilt hier genau dasselbe, was für die Hennenschar hervorgehoben wurde.)

Es erwies sich bei den verschiedensten und eingehendsten Experimenten, daſs die Neigung zur sozialen Gliederung den Hühnern im Blute liegt, und daſs sie sich — wenn die Zeit ihrer Entfaltung gekommen ist — bei den jungen Tieren zeigt, gleichgültig, ob diese ihr ganzes Leben abgesondert von den älteren waren oder nicht. Diese Neigung beruht mit anderen Worten auf Vererbung, nicht auf Nachahmung.

Beschreibung der Kämpfe.

Gleich ihren Verwandten, den Auerhühnern und Birkhühnern im Walde, führen auch die zahmen Hofhühner nicht selten heftige Fehden; doch stehen diese nicht in so naher Verbindung mit dem Geschlechtsleben wie bei den anderen Vögeln. Die Kämpfe der Hühner, die man meist als harmlos auffaſst, sind es ganz gewiſs nicht, und entstehen nicht aus einer augenblicklichen Laune. Gewinnt die Henne im Kampf, so wagt ihre Gegnerin nicht, sie weiter anzurühren (wenigstens nicht fürs erste). Dadurch entgeht sie der Störung, wenn sie auf dem Nest liegt, das Futter wird ihr nicht weggeschnappt, während sie selber dieses bei anderen tun darf. Das Resultat des Kampfes wird· also später von groſser Bedeutung für die Hennen, und ihr Instinkt sagt ihnen wohl das. Sie setzen oft viel, manchmal das Leben aufs Spiel, um zu gewinnen. Ein anderer Faktor, der hier eine Rolle spielen könnte, ist der spielerische Trieb, seine Kräfte mit einem bekannten oder unbekannten Gegner zu erproben. Mit diesem Faktor ist aber hier nicht zu rechnen; denn erstens haben die Kämpfe stets einen bösartigen Charakter, und zweitens sollte man erwarten, daſs sie, wenn sie aus Spielsucht hervorgingen, häufiger bei ganz jungen Geschöpfen als bei älteren auftreten würden — das ist aber nicht der Fall.

Zu sagen, daſs die Hühner in der Mauserzeit (Sept.-Nov.) friedlicher sind, ist streng genommen nicht richtig, selbst wenn sie sich da nicht so viel bewegen. Friedlich sind Hühner nämlich nie. Sicher ist, daſs, wenn die Mauserzeit vorüber ist und im Winter milde Tage vorkommen, das Hühnerblut leicht in Wallung gerät. Bei manchen von den jungen Hennen können Kamm und Lappen, die seit dem Herbst eingetrocknet waren, in ein paar Tagen anschwellen. Der Kamm kann 2—3 mal so dick und 3 mal so lang werden, indem er wieder den Sommerumfang bekommt. In solchen Tagen können die Hennen besonders kampflustig sein; durch den stark entwickelten Kopfschmuck werden ihre Gesichter so verändert, daſs sie sich gegenseitig kaum erkennen. Es kann ja auch sein, daſs die erhöhte Kampfeslust in diesem Zeitpunkt mit derselben Ursache zusammenhängt, die das Anschwellen des Kammes und der Lappen bewirkt. In dieser Zeit habe ich beobachtet, wie von 2 Hennen, die auf einander gerieten, die eine auf dem Wahlplatz blieb, um nie wieder aufzustehen.

Im allgemeinen nützt es nichts, zwei kämpfende Hühner zu trennen. Tut man es, so werden sie jederzeit wieder aufeinander losgehen, wenn sie blofs Gelegenheit dazu haben. Sie wollen ihre Sache abgemacht haben, wollen entscheiden, wer Despot sein soll. Einzelne Hähne mischten sich in den Kampf der Hennen und sprangen gerade zwischen sie, so daſs sie voneinander ablassen muſsten, wenn aber der Hahn weggegangen war, fingen sie meistens wieder an zu kämpfen. Von diesen friedensliebenden Hähnen sind einige sehr parteiisch. 1905 beobachtete ich einen weiſsen Italienerhahn, der einmal dazwischen fuhr, als sich zwei Hennen (Lvk und Bl) schlugen. Es war sonderbar zu sehen, wie der Hahn, als er die kämpfenden getrennt hatte, der Lvk einen scharfen Hack auf den Kopf gab, während er sich augenblicklich mit der Bl paarte. Dasselbe wiederholte sich mehrere Male, wenn diese zwei Hennen kämpften und der Hahn dazu kam. Die Ursache hierzu kann nicht gewesen sein, daſs nur Bl für die Paarung gebaut war, während es Lᴠk nicht war, denn andere Hähne paarten sich gern mit Lᴠk. Daſs der Hahn die eine vorzog[1] und ärgerlich war, wenn er glaubte, daſs ihr ein Leid zugefügt wurde, ist ganz klar. Und doch waren beide Hennen von derselben Rasse, beide weiſs, von derselben Gröſse und gleichem Alter; die Gesichter waren allerdings sehr verschieden, und das war es wohl, was für den Hahn das verschiedene Verhalten bedingte. — Wenn zwei Hähne kämpften, konnte sich auch ein anderer Hahn, einer, der den beiden über war, in den Streit mischen; aber nur um entweder beiden Hacke zu versetzen oder um ausschlieſslich den zu hacken, gegen den er sonst in seinem Despotismus am strengsten war.

Bei den Kämpfen haben die Hennen ebenso wie die Hähne ein ganz verändertes und eigentümliches Aussehen. Wenn sich die Hühner zum Kampf bereit machen, halten sie den Oberkörper schief und dreieckig, schieben die eine Schulter scharf hervor und neigen gern den Kopf zur Erde. Dann entfalten sie ganz oder halb den einen Flügel, die Schwanzfedern spreizen und sträuben sich, und die Halsfedern stehen ab, daſs sie wie ein Kragen aussehen. (Angst- und Paradestellung, die bei verschiedenen Vögeln beschrieben ist.) Zuweilen, oder auch die ganze Zeit, lassen die Tiere einen singenden Drohlaut hören; das klingt, als ob zwei Herolde, jeder auf seinem Instrument, zum Kampf blasen. Nun kommt das andere Stadium: Die Hennen gehen im Schritt und nähern sich einander, vielleicht sogar etwas falsch lockend zwischen den drohenden Lauten. Die Halsfedern stehen noch stärker ab. Zwischen dem ersten und zweiten Stadium vergeht ein Augenblick oder 1—2 Minuten. Dann geht es los, während immer noch so viele Hals- und Schwanzfedern wie möglich sich spreizen. Die beiden Kämpfenden suchen den Kamm oder die Lappen des anderen festzubekommen und dabei wechseln die Tiere beständig Platz, wirbeln rundum. Die Füſse werden wenig angewandt beim

---

[1] Ein ähnliches Beispiel berichtet Kᴀʀʟ Gʀᴏᴏs (Die Spiele der Tiere, Jena 1896) von zahmen Vögeln. Zwischen einem bestimmten Männchen und Weibchen kam nie eine Paarung vor, wahrend sich das Männchen häufig mit dem anderen Weibchen paarte. Gʀᴏᴏs hätte untersuchen sollen, ob sich das Weibchen, das sich nicht paarte, überhaupt mit einem Männchen paarte.

Hennenkampf, wohl aber wenn die Hähne kämpfen. Die ersten Hiebe werden meist abgewehrt, Schnabel stöfst an Schnabel, und die Hennen können dabei in ihrem Feuereifer $1/_2$ Meter in die Luft springen. Aber bald gibt es Risse überall am Kopf. Es wird nämlich auf den Kopf gezielt. Sehr merkwürdig ist es, dafs die Augen fast nie beschädigt werden; man könnte fast glauben, dafs ein Übereinkommen bei allen Hennen existiert, dafs die Augen des Gegners geschont werden sollen. Für die Hähne gilt das nicht. Kamm, Ohren, Gesicht, Lappen und Nacken können sehr blutig werden, aber die meisten Kämpfe verlaufen doch mit verhältnismäfsig unbedeutenden Wunden, die gleich wieder zuwachsen. Dauert der Kampf länger, so verliert er seinen verhältnismäfsig unschuldigen Charakter. Er nähert sich dann den bekannten Hahnenkämpfen, die bei Völkern mit blutdürstiger und wilder Phantasie beliebt sind. Kämpfe bis zu 15 Minuten und mehr quälen die Tiere in hohem Grade, ja verursachen ihnen oft ernstliche Leiden. Zuletzt werden die Tiere stark erschöpft und hacken sich gegenseitig grausam, aber apathisch. Die Augen fallen ihnen vor Müdigkeit fast zu, aber keins will aufhören. Nach derartigen Kämpfen kann die Eierproduktion verringert werden oder sogar für kürzere oder längere Zeit ganz aufhören. Statistisch zeigte sich, dafs Kämpfe zwischen Hennen, die sich nicht von früher her kennen, viel häufiger waren als Kämpfe zwischen bekannten Hennen. Das ist auch nicht sonderbar, weil doch die Hennen gerade beim ersten Zusammentreffen abmachen, wer Despot sein soll — und das führt oft zum Kampf.

Änderungen in der Hackordnung.

Wir haben schon flüchtig berührt, dafs die Hackordnung in einer Hühnerschar mit der Zeit Veränderungen erfahren kann. Jetzt wollen wir erklären, wie das zugeht. Es geschieht einzig und allein dadurch, dafs einzelne Hennen sich gegen ihre Despoten empören. Ein derartiger Aufruhr seitens der Unterdrückten findet weit seltener statt zwischen Hennen, die beim ersten Zusammentreffen durch Kampf abgemacht haben, wer der Tyrann sein soll, als zwischen denen, die stillschweigend (d. h. durch die Angst der einen) die Despotismusfrage abgemacht haben.

In letzterem Falle hegt wohl die furchtsame Henne einen sozusagen schwelenden Hang zum Kampf, und deshalb kann er leichter zustande kommen zwischen einer solchen Henne (I) und ihrem Despoten (II), als zwischen zwei Hennen, die schon einmal durch einen wirklichen Kampf festgestellt haben, welche die überlegene ist.

Bei einem solchen Aufruhr ist es stets die unterdrückte Henne, die angreift. Was kann nun der Despot tun? Entweder sich drücken und flüchten, dann rückt I zum Despoten empor (und es findet eine Veränderung in der Hackliste statt), oder II kann wieder hacken. Es ist äußerst selten, daß der Despot ohne weiteres seine Herrscherstellung aufgibt, das Gewöhnliche, daß er zurückhackt. Wozu führt nun dieser zweite Fall? Entweder wird bei I der Mut so gründlich und schnell verringert, daß sie keine weitere Opposition versucht — wenigstens vorläufig nicht; in diesem Falle bleibt die Henne I unterdrückt; mit anderen Worten, der alte Zustand wird aufrecht erhalten, und es geschieht keine Änderung in der Hackliste. Oder die Gegenhacke der II dämpfen die Kampflust der I nicht. Was geschieht nun? Ein Kampf kommt zustande, und wer aus diesem Kampf als Sieger hervorgeht, wird in Zukunft Despot. Es ist oft vorgekommen, daß die Opponierende gewann. Das braucht einen nicht zu wundern, wenn man bedenkt, unter welchen Verhältnissen II Tyrann über die I wurde; aus Mangel an Mut hat sich I ja seinerzeit nicht in einen Kampf mit II eingelassen. Gewinnt jetzt I, so führt das natürlich zu Veränderungen in der Hackliste.

Es ist mir aufgefallen, daß Hennen, die den Platz der unterdrückten I einnahmen, sich während des Aufruhrs nicht mit allen Kräften in den Kampf gegen ihre Despoten legten, sondern ihn bisweilen mit größerer Schlaffheit führten, als man erwarten sollte, ja merkbar schlaffer, als die betreffenden Hennen sonst kämpfen konnten, und daß demzufolge die Aussichten für die I, den Kampf zu gewinnen, verringert wurden. Es könnte deshalb folgendes als allgemeine Regel ausgesprochen werden. Ist eine Henne erst einmal in eine unterdrückte Stellung geraten, so ist es (wahrscheinlich wegen des gewohnten Zustandes) für sie schwieriger, durch Aufruhr Despot

16

über die andere zu werden, als wenn sie gleich, als sie die andere, damals unbekannte Henne traf, mit dieser Henne gekämpft hätte.

Oppositionen wie die eben geschilderten können auch eintreten von seiten der Unterdrückten gegenüber den Hennen, von denen sie seit der Küchelzeit an gehunzt worden sind, wie auch gegen eine oder mehrere Hennen, die beim ersten Zusammentreffen durch ihr Auftreten und ihren Überfall sich eine vorläufige Oberherrschaft erzwungen haben.

Statistik der Kämpfe.

Ich habe 10 Jahre lang eine Statistik darüber geführt, wie häufig beim Zusammentreffen einer beheimateten und einer fremden Henne von der einen oder der anderen der Kampf gewollt oder nicht gewollt wird und erhielt folgende Zahlen:

Beobachtetes erstmaliges Zusammentreffen.	Anzahl Zusammentreffen, wo die Beheimatete den Kampf wollte.	Anzahl Treffen, wo die Beheimatete den Kampf nicht wollte.	Anzahl Treffen, wo die Fremde den Kampf wollte.	Anzahl Treffen, wo die Fremde den Kampf nicht wollte.
1428	1330	98	458	970

Aus der Übersicht über die 1428 beobachteten Zusammentreffen geht hervor, daſs die Anzahl Zusammenstöſse, wo die beheimatete Henne den Kampf wollte, mehr als 13 mal g r ö ſ s e r war als die Anzahl Zusammentreffen, wo die beheimatete Henne den Kampf nicht wollte. Gleichzeitig war die Anzahl Begegnungen, wo die fremde Henne den Kampf nicht wollte, etwa 2 mal so groſs als die Anzahl Begegnungen, wo die fremde Henne zum Kampf aufgelegt war.

Diese Tatsachen legen deutlich die Stimmung der Hennen im Begegnungsaugenblick zutage. Die beheimateten Hennen fühlen sich in der Regel sicher, bauen auf ihre eigene Stärke und sind erbittert auf fremde Hennen, die in ihr Gebiet kommen. Das macht die beheimateten Hennen durchweg dreist und aggressiv. Die fremden Hennen dagegen, die die Lokalitäten nicht kennen, finden sich schwer zurecht, sind durchweg scheu und möchten den Kampf vermeiden.

Wer hat bei diesen Kämpfen die gröfsere Aussicht zu gewinnen, die beheimateten oder die fremden Hennen?

Beobachtete erst- malige Kämpfe.	Anzahl Kämpfe, die die Beheimatete gewann.	Anzahl Kämpfe, in denen die Fremde siegte.
476	294	182

Es zeigt sich, dafs von den Kämpfen 1,6 mal so viel von den
beheimateten als von den fremden Hennen gewonnen werden.
Das spricht auch dafür, dafs der Mut der neuen Hennen so
verringert ist in der unbekannten Umgebung, dafs sie durch-
schnittlich nicht mit so grofser Kraft kämpfen, wie sie in
Wirklichkeit könnten.

Es sei bemerkt, dafs bei den Begegnungen, die unserer
Statistik zugrunde liegen, natürlich jedesmal e i n e fremde
Henne mit e i n e r beheimateten zusammen geführt wurde.
Eine andere Vorsichtsmafsregel war die, die verschiedenen
Begegnungen nicht zu schnell aufeinander folgen zu lassen,
da sie mit ihren Kämpfen Einflufs auf die Stimmung und
die Kräfte der Versuchstiere und damit auf den Ausfall der
Versuche haben könnten.

α-H e n n e n i m K a m p f. Wir haben vorher erwähnt,
dafs wenn 2 α-Hennen aneinander geraten, der Kampf oft
sehr hart wird. Der Grund liegt sicher darin, dafs eine α-Henne
nicht die geringste Spur von Knechtschaft gewohnt ist. Es
zeigte sich ferner, dafs fremde Hennen zuweilen beim Zwei-
kampf über eine α-Henne siegen — und dadurch ihr Despot
werden konnten — während sie gleichzeitig den Kampf mit
anderen Bewohnern des Hühnerhofes verlieren konnten. Aber
das ist ja gar nicht sonderbar, wenn man bedenkt, was früher
hervorgehoben wurde: dafs eine α-Henne nicht die stärkste
Henne in einer gewissen Hühnerschar zu sein b r a u c h t,
sondern dafs sie oft ihre hohe Stellung durch Macht der Gewohnheit oder durch besonderes Glück erhalten hat.

D e r E i n f l u f s d e s A l t e r s a u f d i e K ä m p f e. Aus
vielen Beobachtungen geht hervor, dafs ein gewisses Alter am
günstigsten ist, damit sich die Hennen in den Kampf einlassen und diesen auch gewinnen. Das mittlere Alter der

16*

Henne (1 ¹/₂ bis höchstens 5 Jahre) ist das beste. Ganz junge
Hennen (unter 1 Jahr) haben durchweg wenig Mut, wenn sie
mit anderen auch bekannten Hennen zusammentreffen. Sehr
alte Hennen können sich oft vor anderen Individuen auffallend
erschrecken und geradezu jeden Kampf verweigern. Ich fand
mehrmals, daſs Hennen, die in jüngeren Tagen kampfeifrig
gewesen waren, sich aber später vom Kampf zurückzogen, nur
noch eine geringe Sehkraft hatten, und dieser Umstand scheint
nicht selten eine Rolle zu spielen, wenn ältere Hennen geringe
oder keine Lust haben, sich in den Kampf einzulassen. Seinen
Gegner undeutlich zu sehen, kann auch kaum die Kampflust
erhöhen. Der Grund für die Unlust zum Kampf kann mög-
licherweise auf Degeneration der Geschlechtsorgane und den
damit verbundenen Veränderungen in dem psychischen Zustand beruhen.

Experiment mit Drahtnetz.

Mehrere Male habe ich folgendes Experiment vorgenommen,
um die Natur des Hackgesetzes näher zu untersuchen: 2 Hennenscharen, die einander ganz unbekannt waren, wurden nicht
zusammen geführt, sondern je in ihren Hühnerhof gebracht.
Diese Höfe waren nur durch ein Drahtnetz mit viereckigen
Maschen geschieden. Die Hennen in beiden Scharen konnten
einander deutlich sehen, aber keine der Hennen konnte zu der
anderen Schar hineinkommen.

Was geschah nun? Einzelne Hennen beider Hühnerhöfe
stellten sich in Kampfstellung gegenüber Hennen, die sich im
anderen Hof befanden, auf und stieſsen Drohlaute aus. Mehrere
Hennen von beiden Seiten gingen aufeinander los, und zwischen
einzelnen Paaren (eine Henne von jedem Hühnerhof) konnten
Kämpfe entstehen, trotzdem das Drahtnetz die zwei Parteien
trennte. Die kämpfenden Hennen hackten sich gegenseitig
durch die Maschen des Drahtnetzes. Die Hacke waren oft
gut gezielt; die Hennen sprangen zuweilen in die Höhe, und
häufig sah man Blut flieſsen. Nichtsdestoweniger konnten
diese Hennen, die mit dem Netz zwischen sich kämpften, sich
nicht richtig packen, und die meisten derartigen Kämpfe
blieben unentschieden, weil die Streitenden nach einer
Weile des Kampfes überdrüssig wurden und dann weg-

gingen — ohne dafs eine die andere zum Aufhören geängstigt
hatte. Sie entfernten sich oft sogar drohend voneinander. Bisweilen konnten sie dann nach einigen Stunden oder Tagen
aufs neue mit ihrem Kampf beginnen, jede von ihrer Seite
des Netzes aus. Wenn es den Hennen allmählich klar wurde,
dafs sie sich durch das Netz nicht ordentlich hacken konnten,
konnten sie sogar mit den drohenden Lauten und Bewegungen
aufhören, indem sie weiter keine grofse Notiz voneinander
nahmen. — Weiter gab es Hennen, die überhaupt in keinen
Kampf durch das Netz gerieten, bei ihnen trat das Gleich-
gültigkeitsstadium gegenüber den Hennen im Nachbarhof bald
ein. — Es ist zu bemerken, dafs augenscheinlich einzelne der
Hennen in dem einen Hühnerhof durch ihr Äufseres oder ihr
Benehmen auf gewisse Hennen in dem anderen Hühnerhof
irritierender oder kampfentflammender wirkten. Nr. 3 führte
5 Kämpfe, Nr. 1 4 Kämpfe, Nr. 2, 7, 9, VI, IX und X jede
2 und Nr. 4, 6, 12, II, VII, XI und XII jede 1. Das deutet
daraufhin, dafs Nr. 3 die Henne war, die die Tiere im Hühnerhof a am meisten irritierte.

Fig. 4 soll das Benehmen der Tiere in 2 derartigen voneinander isolierten Hühnerhöfen mit eingefügtem Drahtnetz
veranschaulichen. Die Tiere im Hühnerhof a (links) sind mit
römischen Ziffern, die im Hühnerhof b (rechts) mit gewöhn-
lichen Ziffern bezeichnet. Ich beobachtete die Hennen 4 Tage
lang den ganzen Tag, jedesmal, wenn ich von den Hennen
fortging (zu meinen Mahlzeiten und zur Nacht), schlofs ich
die Tiere in b in ein Hühnerhaus rechts vom Hofe b ein, damit sie nicht einen Augenblick unbeobachtet in ihrem Verhalten
zu den Hennen in a blieben. In dem Hühnerhaus konnten
die Hennen von b weder die Hennen in a sehen, noch von
ihnen gesehen werden. — Die 2 Höfe a und b hatten zirka
20 qm Flächeninhalt; bei der Wahl dieses verhältnismäfsig
kleinen Areals war es darauf abgesehen, dafs sich alle Hennen
leicht sehen konnten.

Die Hennen, ·deren Ziffern in Fig. 4 miteinander durch
eine gestrichelte Linie verbunden sind, gerieten im Laufe der
Beobachtungszeit in Kampf. Man sieht, dafs im ganzen 14
Kämpfe stattgefunden haben. Theoretisch hätten höchstens
144 Kämpfe vorkommen können, falls jede Henne des einen

Hühnerhofes mit jedem Individuum des anderen gekämpft hätte. Diese hohe Kampfzahl war zwar an sich nicht zu erwarten, da wir ja früher gehört haben, daſs sich Hennen oft dem Kampf entziehen. Aber es wären sicher mehr als 14 Kämpfe eingetreten, falls 12 Hennen 12 ihnen unbekannten Hennen o h n e trennenden Drahtzaun gegenübergestanden hätten. Das Netz hat also hemmend auf die Kämpfe gewirkt.

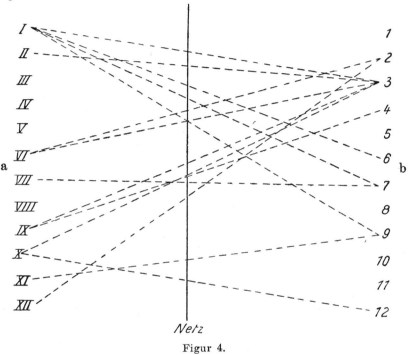

Figur 4.

Bemerkt sei, daſs von den 14 Kämpfen 12 am ersten Tage ausgefochten wurden, 2 am zweiten Tage, während der dritte und vierte Tag kampflos waren. Die Kämpfe waren also im Anfang am häufigsten, während sich die Hennen bald ziemlich gleichgültig wurden. Wurde nun einer von den 14 Kämpfen entschieden? Nein, sie verblieben alle ohne Resultat, und das Kampfstadium ging bei den Hennen, die sich gestritten hatten, in ein passiveres Stadium über, wobei die 2 Gegner gegenseitig Drohlaute ausstieſsen.

Die anderen Hennen, die nicht kämpften, verhielten sich paarweise folgendermaſsen: 1. E n t w e d e r drohten sich beide gegenseitig — mehr oder weniger — mit Stimme und Bewegungen, doch ohne daſs es zum Kampf kam, 2. o d e r die eine drohte der anderen, aber nicht umgekehrt, 3. o d e r keine von beiden drohte der anderen.

Nachdem die 4 Tage zu Ende waren, wurde das Netz ganz weggenommen und die Hennen paarweise zusammengeführt, also jedesmal 1 Henne von dem einen Hof und eine von dem anderen. Daſs die Tiere d u r c h d a s N e t z ihren Streit nicht entschieden hatten, keine O r d n u n g i n d i e G e - s e l l s c h a f t b e k o m m e n h a t t e n, wird nun zur Genüge bestätigt; es kamen einige furchtbare Kämpfe zustande und viele leichtere. Erstens kämpften nun alle die Hennen, die sich unentschieden durch das Netz geschlagen hatten; j e t z t k ä m p f t e n s i e b i s z u r E n t s c h e i d u n g u n d h ö r t e n d a n n a u f. Zweitens wurden Kämpfe von denen ausgefochten, die sich durch das Netz gegenseitig gedroht hatten. Zwischen den Hennen, von denen die eine der anderen gedroht hatte, ohne daſs diese der ersten gedroht hatte, wurde nun die, die gedroht hatte, Despot über die andere o h n e K a m p f, mit einer Ausnahme, wo es zum Kampf kam. Von den Hennen- paaren, die sich untereinander nicht gedroht hatten, wurde jedesmal eins der Individuen nach kurzer Zeit Despot, und zwar o h n e K a m p f. Wir kommen also zu folgenden Er- gebnissen:

1. Solange die Hennen durch das Netz getrennt sind, können wohl Kämpfe entstehen, aber diese verbleiben unent- schieden, und die betreffenden Hennen geben es auf die Rang- ordnung festzustellen.

2. Daſs die genannten Kämpfe wirklich nicht entschieden worden sind, wird dadurch bewiesen, daſs die Hennen, die sich, mit dem Netz zwischen sich, geschlagen haben und dann zur Ruhe gekommen sind, sofort den Kampf wieder anfangen, wenn das Netz entfernt wird und nicht aufhören, bis einer der Gegner besiegt worden ist.

3. Die Hennen, die keinen Kampf durch den Zaun an- gefangen haben, bestätigen durch ihr Benehmen nach der Ent- fernung des Zauns, daſs die Rang- und Gesellschaftsordnung

nicht hatte geordnet werden können, als sich die Tiere sehen, aber ihre Kräfte nicht erproben konnten. Die Kraftprobe ist also entscheidend für die Rangordnung im Hühnerstaat.

4. Die Hennen, die sich gegenseitig nur durch das Netz drohen, kommen nach der Entfernung des Netzes in Kampf und benehmen sich wie die Hennen, die schon durch das Netz in Kampf geraten waren.

5. Die Hennen, welche anderen ohne Erwiderung gedroht haben, werden (mit einer Ausnahme[1]) Despot über die anderen.

6. Von den Hennen, die durch das Netz von Anfang an keine Notiz voneinander nehmen, werden die einen ohne Kampf Despot über die anderen.

Diese verschiedenen Punkte, die sich aus obigem Experiment ergeben haben, wurden durch erneute Versuche immer wieder bestätigt. Die erwähnten Punkte scheinen generell zu gelten, auch in bezug auf Hähnescharen.

Es ist erwähnt worden, daſs selten ein Kampf entschieden wird, wenn ein Drahtnetz die 2 Hühnerhöfe trennt. Ich beobachtete aber, daſs es doch, besonders bei Hähnen vorkommen konnte, nämlich wenn der eine der Kämpfenden den anderen durch das Netz so gut packte, oder ihm einen so heftigen, schmerzenden Hack erteilte, daſs jener Angst bekam und sich als besiegt ansah. Wenn sich dann die beiden nach Entfernung des Zauns trafen, ging der durch das Netz überwundene dem anderen ohne Kampf aus dem Wege und betrachtete den anderen als Despot. — Diese Beobachtungen bestätigen übrigens nur die Behauptung, daſs Hühner nicht in demselben Hühnerhof leben können, ohne eine bestimmte Rangordnung zu stiften.

3. Kapitel.
Die Gluckperiode und ihr Einfluſs.

Wirkt der Gluckzustand, dieses wichtige Stadium im Leben der Henne, auf die gesellschaftliche Stellung der

---

[1] Diese Ausnahme kann als ein plötzlicher Aufruhrdrang bei dem Individuum erklärt werden, das sich eigentlich ohne weiteres unterordnen sollte.

glucken Henne? Verbleibt diese unverändert oder nicht? Auf diese Frage kann keine allgemeine Antwort gegeben werden. Viele Umstände üben ihren Einfluſs hierbei aus. Fängt eine Henne, die in einer Hühnerschar zusammen mit anderen lebt, an zu glucken, so äuſsert sich das zuerst dadurch, daſs sie jedesmal nach dem Legen das Nest ungern verläſst und lange darauf liegen bleibt. Kommt die Henne in den Hühnerhof, so gluckt sie vielleicht etwas, aber nicht viel. Der Glucklaut nimmt schnell an Häufigkeit zu und nicht lange (4 Tage) danach will die Henne das Nest nicht mehr verlassen, auſser 5—10 Minuten 1- oder 2 mal täglich. Sogar nachts bleibt sie auf dem Nest liegen. Sowohl im Nest wie im Hühnerhof ist die Gluckhenne etwas reizbarer als sonst. Erstens ist sie etwas reizbarer gegen die Hennen, über die sie Despot ist, und zweitens neigt sie oft mehr dazu, Aufruhr gegen die zu machen, die sie hunzen.

Im Freien ist die Gluckhenne unruhig und nervös, hackt das Futter mit fieberhaftem Eifer und badet sich eifrig, aber kurz im Sande. Daſs die Gluckhenne so rastlos ist, sich sozusagen zu nichts ordentlich Zeit läſst, beruht ohne Zweifel darauf, daſs sie Sehnsucht danach hat, rechtzeitig zum Nest zurückzukommen. Sie eilt dann wieder dorthin, oft im Laufschritt (oft s t ü r z t die Gluckhenne von ihrem Sandbad weg, als ob sie sich plötzlich an etwas erinnert hätte), geht auf das Nest und fängt wieder an zu brüten. Vom Nest läſst sich die glucke Henne durch starke Hacke seitens ihrer Despoten vertreiben, sie hält aber zäher stand als sonst, indem sie sich mehr hacken läſst, ehe sie von ihrem Platz weicht. Einige Gluckhennen lassen sich aber, ohne die Hacke ihrer Despoten zu erwidern, nicht durch diese vertreiben, und wenn sie noch so gewaltig sind; die Gluckhenne schützt blofs den Kopf so gut es geht und rührt sich nicht vom Nest. Der Despot stellt schlieſslich das Hacken ein und setzt sich zum Eierlegen entweder an die Seite der Gluckhenne ins Nest oder legt sein Ei auſserhalb davon oder in einem anderen Nest.

Ihren Despoten gegenüber unterscheiden sich die Gluckhennen individuell sehr stark. Zuweilen kann — wie gesagt — eine Gluckhenne drauſsen im Freien oder im Nest Aufruhr gegen einen oder mehrere von ihren Tyrannen machen. Es kommt dann selbstverständlich zum Kampf, und es ist gar nicht immer sicher, daſs die Gluckhenne gewinnt. Oft konnte der Gegner nach kurzer Zeit die Gluckhenne besiegen, die dann mit allen Zeichen von Mutlosigkeit die Flucht ergriff.

Das bisher Gesagte gilt für die Hennen, die nachdem sie gluck geworden sind, ohne Unterbrechung in ihrem gewohnten Hühnerhof und Hühnerhaus bleiben und sich nach Belieben hier einrichten können.

Ganz anders wird sich jedoch die Gluckhenne benehmen, wenn man sie aus der gewohnten Umgebung entfernt und ihr ein N e s t mit Eiern und e i n e n R a u m g a n z f ü r s i c h gibt, wo sie einige Tage lang (4 Tage können oft genügen) keine Gelegenheit hat, die übrigen Bewohner des Hühnerhofes zu treffen. Führt man dann die Gluckhenne mit einer nach der anderen von ihren früheren Despoten zusammen, so stellt sich die Gluckhenne fast ohne Ausnahme augenblicklich kampf-bereit. Dann entsteht im allgemeinen ein heftiger Kampf, da der Despot absolut nicht dazu geneigt ist, die Herrschaft, die er bisher über die Gluckhenne hatte, aufzugeben. Gleichzeitig ist die letztere sehr unerschrocken. Entweder greifen nun beide Gegner gleichzeitig an, oder die Gluckhenne wirft sich in vollem Lauf rasend über die andere.

Die Gluckhenne gewinnt nicht immer, aber doch meistens bei solcher Gelegenheit. Wenn der Gegner der Gluckhenne nicht den sog. „schweren" Rassen angehört, während die Gluckhenne einer der leichten Rassen entstammt, verliert die Gluckhenne sehr selten. Gewinnt die Gluckhenne, so ist sie leicht sehr erbittert auf ihren Kampfgenossen, auch nachdem er sich ergeben hat und verfolgt so lange wie möglich den früheren Despoten, der Hals über Kopf flieht. Die geschlagene Henne muſs dann, um nicht zu arg zugerichtet zu werden, entfernt werden. — Verliert dagegen die Gluckhenne den Kampf, so wird sie sehr demütig, senkt den Kopf tief und flieht vor der anderen, die ihr wohl im allgemeinen nachläuft, aber in ihrem Benehmen doch gemäſsigter ist als sie selbst.

Bringt man eine isolierte Gluckhenne mit einer von den Hennen zusammen, über die sie selbst früher Despot war, so macht sich die Gluckhenne o h n e A u s n a h m e zum Kampf bereit, und o h n e A u s n a h m e nimmt sie den Kampf mit den genannten Hennen auf, falls sie kämpfen wollen. Im allgemeinen gehen aber diese Hennen (wir wollen sie als x-Hennen bezeichnen) der Gluckhenne aus dem Wege, weil sie ihren alten Tyrannen noch nicht ganz vergessen haben. Nur selten lassen sich x-Hennen in einen Kampf mit der Gluckhenne ein. Dafs sich x-Hennen überhaupt in einen Kampf mit der Gluckhenne einzulassen wagen, mufs wohl darauf beruhen, dafs sie dieselbe nicht sicher wiedererkennen, sondern glauben, eine fremde Henne vor sich zu haben, mit der sie ihre Kraft prüfen wollen.

Sehr selten gewann eine x-Henne über eine Gluckhenne. Von 305 untersuchten Fällen gewann die Gluckhenne 303. Das hat sicher einen psychologischen Grund. Die x-Hennen, die sich in einen Kampf mit der Gluckhenne einlassen, fühlen vielleicht, wenn sie sich auch nicht genau darauf besinnen, dafs die Gluckhenne früher ihr Herrscher war, unklar, dafs es ihnen im Kampf schlecht ergehen wird, und halten es deshalb nicht der Mühe wert, alle Kräfte einzusetzen. Weiter mufs man bedenken, dafs die Gluckhenne alle diese Kämpfe mit grofser Kraft und Ausdauer führt. Mit ganz fremden Hennen läfst sich eine normale, ausgewachsene Gluckhenne, sofern sie nicht in unbekannte Umgebung versetzt wird, fast immer in den Kampf ein, was zahlreiche Untersuchungen zeigten. Sogar wenn Gluckhennen in ganz fremde Umgebung gebracht werden, können sie, wenn sie ein kriegerisches Temperament haben, den Kampf mit ganz fremden Hennen aufnehmen.

Die beste Bedingung, unter der sich die Kampflust der isolierten Gluckhenne bei der Begegnung entfalten kann, ist, dafs sie sich auf dem Nest oder in dessen Nähe befindet. Ist die Gluckhenne vom Nest entfernt, so soll man sofort die andere Henne zur Gluckhenne lassen, da einzelne Gluckhennen selbst bei kurzer Abwesenheit vom Nest unruhig werden, sich nur danach zurücksehnen und den Mut derartig verlieren können, dafs sie den Kampf verweigern, selbst wenn sie angegriffen werden.

Hat die isolierte Gluckhenne so lange gebrütet, dafs die Küchel ausgekrochen sind, so ist sie weiterhin sehr streitbar, wenn sie die Küchel um sich hat. Nimmt man ihr aber die

Küchel, wenn auch nur für eine kurze Weile, z. B. für ein paar Minuten fort, so verlieren viele Gluckhennen ganz den Mut, sehnen sich nur zurück nach den Kücheln und klagen oft laut. Von den Kücheln entfernt kann also manche Gluckhenne von anderen Hennen überfallen werden, ohne daſs die Gluckhenne es ihnen wiedergibt. Dann werden die anderen zu Despoten.

Ist eine Gluckhenne vom Nest oder von den Kücheln entfernt worden und in diesem Zustand anderen Hennen ausgewichen, oder hat sich widerstandslos von ihnen hacken lassen, so bleiben diese Hennen, wenn die Gluckhenne wieder aufs Nest oder zu den Kücheln zurückgebracht wird, fast ohne Ausnahme Despoten über die Gluckhenne. Spazierte die Gluckhenne mit den Kücheln umher, so war sie gewöhnlich auch vor letzteren Hennen ängstlich, ging ihnen aus dem Wege oder floh vor ihnen, auf den Fersen von den kleinen Kücheln gefolgt. Selbst wenn die feindlichen Hennen die Kücheln anfielen, verteidigte sie diese nicht, was sie hingegen mit Raserei tat, wenn Hennen, mit denen sie während ihres nervösen Zustandes nicht zusammen gewesen war, die Kleinen angriffen. Wieder ein Beweis, daſs das Hacken der Hennen sehr von der Gewohnheit abhängt.

Bei allen diesen Experimenten ist Voraussetzung, daſs die Tiere zahm sind, daſs der Experimentator es versteht, mit ihnen umzugehen, daſs er seine vollständige Ruhe bewahrt, daſs er zur rechten Zeit eingreift usw., alles Dinge, die man nur durch Erfahrung lernen kann.

1

Contributions to
the Social Psychology
of the Domestic Chicken

This article was translated expressly for this Benchmark volume by Monika Schleidt, and Wolfgang M. Schleidt, University of Maryland, from Z. Psychol., 88, 225–252 (1922).

Chapter 1

THE SOCIAL POSITION OF THE INDIVIDUAL

Introductory Remarks. On entering a chicken yard to observe its inhabitants, one finds the chickens either resting or moving about. Some animals stand still or stretch out lazily on the ground while others scratch the ground in search of food, dust bathe in the sand, run back and forth along the fence, or chase each other from the water container.

Chickens that are allowed to move about freely (i.e., not in a fenced-in yard) also behave in the above described manner. Since in an open field it is more difficult to observe all individuals at the same time, it is advisable to keep them in an enclosed yard of moderately limited size; then, the behavior of the animals towards each other may be observed with greater ease. During my studies I observed the animals mainly in a chicken yard, but some observations were made in open fields. I studied small groups (2-25) as well as large groups (25-150).

To avoid misunderstandings, I will use the term "hen" and "rooster" separately, in contrast to "chicken" (which means both hens and roosters).

Means of Defense and Aggression in the Hen. If one throws feed to the hens, they come running from all sides in order to ingest the greatest possible amount. Even the most phlegmatic hens which might have been resting get up and come running to get their share, and if possible, more than their share. This passion for food is natural in all hens and has its roots in the drive for self-preservation. But does the meal procede quietly? No, there is a constant quarrel over the food; each hen shows clearly that she wants to be first to eat.

In order to get the greatest amount of feed during such a meal, the hen has to do the following: 1) defend the morsel of food which she is still carrying in her beak, 2) snatch from the other hens what they have already gotten but are still carrying in their beaks, 3) chase the other hens as far away from the feeder as possible and eat as quickly as she can. These rules are a constant cause for quarrels and fights. How does the hen try to carry out these measures? How does she chase her fellow chickens from the feeder? The answer is: by administering or threatening to administer such strong punishment that others will give up part or all of their food in order to avoid being punished.

Editor's Note: Figures for this article may be found in the preceding original paper.

Defense and aggression in the hen is accomplished by pecking with the beak. Many pecks are delivered during such a meal; hens that are pecked utter short cries or moans. (To a human the pecks feel like a strong pinch or a scratch.) Pecks occasionally cause a feather to be pulled out or, if the face or comb are hit, there might be a scratch and even bleeding. Before a hen pecks she sometimes utters a peculiar "threatening" sound which, together with the short cries and moans, make a meal anything but silent. Pecking is frequently observed even when the hens are not feeding but are just standing around in the yard, or are on or close to their nests or roosts at nightfall.

The Peck Order and Its Peculiarities. We may now ask: are all hens exposed to the same aggression? One would assume that the pecking is mutual (which means that hen A pecks at hen B and the next time hen B pecks at hen A and so forth) or that the strongest hens peck and chase the weaker ones.

The first assumption is definitely wrong. I have observed in more than 1900 cases that if hens A and B live together for some time and hen A pecks at hen B, then hen B will not peck at A; instead B is afraid of A and avoids her. After some time though hen B may oppose hen A and try to peck her. This leads instantly to a fight because A definitely does not want to give up her dominance over B. Such a fight will usually end with the defeat of B and the original situation remains: A chases B. If B wins she will become the despot, possibly forever but in any case for the time being until A wins another fight at some later time. Usually, though, a conflict like the one described above does not occur and many hens do not try to challenge for years. The second assumption, that the strongest individuals peck the weaker ones, sounds likely but it does not always coincide with reality. Later in this paper I will show that this assumption is unwarranted.

Let me now list several "peck orders" that are based on long-term observations.* The letters are abbreviations for the names of individual chickens.

Peck Order 1, seven hens (Hvl, Pl, Oml, Byl, Afl, FrSl).
        Hvl pecks all others.
        Pl pecks Oml, Byl, Sml, Afl.
        Oml pecks Byl, Sml, Afl, FrSl.
        FrSl pecks Pl, Byl, Sml, Afl.
        Byl pecks Sml, Afl.
        Sml pecks Afl.
        Afl pecks none.

It can be seen from this list that hen Hvl is definitely the most fortunate since she pecks all the others, which means that they all avoid her and are afraid of her freely administered pecks. No hen dares to take food away from her and she rules the roost.

---

*Such peck orders are not revealed very quickly since they require many hours with the animals and many consecutive days of observation (for 15 hens at least 14 days, all day long). To discern peck orders, it is advisable to pay particular attention during feeding time or when they roost at night, since pecking is most frequently observed at these times.

Thorleif Schjelderup-Ebbe

Hen Af1 is in the worst position, daring not to peck anyone. The positions in between are taken by the other five hens, of which P1, Om1 and FrS1 are considerably better off than By1, who is turn is in a better position than Sm1.

It might be assumed that hens lead a more or less worry-free existence according to their position in the peck order, since it is probable that a hen that is pecked by many others has a less pleasant life than one that is not attacked but instead pecks most others. And such is the case. I noticed for example that in the above flock, Af1 got the least amount of food and was very nervous because of the number of pecks she received. I had the impression that she tired herself out in a constant attempt to avoid punishment and to get enough food. I observed similar behavior in other hens which were low in the peck order. In contrast, the despot Hv1, who always got her way, who chased others from the food or the nest and who was never bothered by anybody, seemed to feel very well.

Pecking Order 1 is not continuous. If it were, the most powerful hen would have been despot over 6 others, the next over 5, the next over 4, the following over 3, the next over 2, the next over only 1 and the last in line over none. Actually, though, P1, Om1 and FrS1 caused deviations: P1 dominates Om1 and Om1 dominates FrS1; it would therefore be expected that P1 is also despot over FrS1, but instead FrS1 is despot over P1.

In this one can see proof that the hens are not arranged in a peck order exclusively according to their individual power.

It would be presumptuous to infer that the cause of Hv1's despotism over all other six hens lies in the fact that she is so much stronger than the others. It could be her strength, but there could also be other reasons to explain her superior social position. We will come back to this point later, but let us now look at another peck order.

Peck Order 2, 10 hens: Om2, Att1, Sp1, Gam1, P2, Lvk1, Sm2, Bl1, Gr, Hv2.
          Om2 pecks all but Sp1.
          Att1 pecks all but Om2.
          Sp1 pecks all but Att1.
          Gam1 pecks P2, Lvk1, Sm2, Bl1, Gr, Hv2.
          P2 pecks Lvk1, Sm2, Bl1, Gr, Hv2.
          Lvk1 pecks Sm2, Bl1, Gr, Hv2.
          Sm2 pecks Bl1, Gr, Hv2.
          Bl1 pecks Gr, Hv2.
          Gr pecks Hv2.
          Hv2 pecks none.

Looking at this peck order, it is apparent that there is no hen despot over all the others. This is in contrast to what we observed in Peck Order 1. Three hens in Peck Order 2 (Om2, Att1 and Sp1) are in the same favorable position, each pecking 9 other hens. Among themselves the 3 hens behave in the same way as hens P1, Om1 and FrS1 in Peck Order 1, that is Om2 pecks at Att1, which in turn pecks at Sp1 while Sp1 pecks at Om2.* These 3 hens peck each other in a

Figure 1

---

*Another example that hens do not always peck according to their strength.

37

triangular fashion, as shown in Figure 1 where the arrows point in the direction in which pecking takes place.

It should be noted that there is a variety of different pecking combinations which leads to deviations in the continuous sequence of the peck order in other flocks.

If one disregards the 3 hens which form the triangle, the pecking of the other 7 hens in Peck Order 2 is absolutely continuous and could be pictured as a straight line, as in Figure 2.

Figure 2

The arrows indicate the direction of pecking. Along the straight line from left to right are the names of the 7 hens; each hen is despot over all hens whose names are printed to her right. This line may be called the base line. The simplest (a perfectly continuous) pecking scheme is one which consists of only such a base line. But the peck order in a flock of chickens is rarely so simple. Quite frequently one finds triangles, squares, etc. which have to be taken into consideration.

Considering the intricacy of some peck orders, it is not surprising that on superficial observation of the pecking in a flock of hens one gains the false impression that they all peck each other.

SUMMARY OF THE PECKING IN A FLOCK OF HENS

In reviewing the pecking in a flock of hens one may conclude, on the basis of the above listed peck orders and many others, the following:

1. In a flock of hens, one hen assumes a special position and is not pecked by any of her fellow hens; she is despot over all of them.

2. There are flocks, however, in which no one hen is the despot over all the other hens in the chicken yard.

3. Occasionally there is one hen in a flock who is so to speak Cinderella, since she is pecked by all other hens.

4. In some flocks, however, there is no hen that is pecked by all others. This relationship occurs rather frequently, but nevertheless I have observed the following extreme: out of 21 hens, hen Z pecked 19; all of these 19 pecked hen Y, but hen Y pecked hen Z who was very much afraid of her.

5. In a flock of more than 10 hens, one rarely finds a continuous peck order.

6. Over a period of time changes can occur in the peck order. How this comes about is easiest to understand in connection with what is explained below in Chapter 2, and therefore we will come back to this point later. It should be mentioned here that many experiments have demonstrated that the development of a peck order among animals hatched and raised in isolation is the same as among those hatched and raised in groups.

In terms of social psychology, the following must be mentioned: when a hen (let us call her an alpha hen) pecks all others in a flock, her behavior is modified in that she becomes less aggressive towards the others. In fact, an alpha hen may assume some kind of leader position, especially if she is somewhat older or if there is no rooster in the flock. The alpha hen may even court one of the other hens. It can not be denied that in some cases courtship may result from changes in the organs of procreation, but it must be pointed out that even heavily laying alpha hens may behave in the above described manner.

Generally it is not difficult to tell from the behavior whether the hen is feeling secure in her position. Rarely do we observe a more embittered fight than an encounter between the alpha hens of two different flocks.

Evidence shows that a hen with a low position on the peck order is crueller towards the few she can peck than a hen in a high position is towards the many others over whom she is despot. The following illustrates this point: if a hen with a low position in her flock is transferred to another flock in which she achieves a higher position, her behavior towards the many hens she is able to peck in the new flock will almost invariably be more moderate compared to what it was toward the few she pecked in her old flock. A typical example is hen Gr in Peck Order 2 where, as one can see, she pecked only one other hen (Hv2). The next year she was put in a different flock of 11 hens, out of which she pecked 9. Although she had been very unpleasant and nasty towards Hv2, she became very moderate towards all the hens she was able to peck in the new flock.

THE PECKING IN A FLOCK OF ROOSTERS

Pecking in a flock of roosters is the same as the pecking in a flock of hens, with the exception that the roosters are fiercer.

Chapter 2

FIGHTS AMONG CHICKENS

Fighting and Pecking Rules. Anyone who thinks the inhabitants of a chicken yard are thoughtless happy creatures with a daily life of undisturbed pleasure, at peace with each other and, without a worry about anything, crow, lay eggs and eat, is thoroughly mistaken. A grave seriousness lies over the chicken yard and hens exhibit much anger and fear. Even the liveliness demonstrated during feeding time is not in fact happy liveliness, since it is all based on the effort of each hen to consume as much as possible herself and to drive away the others as much as possible.

One of the most serious, meaningful and important events in the life of a chicken society is a fight, especially a sudden, sometimes terrible fight between two animals. Most of the time, but not always, the fight decides which of the two hens is going to be despot over the other.

We now want to look more closely at fights between chickens. One remark I would like to make at the outset: from my observations it is clear that there are no two hens living together which have not decided whether A is despot over B or B is despot over A. At what time is this decided? It happens quite early in their relationship.

When two strange hens meet each other, it is decided very quickly which of them will be despot. The two animals eye each other and the following happens:

1. Either one of them is startled at the sight of the other and the other, noticing it, approaches and threatens. The case is then decided without a fight. The frightened hen will be mistreated by the other one, either forever or at least for the time being. This is quite a frequent occurrence when two hens meet. It seems that the frightened hen is so thoroughly afraid of losing that it surrenders rather than fights.

2. <u>Or</u> both hens are frightened. <u>The first to overcome her fear becomes despot</u>.

3. <u>Or</u>, quite commonly, neither is afraid and they will start to fight. This leads rapidly into a regular battle; the victor becomes dominant over the loser, forever or at least for the time being, and can peck and chase her without fear of a counter attack. Frequently, momentary courage or luck is decisive in establishing dominance relationships.

<u>The above three rules constitute the basis for the peck law in chickens</u>.* We should note the following. It can be determined indirectly whether or not a hen, F, is really stronger than another hen, H: hen F is put together with a strange hen which previously won a fight over hen H. F and the stranger fight and F wins easily. Considering the three rules stated above one can now understand how a peck triangle can develop, even if A is really stronger than B and B stronger than C (where one should therefore expect continuity). The irregularity in the following triangle (C pecks A, though C is weaker than A) could be the result of A being frightened when she first met C and C using the opportunity to become despot over A in spite of the fact that she is weaker than A.

Figure 3

I have observed such situations many times and triangles, squares, etc. develop without doubt out of these or similar circumstances.

The hen that is lucky enough to be "tough" may be able to keep her position of leadership over another for a month, possibly for life, whereas the other has to put up with being mistreated and chased from the feeder as well as from the nest. Quite frequently one can observe a large strong hen F being pecked every day by a small or weaker hen H. If F would fight back she would surely win the fight and so become superior to H; <u>but F does not dream of fighting back</u>.**

Habituation surely plays an important role; F is habituated to avoid H and so it does not occur to her to rebel. We may now ask: how did it happen in the beginning that the strong hen F let the weaker hen H mistreat her? There are several different answers:

a. As is usually the case, adult hens (except broody hens) terrorize young chicks. The fear which growing chicks have of adult hens usually remains when they grow up themselves. In this way the oldest hens become despots, regardless of whether or not they or the younger hens are really stronger. This is repeated with every new generation of chicks. The youngest animals usually do not lose their fear of the older ones even when they themselves grow up.

b. Hens newly arrived in the chicken yard have usually been traveling for some time, are frightened, tired or hungry (or all of these) at their first encounter with the other hens, so that the latter hens have no difficulty dominating the newcomers and the relationship persists; this phenomenon is striking. If hens become sick and, in their state of weakness are attacked by healthy ones, the reign of terror of the healthy hens continues frequently even after

---

*The presence of a similar law has been observed in several wild birds, e.g. sparrows (<u>Passer</u> <u>domesticus</u>), and Kingfishers (<u>Alcedo</u> <u>ispida</u>).
**Analogy: The old Arab proverb: "Woe to the rider if the horse feels its strength."

Thorleif Schjelderup-Ebbe

the sick have completely recovered. There are many causes which make hens in-
disposed to fights, and other hens then take advantage of the indisposition.
At such a time it is easy for them to attain superiority, which they then retain.

  c. One hen may be simultaneously attacked by several weaker hens (or several strong plus weaker ones); their combined strength makes a victory possible.
From then on the one hen is afraid of each member of that group.

  The above three cases, and especially the first two, occur rather
frequently.

  We have now pointed out that pecking among chickens is not as arbitrary as
it may seem on first glance; it is dependent upon certain laws that are obeyed
by all hens. There are no two hens in the same group that do not know exactly
who is "over" and who is "under". The peck order does not so much depend upon
the strength of the hens but rather on habituation and certain circumstances
that occurred at the first encounter of the hens. (The same is true for a
flock of roosters.)

  It has been shown in a number of thorough experiments that the tendency to
social structure is in the chicken's blood; it is exhibited by young animals
when they grow up, regardless of whether or not they have been raised apart
from the older chickens. In other words, the tendency to social structure is
inherited rather than learned.

DESCRIPTION OF THE FIGHTS

  Like their relatives, Capercaillie and Black Grouse in the forest, domesticated chickens have serious feuds, but unlike the other birds, the feuds are
not so closely related to their sexual life. Fights among chickens, which are
usually considered to be quite harmless, are certainly not so and do not result
from a momentary whim. If one hen wins a fight her opponent does not dare
touch her (at least not for the moment). Thus the winner is not disturbed
while sitting on the nest, food is not taken away from her but she may do all
these things to others. Thus the result of a fight is of great importance to
the hens at a later time and their instinct apparently tells them that. They
put a lot at stake, sometimes even their lives, in order to win. Another factor which could possibly play an important role is the playful drive to measure
one's strength against a known or unknown opponent. But this factor can be
disregarded here: first of all, chicken fights are always of a vicious character and secondly, if they were merely play one would expect them to occur more
frequently in young animals rather than in older ones, which is not the case.

  To claim that chickens are more peaceful during molting season is, strictly
speaking, not true, even if they do seem to be less active at that time.
Chickens are never peaceful. Indeed, after the molt is over and there are a
few mild days in winter, chicken blood starts to boil. In some young hens the
comb and wattles, which have been dried up since fall, can enlarge in a few
days. The comb can resume its summer dimensions, some 2 to 3 times as thick
and 3 times as long as it is in the winter. On such days the hens can be extremely pugnacious; their highly developed head gear lets their faces appear
so changed that they have a hard time recognizing each other. It is also possible that the greater pugnacity at this time is related to the factor which
causes the enlargement of the comb and wattles. At this time of the year I
have observed two hens fight with each other with the net result that one of
them was left on the place of fighting, never to get up again.

Generally there is no use in separating two fighting hens. If one does separate them they will resume their attacks the moment they have an opportunity. They want to have the case decided, want to know who is the despot. On some occasions roosters interfered in fights between hens and separated them, but in most cases as soon as the roosters left the hens resumed their fight. Some of these peace-loving roosters are very partial. In 1905 I observed a white Italian rooster interfering in a fight between two hens (Lvk and Bl). It was interesting that the rooster gave Lvk a sharp peck on the head after he had separated the two hens and immediately copulated with Bl. The same was repeated several times whenever these two hens had a fight and the rooster was nearby. The reason for this behavior could not have been that Bl was better equiped for mating than Lvk because other roosters frequently mated with Lvk. It was quite obvious that the rooster preferred Bl* and became angry whenever he thought that harm was befalling her. Both hens were the same breed, the same size, both were white and about the same age; however their faces were very different and that must have caused the rooster to treat them so differently. In a fight between two roosters, a third one (superior to both of the fighters) might interfere, but he would do so only to either peck at both of them or peck only at the one who is lower in the peck order.

Hens as well as roosters assume a changed and peculiar appearance during fighting. When chickens get ready to fight, they hold their trunk in a diagonal and triangular position, push one shoulder forward and lower the head to the ground. Then they partially or completely unfold one wing, spread their tail feathers, and erect the neck feathers so that they resemble a collar (Fear- and Parade-position, as has been described in different birds). Often, the animals emit a singing threat sound, as if two heralds, each on his instrument, were sounding the charge. In the next stage, the hens take a few steps and approach each other, possibly trying to attack the other one during the threatening sounds. The neck feathers stick out even further. Anywhere from two minutes passes between the first and the second stage. Then the fight starts, with as many neck and tail feathers as possible being spread out. The two fighting animals try to get hold of the comb or wattles of the other one, while they constantly change their places going round and round. The legs are used little in a fight between hens, but very much so in fights between roosters. The first thrusts are usually diverted, but beak meets beak and the hens in their excitement may often jump as much as a half meter into the air. Soon there are cuts all over the head; the head is the aim. The strange thing is that the eyes are hardly ever hurt. One can almost believe that there is an agreement among the fighting hens that the eyes of the opponent are to be spared. This does not hold true for roosters. Comb, ears, face, wattles and neck can become quite bloody, but most fights result in rather small, unimportant wounds which heal quickly. If a fight lasts for a longer time it loses its rather innocent character; it then approaches the well-known cock fight, which is so popular among people with bloodthirsty and wild fantasies. Fights

---

*A similar example is told by Karl Groos (Spiele Der Tiere, Jena 1896) about tame birds. No mating took place between a certain male and female, but the male mated frequently with other females. Groos should have tried to find out whether the female, unmated by this male, ever mated with any other males.

of 15 minutes or more torture the animal to a high degree and sometimes result
in serious injury. Towards the end the animals are thoroughly exhausted and
peck at each other in a cruel way, but rather apathetically. They are so tired
that their eyes almost fall shut, but neither wants to stop. Statistically it
has been shown that fights among hens which did not previously know each other
are more frequent than fights among acquainted hens. This is not surprising,
since acquainted hens have already had their fight at first meeting, when they
decided who is going to be despot.

CHANGES IN THE PECK ORDER

We have already mentioned that the peck order in a flock of chickens can
change over a period of time. Now we want to explain how this happens. The
only way this can happen is if one hen rebels against her despot. Such an uprising of the oppressed occurs much less frequently among hens that settled the
problem of despotism at their first meeting by a fight; rebellion is more common among hens that have quietly decided who is the despot (i.e., through the
fear of one of them).

In the latter case the fearful hen may harbor a smoldering desire for a
fight and for this reason it is more likely to occur between such a hen (I) and
her despot (II) rather than between two hens which have already decided who is
the superior by a real fight.

In such an uprising it is always the oppressed hen which attacks. What can
the despot do? Either hide and flee, in which case I becomes despot (and a
change in the pecking order takes place) or II can fight back. It is extremely
rare that the despot freely gives up her leader position; usually she fights
back. Where does the second choice lead to? Either I loses her courage so
rapidly and so thoroughly that she ceases any further opposition, at least for
the time being. In this case Hen I remains the oppressed. In other words, the
situation remains as it was and no change in the peck order takes place. Or
the pecking of II does not dampen the aggression of I. What happens now? A
fight develops and whoever is winner of this fight will be despot in the future.
It frequently happens that the opposing hen wins such a fight. This is not surprising if one remembers under what circumstances II came to be tyrant over I.
It was a lack of courage which prevented I from starting a fight with II. If
I now wins, there will naturally be a change in the peck order.

I've noticed that during such an uprising hens in the place of the oppressed
I do not fight against their despot with full strength but rather exhibit an un-
expected amount of slackness, definitely more slackness than they show in other
fights; thus the chances for I to win the fight decrease. The following could
be set as a general rule: once a hen has arrived at a lower position in the
peck order, it is much harder (probably due to the condition that she has become accustomed to) for her to become despot over the other through rebellion
than it would have been at the beginning, i.e., if she had gotten into a fight
with the then unknown hen.

Rebellions such as the one described above can also come from the oppressed
against those hens which mistreated them from the time they were chicks or
against one or several hens which, at their first contact, won temporary leadership through their courage and aggression.

43

SOCIAL PSYCHOLOGY OF THE DOMESTIC CHICKEN

STATISTICS OF FIGHTS

For ten years I have kept records on the following: in their initial encounter, how often is a fight desired or not desired by either the resident hen or the introduced stranger? Following are the data:

Table 1

Observed first encounter.	Number of encounters in which the native desired the fight.	Number of encounters in which the native did not desire the fight.	Number of encounters in which the strange hen desired the fight.	Number of encounters in which the strange hen did not desire the fight.
1428	1330	98	458	970

From the 1428 observed encounters one can see that the number of occasions wherein the resident hen wanted to fight was more than 13 times that of the number of occasions in which she did not want to fight. At the same time, the number of occasions wherein the strange hen did not want to fight was twice that of the number of encounters wherein she felt like fighting.

These facts show clearly the mood of the hens at the moment of encounter. Resident birds as a rule feel secure, count on their own strength and are angry at strange hens coming into their own territory. This makes the resident hens generally bold and aggressive. On the other hand, strange hens do not know the locality, find it hard to adjust, are generally shy, and try to avoid the fight. Who is more likely to win these fights, the resident or the strange hen?

Table 2

Observed initial fights	Number of fights won by native hen	Number of fights won by strange hen
476	294	182

The data show that resident hens win 1.6 times more fights than strange hens. They also suggest that the courage of the strange hens is so diminished in the new surroundings that they usually fight with less strength than they really could.

It should be mentioned that our data are based on encounters wherein one strange hen was introduced to one resident hen. Another precautionary measure was not to stage the encounters in too close succession, since fighting itself influences the mood and strength of the experimental animals and thus affects the outcome of a fight.

Thorleif Schjelderup-Ebbe

Alpha Hens In a Fight. We have mentioned before that a fight between two alpha hens is very serious. The reason for this is that an alpha hen is not accustomed to the slightest amount of slavery. It has further been shown that strange hens occasionally win a fight against an alpha hen, thus becoming their despot, while at the same time they might lose fights against other inhabitants of the chicken yard. But this is not surprising if one considers what we have mentioned before: that an alpha hen may have achieved high position through habit or through special luck and therefore does not necessarily have to be the strongest hen in the flock.

The Influence of Age on Fights. Many observations have shown that particular ages are more advantageous for hens in starting and winning fights; the middle ages (from 1½ to at most 5 years) are the best. Very young hens (under 1 year of age) are generally less courageous even when meeting hens they know. Very old hens sometimes are frightened by other individuals and literally refuse to fight. I have found on several occasions that hens which at younger ages were quite pugnacious but later in life refused to fight, had impaired visual abilities; this seems to be one of the major factors if an older hen shows little or no desire to enter into a fight. To see one's opponent fuzzily cannot possibly heighten the desire to fight. The disinclination to fight can also possibly be caused by the degeneration of the reproductive organs and the change of the psychic condition which goes along with such degeneration.

EXPERIMENT WITH WIRE MESH

On several occasions I have carried out the following experiment in order to further investigate the nature of the peck order: two flocks of hens, unacquainted with each other, were not mixed together but instead put into two separate pens that were separated by a chicken wire fence. The hens in both flocks were able to see each other clearly, but no hen could get into the other pen.

What happened now? A few of the hens assumed fighting posture towards the strangers on the other side of the fence and emitted threat sounds. Several hens from both sides attacked each other and some fights broke out between pairs (one hen from each pen) even though the chicken wire fence separated the two parties. The fighting hens pecked at each other through the mesh of the chicken wire fence. Some of the pecks were very well aimed; the hens jumped up and blood frequently flowed. Nevertheless, these hens, fighting with the wire fence between them, could not really get hold of each other and most fights remained undecided, because after a while the opponents tired and walked away without either one of them having forced the other to stop the fight. Frequently, they even threatened the other as they walked away. Occasionally they would start the fight over again, after several hours or even days. Once the hens realized that they could not peck properly through the wire fence, they stopped their threatening sounds and gestures and took no great notice of each other. There were also some hens that never even started a fight through the fence; they were indifferent to the hens in the adjacent pen. It was noticed that apparently some hens, through their appearance or behavior, had a more irritating and fight-provoking effect on hens in the other chicken yard. Hen No. 3 entered into five fights, No. 1 into four fights, Nos. 2, 7, 9, VI, IX and X had two fights each, and Nos. 4, 6, 12, II, VII, XI and XII had one fight each. This shows clearly that No. 3 was the one hen that irritated most animals in chicken yard a.

Figure 4 illustrates the behavior of the animals in two such pens separated by a fence of chicken wire. The animals in pen a (left) are numbered with Roman numerals and the animals in pen b (right) with Arabic numerals. I observed the animals for four days all day long and every time I left the yard (for meals and at night) I locked the chickens from pen b into a chicken house to the right of pen b so that their behavior toward the hens of pen a would not be unobserved for even a moment. Hens in the chicken house could not see the hens in pen a nor could they be seen by pen a birds. The two pens, a and b, covered an area of about 20 square meters. This relatively small size was chosen to make sure that all hens could see all others at all times.

The hens whose numbers in Figure 4 are connected by dotted lines entered into a fight during the time of observation. One can see that in the whole time, 14 fights occurred. Theoretically, a maximum of 144 fights would have been possible if each hen in one pen started a fight with every hen in the other pen. However, this high rate of fighting was not expected since we have already mentioned that some hens avoid fights, but more than 14 fights would certainly have occurred if 12 hens would have met 12 strange hens without the dividing chicken wire fence between them. <u>The fence seems to have had an inhibiting influence on the fights</u>.

Figure 4

Of the 14 fights, 12 took place on the first day, 2 on the second day, and on the third and fourth day there were no fights at all. Fights were most frequent at the outset but after a relatively short time the hens became rather indifferent towards each other. Were any of the 14 fights decided? No, they all remained undecided and among the hens which did fight, the fighting stage passed into a more passive stage during which the two opponents uttered threat sounds. The other hens that did not fight behaved in the following manner: (1) <u>Either</u> they more or less threatened one another with sound or gestures but without actually entering into a fight, <u>or</u> (2) one of them threatened the other but not the reverse, <u>or</u> (3) neither of them threatened.

After the four days had passed, the chicken wire fence was removed and the hens were brought together in pairs, one from each pen. It became evident that the animals did not decide their fights <u>through the fence</u> and <u>no order in their society had been established</u>. A few terrible and many minor fights took place. Now all the hens that had had an undecided battle across the fence started fights; <u>they fought all the way to the end and then stopped</u>. Fights were also carried out between animals which only threatened each other through the fence. In those situations where one hen threatened another who did not respond to the threat, the first hen (the one that threatened) became despot over the other without a fight, with one exception where a fight did occur. Among the chicken pairs which did not threaten each other at all, one individual became despot in a very short time <u>without a fight</u> taking place. We therefore conclude:

1. As long as the hens are separated by a wire fence, fights can occur but they remain undecided and the hens stop trying to establish a peck order.

2. That the fights which occur through a fence remain undecided as proven by the fact that the hens taking part in those fights at the outset (and subse-

quently settling down), take up the fight immediately after the fence has been removed and continue fighting until one of the opponents is defeated.

3. Hens which did not start a fight through the fence prove through their behavior after removal of the fence that rank and social order was settled merely by the animals seeing each other without any chance for physical inter-action. The trial of strength is thus the decisive factor in establishing a rank order in a society of chickens.

4. Hens which only threaten each other through the fence enter into a real fight after removal of the fence and thus behave like hens which already fought through the fence.

5. Hens which threaten others through the fence without receiving a response (with one exception*) become despot over the others.

6. Of the hens which ignored each other when the fence was present, one becomes despot over the other without a fight when the fence is removed.

These different points, which are the result of the above experiment, have been confirmed over and over again by repeated experiments. They seem to be valid quite generally and also apply to flocks of roosters.

It is suggested above that a fight is rarely decided if a fence separates the two chicken pens. However, I have observed that a fight can be decided, especially among roosters, if one of them either gets a good grip on the other through the fence or pecks him so hard and so painfully that the other becomes frightened and considers himself defeated. If they meet again after the fence has been removed the defeated bird avoids the other one without a fight and considers him to be the despot. These observations confirm the statement that chickens cannot live in the same chicken yard together without establishing a rank order.

Chapter 3

THE PERIOD OF BROODINESS AND ITS INFLUENCE

Does the state of broodiness, which is the most important time in the life of a hen, have an influence on the social position of the brooding hen? Does the social position remain unchanged or not? These questions cannot be answered broadly; many circumstances influence the answers. When a hen which lives to-gether with others in a flock becomes broody, the first thing that becomes noticeable is the fact that she remains on the nest for some time after laying an egg and does not like leaving the nest. She may cluck somewhat when entering the chicken yard, but not much at the outset. The amount of clucking increases steadily and shortly thereafter (about 4 days) the hen does not leave the nest except for 5 to 10 minutes once or twice a day. Even at night she remains on the nest. The broody hen is somewhat more irritable than usual when she is on the nest or in the chicken yard. She is more irritable towards the hens that she dominates and at the same time more likely to rebel against hens that mis-treat her.

The broody hen is restless and nervous out of doors; she pecks at food with great fervor and eagerly bathes in the soil but only for a short period of time.

---

*This exception can be explained as a sudden desire to revolt in an individual which should have been subordinate.

SOCIAL PSYCHOLOGY OF THE DOMESTIC CHICKEN

The reason for the restlessness of the broody hen, for not taking time for anything is without a doubt her desire to get back to her nest. She hurries back
to the nest, sometimes running (sometimes the hen dashes away from dust bathing
as if she has suddenly remembered something), gets back on the nest and resumes
incubation. The broody hen can be displaced from the nest by much hard pecking
from her despot, but she holds out much more steadfastly than usual and lets
herself be pecked much more before she retreats from the nest. Some hens will
not let their despots chase them from the nest. They will not fight back no
matter how hard they are being pecked but try only to protect their heads as
well as possible and to remain quietly on the nest. The despot finally stops
the pecking and sits down next to the subordinate to lay her egg or lays it
next to the nest or in some other nest.

There are strong individual differences in broody hens in relation to their
despots. Occasionally, as we have mentioned before, a broody hen may rebel
against one or more of her despots. This results inevitably in a fight and it
is not always certain that the broody hen will win the fight. In many cases
the opponent defeats the broody hen after a short time and the broody hen retreats full of dejection.

What we have said so far is true for hens that become broody while remaining
in the chicken yard and house that they are accustomed to and where they can
arrange everything to their liking.

The broody hen behaves quite differently though if one removes her from her
familiar surroundings and puts her into a nest with eggs in a room all by herself, where for several days (four is usually sufficient) she does not have an
opportunity to meet other inhabitants of the chicken yard. If she is then put
together with one or another of her earlier despots, she will almost without
exception be ready for a fight. There usually ensues a major fight because the
despot is clearly unwilling to give up the control she had over the broody hen.
At the same time the latter is usually quite fearless. Either both opponents
attack now or the broody hen throws herself at her opponent with full force.

The broody hen does not always win on such occasions but she does so in
most instances. If the opponent is not one of the "heavy" breeds of chickens
and the broody hen is of the lighter breeds, only rarely does she lose. If she
wins the fight she is usually quite angry at her opponent and follows the former
despot as long as possible even after it has given up the fight and flees head
over heels. The defeated hen must then be removed in order not to be damaged
too much. If the broody hen is defeated in the fight, she becomes very submissive, lowers her head and flees from her opponent who follows her but is
usually much more civilized than she herself would be.

If one puts an isolated broody hen together with hens that she formerly
dominated, she will without exception get ready to fight and without exception
fight if the others are willing to do so. Generally these are the hens (which
I will call X hens here) that avoid the broody hens because they haven't quite
forgotten their old tyrant. Only very rarely will such X hens get into a fight
with a broody hen. The fact that occasionally X hens dare to enter into a fight
with a broody hen must mean that they don't recognize her with certainty but
think they are confronted with a strange hen with whom they want to measure
their strength.

Very rarely does an X hen win over a broody hen. Out of 305 observed
contests, the broody hen won 303. This must certainly have a psychological
reason. The X hens who enter into a fight with a broody hen feel perhaps un-
clearly, even if they don't really know it, that the broody hen was their former

despot, that they are probably going to lose the fight and therefore don't bother to use their total strength. Further, one has to consider that the broody hen enters all these fights with great strength and endurance. Many observations have shown that a broody hen will almost always enter into a fight with a totally strange hen if she is not placed in a strange surrounding. But even if they are placed in a totally unfamiliar surrounding, broody hens with a violent temperament start fights with total strangers.

The broody hen is most likely to enter into a fight if she is on or in close proximity to the nest. If the broody hen is some distance from the nest, one should bring the other hen to her immediately since some broody hens become very restless after a short absence from the nest; they wish only to get back to the nest and lose their courage to such an extent that they refuse to fight even when they are being attacked.

After the isolated broody hen has been sitting on the nest long enough for the chicks to have hatched, she is still very pugnacious if she has her chicks with her. However, if one takes the chicks away from her for even a short period of time (a few minutes, for example), some hens become very upset, long to get back to their chicks and complain bitterly. Thus, a hen removed from her chicks can be attacked by others without defending herself. In this case the others will become despot over her.

If a hen is removed from her chicks or from the nest and in this state avoids or is attacked by others without defending herself, the other hens almost invariably remain despot over the broody hen when she is brought back to the nest or to her chicks. She is usually fearful of these hens when she walks her chicks about, and is closely followed by her chicks as she avoids or flees from the other hens. She does not even defend her chicks if one of these enemies attacks them, something she would have done vigorously if they would have been attacked by hens she did not meet during her nervous period. This is another proof that the pecking among hens depends to a large extent on habit.

One of the prerequisites in all of these experiments is that the chickens are tame, that the experimenter knows how to handle them, is able to keep calm and to interfere only at the right moment; all these things can be learned only through experience.

2

Reprinted from *Auk*, **51**(3), 306–327 (1934)

THE SOCIAL ORDER IN FLOCKS OF THE COMMON CHICKEN AND THE PIGEON.[1]

BY RALPH H. MASURE AND W. C. ALLEE.

THE problems centering about the organization of flocks of birds are by no means solved despite the attention they have received. A portion of the pertinent literature to date has been summarized by Allee (1931, 1934) both with regard to the observed facts and their general social implications; hence no general literature survey will be undertaken now. It will suffice to call attention to certain work which is directly antecedent to ours. Schjelderup-Ebbe (1922) has analyzed the organization of flocks of Domestic Chickens and of both wild and tame Ducks (1923). In all of these he found a more or less definite organization revealed by the way in which the birds reacted in contact situations. He recognized a so-called peck-order in which the animal highest in the order pecks and is not pecked in return while that at the extreme bottom of the order is pecked without pecking in return. Throughout the entire order, any individual with the peck-right over another remains steadily dominant over it until by a combat their positions are reversed.

More recently (1931) Schjelderup-Ebbe has extended his observations to include a large number of different sorts of birds both in nature and in various kinds of confinement; he finds that when two birds of one species are together, one is despot and the other is subservient. Schjelderup-Ebbe believes that this sort of despotism is one of the fundamental principles of biology.

The older bird of a flock is usually despot because her matured body gives her strength which the young, partially developed birds lack; even after the latter attain their full size and strength, if of the same sex, the older individual maintains her despotic rights. Between the sexes, the larger males are usually despots over the females. When the two sexes are alike in size and strength and the

[1] The work upon which this report is based has been supported in part by a grant from the Rockefeller Foundation to aid investigations in the biological sciences at the University of Chicago. We are indebted to Dr. L. V. Domm for permission to study the group organization of Chickens in certain of the pens under his control and for two post-mortem examinations.

male possesses ornamentation, he is despot; otherwise either may be despot. Often males put on and lose their despotic rights with the assumption and loss of breeding plumage.

We have repeated Schjelderup-Ebbe's observations first on Brown Leghorn Chickens, both males and females but with sexes separate; then with flocks of Pigeons, first with sexes segregated, then when mated, and again with the sexes together. The observations were the result of joint planning and while the senior author was in very close touch with the work throughout, the actual observations were all made and the preliminary summary prepared by the junior author.

Two flocks of Brown Leghorn Chickens were observed in their Whitman Laboratory quarters during January, February and March, 1932. They were housed in pens 14.5 feet long, 5 feet wide and 8 feet high. These pens were in a heated house that was kept between 60° and 70° F. They were each provided with raised roosts which occupied a corner space of 3 by 4 feet. The floor below the roosts was protected from droppings. The cement floor was covered with straw litter which was changed weekly. On a few warm days, the Chickens were given access to separate outside runs which were about 5 feet long and were floored by wire mesh about a foot from the ground to prevent attacks by rats.

All the Chickens were about 10 months old when observations were begun. They had been raised together in the Whitman pens since hatching. One pen contained 11 cockerels and the other at the start contained 26 pullets. Toward the end of January, 12 days after the observations were begun, half of the pullets were removed and on February 27 a Barred Plymouth Rock rooster was added. The flock of cockerels was disturbed only by the experiences of one bird. On February 29, RB was removed because his eye was injured in fighting; he was returned to the flock on March 14 when his eye was completely healed and sight was normal; he was again removed five days later in a much battered condition.

All these Chickens were fed twice daily with mixed grain and in addition the cockerels were given a small amount of grain when the observations were being made, since relatively little pecking was done in the absence of food. In addition to the grain, the pullets were provided with a hopper of mash which remained in the pen constantly. The cockerels received no mash since by omitting it most of their fighting was avoided.

The majority of the observations on contact reactions were made in the afternoon. The observer sat quietly in an adjoining pen, note pad on knee, and was separated from the Chickens under observation only by ordinary chicken-wire netting. When two animals came into contact with each other and one was pecked and retreated, this was considered to be the subservient member of that particular contact pair. Many times two Chickens would meet without showing any signs of pecking or being pecked; such neutral encounters were not considered in deciding on the peck-order. Often such contacts would start a battle, more often with the cockerels than with the pullets; pecking would be mutual. In these cases the Chicken that gave in and retreated was considered to be subservient to the other. The majority of the pecking contacts were observed over and over. Colored celluloid leg bands furnished a ready means of individual identification by the observer; the birds themselves appeared to recognize other individuals by means that were not always apparent to the human observer.

SOCIAL ORDER AMONG THE PULLETS.

The social order in the flock of 26 pullets had not been determined when, twelve days after the observations began, the flock was reduced to 13. The definite status of any one hen had not been established but the evidence at hand allows a fair approximation of the birds in the upper half of the flock and extreme lower part as follows:

1. W	7. GY	13. WW
2. B	8. BY_2	22. Y
3. RW	9. G	23. BB
4. GY_2	10. RY	24(?)BY
5. RY_2	11. RG	25. M
6. RR	12. BG_2	26. A

Although the observations were insufficient to establish the contact reactions between any one bird and all of its associates, the data at hand show that there was no absolute despot. One bird, GG, was not observed to receive pecks from another individual but it pecked only A and YY during these observations. YY ranked below the 13 just given and A ranked at the bottom of the indicated order in the larger flock and at the very bottom of the better tested order of the reduced flock: hence it appears that GG cannot be assigned a place in the upper half of the peck-order.

In all the contacts observed in this crowded pen, there was only one reversal. W was seen to peck RY_2 once and the reverse was observed once also. This indicates a strong and stable social organization among the pullets. Pullets of inferior position always gave way to their superiors at the food hopper or water dish. Superiors were often observed pecking food from the bills of their inferiors. M, A, and BB, individuals low in the peck order, spent much of their time on the roost where there were fewer birds and when they did venture onto the floor, they were alert and avoided many pecks by darting away from an approaching hen.

After the flock was reduced to 13 individuals, it was observed for 60 days and the complete order obtained as shown in Table I.

TABLE I. Showing the Status of Each Individual in a Flock of 13 Brown Leghorn Pullets on the Basis of the Peck-order.

RW pecks all 12: A, BG, BB, M, Y, YY, BG₂, GR, R, GY, RY, RR.
RR pecks 11   : A, BG, BB, M, Y, YY, BG₂, GR, R, GY, RY.
RY pecks 10   : A, BG, BB, M, Y, YY, BG₂, GR, R, GY.
GY pecks 9   : A, BG, BB, M, Y, YY, BG₂, GR, R.
R pecks 8   : A, BG, BB, M, Y, YY, BG₂, GR.
GR pecks 7   : A, BG, BB, M, Y, YY, BG₂.
BG₂ pecks 6   : A, BG, BB, M, Y, YY.

YY pecks 4   : A, BG, BB, M.
M pecks 4   : A, BG, BB,   Y.
Y pecks 4   : A, BG, BB,   YY.

BB pecks 2   : A, BG.
BG pecks 1   : A.
A pecks 0

It is apparent from the data of Table I that RW is the despot of the flock and that down to YY there is a straight line order; then YY, M and Y form a triangle order below which the straight line order continues to A, the lowest member which pecks none. As in the larger flock, superiors often ate food from the bills of their inferiors without resistance from the latter. When a hen was sitting on even one or two eggs, it took much more pecking before she would relinquish her position to the superior hen; even so, no reversals in peck order for this or other reasons were observed in the reduced flock. RR, second in the social order, was much given to pecking her inferiors, more so than RW, the ranking hen. The impression of a social order gained in the observations on the large flock is definitely strengthened. The reduction in flock size did not change any of the observed peck-rights. R, which stood fifth in the social order of the reduced flock, had not been observed in contact with any of this group before the reduction took place; it had been seen to peck G and BR and to be pecked by W and RY_2.

SOCIAL ORDER AMONG THE COCKERELS.

Practically the complete order for the flock of eleven cockerels was determined during the 70 days of observation. The findings are summarized in Table II, which is built on the same plan as Table I, and so allows ready comparison with the social order obtaining among the pullets. In this table the characters in italic indicate that these peck-rights were not settled; those in heavy faced type indicate that there was one reversal of the peck-order observed in each of these cases.

TABLE II.   SHOWING THE SOCIAL ORGANIZATION OF A FLOCK OF 11 BROWN LEGHORN COCKERELS ON THE BASIS OF THEIR PECK-ORDER.

```
BW pecks 9 : W, BY, G, RY, B, BG, Y, R, GY.
BR  pecks 8 : W, BY, G, RY,    BG, Y, R,     BW.
GY  pecks 8 : W, BY, G, RY, B, BG, Y,             BR.
R   pecks 7 : W, BY, G, RY, B, BG,       GY.
Y   pecks 6 : W, BY, G, RY,    BG,    R.
GB  pecks 5 : W, BY, G, RY, B.
B   pecks 4 : W,     G, RY,       Y.
RY  pecks 3 : W, BY, G.
G   pecks 2 : W, BY.
BY  pecks 2 : W,           B.
W   pecks 0.
```

In this order there are six triangle situations as follows:

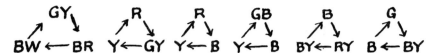

These triangle situations run almost through the whole gamut of the social order. Those high or low in the peck-order are least involved and those in the middle of the order are most concerned. BW, the highest ranking cockerel is involved in one triangle; W, the lowest, in none; B, one of those near mid-rank is involved in four and Y, in three.

The social organization of this flock of cockerels was not as stable as was that of the pullets. After 70 days of observation, peck-rights were still unsettled as indicated in Table II by the letters in italic. They were those of BY–R and BY–G. BY was observed to peck G on six occasions and the reverse was observed on eight occasions. BY was likewise observed to peck R on six occasions and the reverse was true for the same number of contacts. The place of these cockerels must therefore be assigned tentatively and they are placed in their logical positions with reference to other relations which seemed to have been settled.

Four other cases of reversal were observed and are indicated in Table II by heavy faced type. In these the reversal occurred but once in each case. It is of course possible that there may have been an error in observation but the likelihood is no greater here than with the pullets and we are inclined to think that these reversals indicate a less stable and fixed peck-order in the cockerels as compared with the pullets.

On account of a fight which incapacitated BR, other consequences of which will be given later, this individual was removed before the peck-order between him and B had been observed. The other relationships indicate that B would probably have been pecked by BR, but one cannot be certain of this without direct observation.

The cockerels were much more given to fighting than were the pullets. Many times two of them would face each other and start to fight even though the peck-right records showed that the relations between the two were fairly definitely settled. Usually such combats were interrupted by another individual walking between the two combatants, but if the battle was not interrupted, the cockerel which was usually pecked, finally retreated, but not, perhaps, until he had put up quite a fight. The individuals in the lower positions of the social order have a difficult life; they are continually being pecked by their superiors. When food was given, W, which stood lowest in the order, never ventured near those cockerels that were eating but waited until they had left the food and then ate what remained. W also spent most of his time on the roost where the others, if present, seemed to be more tolerant of their inferiors. As a rule very little pecking took place among the cockerels on the roost even though they were somewhat crowded at times. When W did venture on the floor of the pen, he was constantly running to avoid contacts with the other members of the flock. BW, which was first in the order, seemed more inclined to peck his inferiors than did any other cockerel; about half of the observed pecks were delivered by this individual.

The order among cockerels differs from that found among the pullets in the type of organization. That of the pullets is fairly regular while that of the cockerels is built about triangular relationships. For a time there was no despot among the males and in fact BW, the cockerel standing highest, was pecked by BR who stood just below him in rank. On February 29, this latter individual and Y started to fight, as they had many times before, but during this battle BR received a hard peck in the eye which closed the left eye and he retreated. He was then removed to another pen and two weeks later on March 14 he was replaced with the group of cockerels which he had almost dominated. At this time his eye was entirely healed and his sight was apparently normal; he was found to have lost his position in the peck-order completely. In fact he now stood lowest of all and was even pecked by W, which had not been observed to peck a fellow cockerel before. One or all of the following factors may have entered into his loss of status: he had lost an encounter; he had sustained a severe injury and he had been absent from the flock for 14 days. Whatever the reason, he was persecuted so badly that on March 19 his injuries prevented him from standing and he was removed permanently to save his life.

During the five days BR was in the pen after his recovery, he avoided contacts with the others as much as possible and spent much of his time partially hidden under a low shelf upon which the water dish was kept. Not once was he observed to attempt to assert his former high position. After BR was removed no change was observed in the pecking. The absence of BR made BW despot. This change in the relations of BW did not seem to affect his behavior towards the other members of the flock.

OBSERVATIONS ON PIGEONS.

The observations on Pigeons were made upon 14 White King Pigeons, a heavy squab-breeding variety. The birds were equally divided as to sex. They were unmated but sexually mature when obtained from a reliable dealer. They were housed in a large out-of-doors pen about 30 ft. long by 18 ft. wide and 10 ft. high. The pen extended between a laboratory greenhouse and a brick building. It was closed by wire on both sides and on top. A small wooden roost and nesting cote was furnished and both runways and cote were divided into two equal parts by wire netting and boards during the periods when the Pigeons were sex segregated.

The majority of the observations were made at feeding time since the close contacts incident to feeding greatly increased the opportunities for pecking. The group of males was observed for 24 days and the females 30 days. After these observation periods, which ran concurrently, the partition was removed and the two groups were allowed to be together.

Soon after the observations on the sex segregated groups were started, two of the females simulated mating. One, GW, took the part of the male and the other, BW, that of the female.[1] Twelve days after mating both of the birds laid eggs which were removed. During the month that the sexes were together, five pairs mated; three of these matings resulted in fertile eggs that were hatched out before the sexes were again segregated for a second observation period that lasted for 28 days. RY and BY of the females and B and Y of the males did not mate.

OBSERVATIONS ON MALES AND FEMALES WHEN COMBINED.

When the sexes were united into a common flock, five pairs of the

---

[1] A common occurrence according to Whitman, 1919, p. 28.

Pigeons soon mated. Thereafter practically all the pecking was done by RY and BY, the two unmated females. The males were stimulated by the females and chased them almost continuously when they were not on the nests. Under these conditions the mated females were constantly running away from the pursuing males, eliminating most of the contacts between males and females, particularly between those belonging to different pairs. The resulting picture is of a flock composed now of couples, each couple with but few contacts with other couples.

The two unmated males were in about the same social position as were the mated ones. Both were stimulated by the presence of the females and were continually courting them. BY and RY, the two females that did not mate, were killed at the end of the post-mating observations and examined for any abnormalities of gonads which might show them to be intersexes. On gross examination by an experienced worker[1] both of these birds showed normal female gonads.

THE SEX SEGREGATED GROUPS.

Neither the segregated males nor the females showed a definite peck-right such as has just been reported for Chickens and such as Schjelderup-Ebbe reported for Chickens and Pigeons and many other birds as well. In our pens both the sex segregated groups showed a relationship which we shall call peck-dominance. Where two individuals peck back and forth, now one retreating and now the other, the one that is observed to retreat the fewest times is said to have the peck-dominance for that particular contact-pair. In only a few of the relationships observed was there a definite peck-right in which one of the contacting individuals does all the pecking and the other does all the retreating. The more usual relationship with these Pigeons was to have the pecking frequently shifting from one to the other of any given contact-pair of birds. The interval between such reversals varied from a few minutes to several days. It was not uncommon to see one bird being chased by another at the beginning of an hour of observation and itself chasing the former dominant before the end of the hour. To be counted as an actual reversal the temporarily subordinate indi-

---

[1] The examinations were made by Dr. L. V. Domm.

TABLE III. PECK-DOMINANCE IN A FLOCK OF SEVEN FEMALE PIGEONS.

Pigeons	Pre-mating period				Post-mating period			Grand Total
	14 days	20 days	26 days	32 days	16 days	22 days	28 days	
BY:BB	15:11	23:17	30:34	36:38[1]	5:7	5:9	7:11	43:49
BB:GW	10:11	12:19	14:28	15:37	7:22	9:44	10:49	25:86
BB:BW	1:6	3:12	6:26	9:41	11:22	11:25	11:30	20:71
BB:BR	1:7	2:14	3:20	3:26	0:6	1:8.	1:10	4:36
BB:RY	1:6	2:12	7:27	16:47	42:17[2]	48:21	53:22	69:69
BY:GW	9:12	9:21	10:33	17:41	5:20	6:28	7:39	24:80
BY:BW	2:8	4:19	5:35	9:42	0:10	2:16	6:24	15:66
BY:RW	5:10	8:14	9:25	11:37	2:16	4:33	4:50	15:87
GW:BW	1:7	2:8	2:8	4:13	3:9	7:12	8:14	12:27
BR:GW	6:3	7:6	7:7	9:10[1]	2:21	3:21	4:21	13:31
GW:RW	11:9	3:10	5:10	6:12	2:8	3:9	3:10	9:22
BY:BR	1:3	14:15	15:20	17:24[1]	0:16	0:21	0:31	17:55
BR:BW	2:6	2:7	4:8	7:10	1:0[2]	3:1	5:2[3]	12:12
BW:RW	1:2	2:6	4:6	6:8	1:0[2]	2:3	2:6[3]	8:14
RW:RY	1:3	1:4	3:6	10:13	41:0[2]	48:2	53:3	63:16
RW:BR	2:0	1:3	1:3	3:4	0:1	0:1	0:2	3:6
RY:GW	2:0	4:1	5:4	7:10[1]	9:47	11:49	11:49	18:59
RY:RW		3:1	7:4	8:10[1]	0:39	0:46	0:51	8:61
BY:RY	0:2	0:7	2:22	20:33	68:13[3]	76:16	83:18	103:51
BR:RY	0:3	0:3	2:5	8:7[1]	33:1[2]	36:3	40:4	48:11
BB:RW	0:5	0:6	0:18	0:23	1:10	1:11	1:12	1:35

Order of Dominance.

14 days	20 days	26 days	32 days	16 days	22 days	28 days	Grand Total
BB(6)	BB(6)	BY(6)	BY(6)	RY(6)	RY(6)	RY(6)	RY(5½)
BY(4)	BY(5)	BB(5)	BB(5)	BY(5)	BY(5)	BY(5)	BY(5)
GW(4)	GW(4)	GW(3)	RY(3)	BB(4)	BB(4)	BB(4)	BB(4½)
BR(3)	BR, BW,	BR, BW,	GW, BR,	GW, RW(2)	GW, BW(2)	GW, BW(2)	GW(2)
BW, RW(2)	RW(2)	RW(2)	BW(2)	BW, BR(1)	RW, BR(1)	RW, BR(1)	BW, BR (1½)
RY(0)	RY(0)	RY(0)	RW(1)				RW(1)

[1] Peck-dominance changed in pre-mating observations.
[2] Peck-dominance changed during mating period.
[3] Peck-dominance changed in post-mating observations.

vidual must actually retreat from the attack of a bird from which
it had formerly retreated: merely pecking back and forth was not
so considered. Such mutual peckings were frequently observed.
The individual that did the pecking, whether male or female,
usually showed a swelling of the crop, cooing and bowing not unlike the mating behavior.

The actual observations are summarized in Tables III, IV and V.
The longer tables were constructed as follows: As with the Chickens,
the individuals are represented by letters and are arranged at the
left of the table in the order that indicates the relative dominance
of the different individuals at the end of the first period of observation. The more dominant individuals are listed to the left in any
given pair. The successive columns give the number of times of
observed peck-dominance at the end of the indicated number of
days from the beginning of observations. Each table is divided by
the period in which the birds were mated into a pre- and a post-
mating period.

Some of the observed relationships show only slightly greater
peck-dominance on the part of one bird than the other of a given
pair. Obviously great significance cannot be attached to these
cases. Some measure of their significance is given by the regularity
with which the same bird continues to be dominant in its relations
with another individual.

THE FLOCK OF FEMALES.

In the relations between different females which are outlined in
Table III, of the 21 possible contact-pairs 16 showed the same
individual had the greater peck-dominance during the 32 days of
the pre-mating period. During the mating period, 6 reversals of
dominance occurred, two of these, those in which RY was not concerned, showed another reversal before the end of the post-mating
season observations. In all, 11 of the 21 possible contact pairs,
showed at least one reversal during the entire time of observation.
Three of these pairs showed two reversals during this time. The
reversals that occurred during the pre- or post-mating observations
were all shown by contact-pairs in which there was but slight
difference in the peck-dominance; changes during the mating
period occurred in the case of RY even though the peck-dominance
seemed to have been firmly established.

The position of this Pigeon in the social organization shows some of the elasticity of the system. For the first 26 days of observation, RY stood at the bottom of the peck-dominance order. In the last six days before mating, RY became slightly dominant over BR, GW and RW, all of which stood low in the social order. After the month of mating during which RY remained unmated, she was clearly dominant over the whole group. This change in the position of RY accounts for four of the six reversals that took place during the mating interval. The summary of the pre- and post-mating season position of this bird in the flock is of value and is given in Table IV.

TABLE IV. THE PRE- AND POST-MATING SEASON POSITION OF RY WITH RELATION TO THE OTHER MEMBERS OF THE SAME FLOCK.

Pairs	Pre-mating $\overrightarrow{}$ $\overleftarrow{}$	Post-mating	
		16 days $\overrightarrow{}$ $\overleftarrow{}$	28 days $\overrightarrow{}$ $\overleftarrow{}$
RY:BR	7:8	1:33	4:40
RY:BY	33:20	13:68	18:83
RY:RW	8:10	0:39	0:51
RY:GW	11:10	9:47	11:49
RY:BW	13:10	0:41	3:53
RY:BB	47:16	17:42	22:53
RY:ALL	115:74	40:270	58:329

In the 32 days of observation of the pre-mating flock of females, RY was dominant about two-thirds as many times as it was subordinate; in the 28 days of observation of the post-mating flock composed of the same individuals, RY was dominant over five times as frequently as it was subordinate. In each of its pair-contact relationships, in the post-mating flock, it clearly held peck-dominance without having absolute peck-right except with RW.

The other important changes in the peck-dominance order following the mating month were apparently a result of RY's move to power. BY, the other Pigeon which did not mate, had been the most dominant of the lot during the last half of the pre-mating period; it now ranked second to RY. BB likewise sank one place in the social order from second to third.

The habits of the birds may control their position in a social order arranged on the basis of total number of pecks received and administered. Thus in the pre-mating flock BB always seemed to do the pecking when she was at the entrance of the roost: when BY tried to enter she would be pecked and would retreat. On the ground, however, BY was usually dominant. As long as BY tried to enter the roost during the day time, she lost in a majority of her contacts with BB and the latter stood at the head of the social order. Later BY did not try to gain entrance to the roost except at night: rather she stayed on the ground where she was usually dominant. This change in the behavior of BY during the first period of observation, shifted the social dominance from BB at the beginning to BY just before the mating period.

At the beginning of the observations RY was very peaceful: she gradually became more aggressive, and as stated before, became despot of the post-mating period flock. As despot, she usually stood in the pan of food at feeding time and drove the others away between times of taking mouthfuls of food.

The social relationships of the subordinate members of the female flock of seven Pigeons are shown in diagrammatic form in Fig. 1. Fig. 1A brings out some of the intricate relationships that may exist in such a flock. The relationships existing in the post-mating flock (Fig. 1C) and when all the information collected is considered together (Fig. 1B) are more simple.

Individuals differed widely in the number of contacts with other birds. In general, Pigeons low in the social order had fewer encounters with other members of the flock than did those of high social rank. If all the observed contacts in both observation periods are considered, the three birds ranking highest had 1407 observed pair-contacts, while the three ranking lowest had only 822 such contacts. Those high in the social rank met each other more frequently than did those ranking lower. Again using all the observations at hand, RY, BY and BB had 384 pair-contacts with each other while in the same length of time, RW, BW and BR, low in the social scale, were seen to have only 55 such contacts between themselves. BY and RY, leaders respectively in the pre- and post-mating flock, met as a pair 154 times, while RW and BR, low in the social scale, with otherwise equal opportunities, met only 9 times.

The relationships existing between the total number of pair-

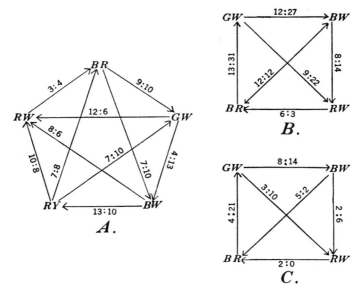

Figure 1.   Relationships among those lowest in the social order in the flock of females.   Arrows point to the subordinate birds.   Numbers show the observed times of dominance and subordination.   Thus BR dominated GW 10 times and was subordinate to GW 9 times in their pair-contacts.

A. The lower five birds in the premating flock: all these were dominated by BY and BB.

B. The lower four birds for the entire period of observation; all of these were dominated by RY, BY and BB.

C. The lower four birds in the post-mating flock; all these were dominated by RY, BY and BB.

contacts and place in the social order is again illustrated by the behavior of RY before and after the mating period.   In the pre-mating flock when she was low in social rank, RY had in all 189 pair-contacts, an average of 5.9 per day.   In the post-mating flock in which she occupied the highest social rank, she engaged in 387 such pair-contacts, an average of slightly less than 12 per day.

THE FLOCK OF MALE PIGEONS.

The pair-contact behavior of the flock of male Pigeons is summarized in Table V in practically the same manner in which the behavior of the females is summarized in Table III. In the 24 days of observation in the pre-mating flock, the same individual retained dominance in 17 of 21 contact-pairs. In one contact-pair, W:B, the few contacts observed were exactly even at each summarizing period. There were five reversals during the mating period; two concerned pairs which had already shown a reversal of peck-dominance and one other was in the case of the pair just mentioned in which there was a slight change in the hitherto balanced relationship. For the entire period of observation, 15 of the 21 contact-pairs of males showed the same individual dominant at each of these summarizing periods.

The general picture is of a more stable flock organization than was found among the females where, however, the social arrangement was upset by the rise of RY. A further evidence of greater social stability is found in the lack of reversal of dominance in individual contact-pairs in two cases in the pre-mating, and six cases in the post-mating flock and two cases for the entire period of observation. Among the females, there was one such case in the pre-mating, one in the post-mating flock and none that held so throughout the observations.

As with the females, the change in peck dominance during the pre-mating period did not concern pairs which had shown a high degree of uniformity in dominance. In the mating period something happened to BL, which had ranked near the top in the pre-mating flock; thereafter, it ranked below Y, YY and G. Its relations with G had not been clearly defined in the pre-mating period; it had definitely dominated YY and had dominated 15 out of 24 observed contacts with Y. This change in status of B1 accounted for three of the five shifts during the mating interval. B and Y remained unmated as did RY and BY of the females. Y's standing in the flock improved following the mating period while B remained at the bottom of the social order.

The peck-dominance which had been somewhat confused in the pre-mating flock (Fig. 2A) became a regular sequence in the post-mating period with Y having peck-dominance over all the rest and with B showing no regularity in dominance in pair-contacts. As with the females, some of the males showed a greater tendency toward dominance in certain spatial positions. YY stood higher

TABLE V. PECK-DOMINANCE IN A FLOCK OF SEVEN MALE PIGEONS.

Pigeons	Pre-mating period.			Post-mating period.		Grand Total	Order of dominance.
	12 days	18 days	24 days	16 days	22 days		
BL:G	10:8	14:12	19:19[1]	15:6[2]	22:9	41:28	
G:W	4:8	4:11	5:14	1:38	1:41	6:55	
G:Y	8:8	12:14	15:20[1]	13:12[2]	21:13	36:33	
G:B	3:8	3:14	4:16	0:5	0:9	4:25	
BL:W	2:8	2:12	3:19	3:9	4:13	7:32	
BL:YY	4:13	7:30	18:30	21:6[2]	28:8	36:38	
BL:Y	4:7	6:11	9:15	22:2[2]	26:4	35:19	
W:YY	13:18	22:20	28:23[1]	2:11	25:3	53:26	
R:W	7:10	7:16	8:22	5:16	2:12	10:34	
YY:R	5:20	7:28	13:45	20:41	8:21	21:66	
YY:G	8:11	11:30	17:47	1:5	24:51	61:98	
R:B	1:15	1:20	2:22	13:28	1:5	3:27	
G:R		12:18	20:28	8:26	16:30	36:58	
BL:R	8:9	10:18	13:26	9:16	8:31	21:57	
Y:YY	5:10	11:19	15:26	0:24	9:21	24:47	
Y:W	2:7	4:9	5:12	0:4	0:10	5:36	
Y:B	7:9	10:12	10:12	7:15	0:4	10:22	
Y:R	12:14	16:19	21:25	0:12[2]	9:26	30:51	
W:B	1:1	1:1	2:2	0:7	0:1	2:3	
BL:B	0:8	0:13	0:14	0:9	0:11	0:25	
YY:B		0:20	0:29		0:11	0:40	

Order of dominance:

	12 days	18 days	24 days	16 days	22 days	Grand Total
	B(5)	G(5), BL(5)	BL(5½)	Y(6)	Y(6)	Y(6)
	Y(4½), G(4½)	YY(4), Y(4)	G(4½)	YY(5)	YY(5)	YY(5)
	R(2)	R(2)	YY(4), Y(4)	G(4)	G(4)	G(4)
	YY(3)	W(½)	R(2)	BL(3)	BL(3)	BL(3)
	W(1½)	B(½)	W(½), B(½)	R(2)	R(2)	R(2)
	B(½)			W(1)	W(1)	W(1)
				B(0)	B(0)	B(0)

[1] Peck-dominance changed in pre-mating observations.   [2] Peck-dominance changed during mating period.

in the social order when near the food pan and W when at the entrance to the roost.

Like the cockerels, the male Pigeons were inclined to fight more than were the females. The Pigeon fights were less severe. At times two males would indulge in a rapid pecking back and forth with neither giving way; such encounters were recorded without showing dominance. It was the individuals high in the social ranking that showed such behavior.

As with the females, the males lowest in the flock order had fewer pair-contacts than did those higher in social rank. B and W, which were consistently at the bottom of the order showed a total of 425 such contacts with other members of the flock while Y and YY, leaders of the post-mating flock showed 797.

In neither the flock of males nor of females, did the social rank necessarily run parallel to the number of pair-contacts. With the females, BY showed more such contacts than did RY although the latter clearly outranked her in the post-mating flock and for the entire period of observations. Similarly among the males, YY which held second place in the post-season flock had a total of 520 observed pair-contacts while Y, the more dominant individual in the latter flock and for the whole period of observations, showed only 338 pair-contacts in all. R, third from the bottom of the post-mating flock showed 404 such contacts.

Despite this irregularity, the general relationship held and the individuals high in the social order showed the greater number of pair contacts. The four ranking males had an average of 298 pair-contacts while the three subordinate ones showed a mean of 252. These three subordinates, R, W and B, had only 86 pair-contacts among themselves while, Y, YY and G, the dominants of the post-mating flock had a total of 299 such contacts. This substantiates the observation with the females that the subordinate members have fewer contacts with each other or with the group as a whole than do those high in the social scale.

The largest number of pair-contacts among the males was between YY and G, a total of 159. These birds ranked second and third in the post-mating flock. The lowest number of such contacts was between B and W, a total of only 5 between these two lowest ranking birds. With the females the similar records for greatest

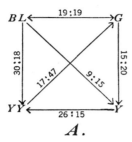 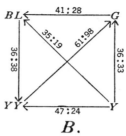

A.   *B.*

Figure 2. Relationships among those highest in the social order in the flock of males. As in Fig. 1 arrows point toward the subordinate birds and numbers show observed times of dominance and subordination.

A. Social order of the four more important males during the pre-mating period. All have peck-dominance over the remaining three.

B. The social order for the same males based on combined results from both periods of observation.

and fewest contacts between two individual birds rested with the highest and two of the lowest ranking birds respectively.

No one of the males showed as extreme a change in social status during the period covered by the observations as did RY among the females; BL came nearest. From being among the first rank in the pre-mating flock he came to occupy definitely the fourth place in the post-mating season observations. His pair-contacts averaged 7.7 per day for the first period and 7.0 for the latter one. The decrease in number of contacts with lowered social status while supporting the observations on RY of the females, is too slight to be certain of significance.

DISCUSSION.

The interrelations between the different phases of this investigation have already been discussed in close connection with the data. There remains the placing of these data in relationship with other published work, particularly that of Schelderup-Ebbe, and the suggesting of some of the more general significance of the facts which have been revealed by such studies.

In making comparisons with Schjelderup-Ebbe's work, it must be remembered that his studies on the flock organization of Domestic Fowls are much more extensive than are our own and that they were made mainly upon Chickens running at large in roomy yards while our observations were made upon animals confined in somewhat crowded pens. This lack of space makes for greater frequency of contacts which is especially important in the case of the more pugnacious males.

Despite this difference, our findings support those of Schjelderup-Ebbe remarkably well and we may conclude that his picture of the flock organization of Chickens is essentially sound. Our observations differ from his in that with the Chickens in our pens, the ranking cockerel or pullet did not necessarily have a less pugnacious disposition than was shown by other high ranking but subordinate members of the flock. RW, the despot of the pullets, did as much or more pecking than did any other, and BW, first in order at all times and later despot of the flock of cockerels, was by far the most vicious of all in his relations to the other cockerels,

Our greatest disagreement with Schjelderup-Ebbe comes in the work with Pigeons, where we have to modify his conception of peck-right to include a situation in which it is the rule for inferiors to peck superiors and for the latter to retreat at times before the attack of an individual which is more usually subservient in its contact-pair relations with that particular bird. At times even after many such encounters, the dominance and subordination relations remain wholly unsettled. Usually, given sufficient time for contact relations and for their observation, the order becomes fairly definitely settled, but in the majority of these cases the sub-servient individual at times successfully attacks the dominant member of the contact-pair and forces it to retreat without, however, causing a permanent reversal of peck-dominance. Under these conditions, social ranking is apparently not determined with a high degree of finality at the first social contact of two individuals but is a matter of gradual development.

The fairly high degree of regularity in the peck-dominance revealed in Tables III and V is evidence that the relationship here revealed is a variation of Schjelderup-Ebbe's principle of despotism rather than representing a different fundamental principle. Absolute despotism is lacking in the relations revealed among these Pigeons since the dominance in a given contact-pair, while usually remaining with one individual in a majority of its contacts with another, is yet subject to certain spatial restrictions in some cases and to temporary reversals for unknown causes in others.

Another difference in our report from that of Schjelderup-Ebbe lies perhaps in the realm of our personalities. We prefer to record our observations objectively without reading into them underlying motives in terms of such human traits as courage, fear, etc. The similarities between the social organization of these flocks of birds and those existing in human social circles are striking and one is constantly tempted to make anthropomorphic interpretations. Experience with other phases of animal behavior, however, compels us to use great caution in interpreting the behavior of one animal as being motivated by forces effective with another.

Unfortunately the working conditions available in our city location do not allow us an opportunity to test the relationship between social dominance of the type we have been discussing and flock leadership in group activities. Fischel (1927) has reported observations with Chickens covering this point. In his studies, groups of hens were observed in a large orchard where the animals moved about, according, in part, to the lay of the land and, in part, according to inner stimuli which were not analyzed. Fischel has evidence that among such groups, the group despot is not necessarily the group leader; in fact the leadership changes readily and only exceptionally is the same individual long at the head of the flock. The group despot, on the contrary, rarely changes.

There is evidence of social coherence in the groups that Fischel observed, since he found that the leader is dependent on followers; she goes little further than the other hens follow.

Neither do we have definite evidence concerning the relationship between position in the peck-order and relative intelligence as measured by ability to learn a given problem. Katz and Toll (1923) found evidence that hens high in the peck-order also stood high in relative intelligence. We have not collected sufficient data to permit a statement on this point.

The general significance of social organization such as we have been discussing deserves brief comment. The similarity to certain aspects of human society is close enough to be immediately apparent. This does not mean that human social life evolved from a bird-like flock organization nor that both human and bird societies have developed from a common social pattern. It does mean, we think, that it is no longer possible to regard human, or avian society for that matter, as something definitely unique. These studies do not support the contentions of Rabaud (1931) that there is nothing to indicate that the animal living in a society has evolved further than the solitary animal since each behaves as if alone! Rabaud states further that there are no collective performances properly so-called among animals other than men; that such unions as are formed come about without the creation of anything which can be called social, since language is lacking.

Unless we are greatly misled, these flocks of Chickens and Pigeons which we have observed represent a condition in which actions speak louder than words and proclaim a social organization similar in some respects to that found among men in which the same principle often holds. We are not prepared to admit the entire lack of voice control in these social groupings of birds, but that is outside the field of our immediate interests and at least for Pigeons has been adequately discussed elsewhere (Craig, 1908).

SUMMARY.

Following leads furnished by the work of Schjelderup-Ebbe, the social organization has been studied as it exists in flocks of Brown Leghorn Chickens and of White King Pigeons. The results obtained are:

1. A social order exists in all the sex-segregated flocks studied. These included Brown Leghorn pullets, Brown Leghorn cockerels, female Pigeons and male Pigeons.

2. The social organization of the cockerels was more complex and was not as definitely organized as was that of the pullets of the same strain and age.

3. When Pigeons were allowed to mate, the resulting picture was of a flock composed mainly of couples, each couple with but few contacts with other couples. The majority of the pecking in such a flock was done by two unmated females.

4. The social order among both male and female Pigeons was based on peck-dominance worked out after many contacts rather than upon an initial combat with one member of any given contact-

pair regularly dominant thereafter. The latter relationship is characteristic for Chickens.

5. The Pigeons standing high in the social order have more social contacts than do those low in the scale.

6. In general our results support those of Schjelderup-Ebbe. The exceptions to this statement and some of the general implications of the work are discussed briefly in the last section of the paper.

REFERENCES.

ALLEE, W. C.

    1931. Animal Aggregations: A study in general sociology. Chicago. University of Chicago Press. 431 pp.

    1934. Recent studies in mass physiology. Biol. Rev. *8:* 1–48.

CRAIG, W.

    1908. The voices of pigeons regarded as a means of social control. Amer. Jour. Sociol. *14:* 86–100.

FISCHEL, W.

    1927. Beiträge zur Sociologie des Haushuhns. Biol. Zentrbl. *47:* 678–695.

KATZ, D. UND A. TOLL

    1923. Die Messung von Charakter- und Begabungsunterschieden bei Tieren. (Versuch mit Hühnern). Zeitschr. f. Psychol. u. Physiol. d. Sinnesorg. Abt. I, *93:* 287–311.

KROH, O.

    1927. Weitere Beiträge zur Psychologie des Haushuhns. Zeitschr. f. Psychol. *103:* 203–227.

RABAUD, E.

    1931. Les origines de la société. Essai sur les sociétés animales. Renaissance du livre. Paris, 1931.

SCHJELDERUP-EBBE T.

    1922. Beiträge zur Sozialpsychologie des Haushuhns. Zeitschr. f. Psychol. *88:* 225–52.

    1923. Das Leben der Wildente (Anas bosehas) in der Zeit der Paarung. Psychol. Forsch. *3:* 12–18.

    1931. Die Despotie im sozialen Leben der Vögel. s. 77–137 in Arbeiten zur Biologischen Grundlegung der Soziologie. Leipzig.

    1931a. Soziale Eigentümlichkeiten bei Hühnern. Kwart. psychol. *2:* 206–212.

TAYLOR, W. S.

    1932. The gregariousness of pigeons. J. Comp. Psychol. *13:* 127–131.

WHITMAN, CHARLES O.

    1919. The behavior of pigeons. (Posthumus works, Vol. III, edited by Harvey A. Carr). Carnegie Institution of Washington, Publ. No. 257.

3

Reprinted from *J. Genet. Psychol.*, **48**, 88–110 (1936)

A STUDY OF A SOCIAL HIERARCHY IN THE LIZARD, *ANOLIS CAROLINENSIS**

From the Biological Laboratories of Harvard University

LLEWELLYN THOMAS EVANS

It is doubtful if a better species of lizard could be chosen for studies of dominance than *Anolis carolinensis*. It is a very hardy animal, lives well in captivity, and the males fight frequently during the sunny days of spring, summer, and fall. It is the latter fact which made this study possible.[1]

Dominance has been reported in a few cases among lizards. Zapf (10) describes the dominance of one male over other males of the same species of Lacerta in the same cage. Noble and Bradley (6) in a splendid monograph on lizard mating behavior describe the same phenomenon among captive *Anolis* but they state that it occurs but rarely. Fighting and dominance (Evans, 1, 2, 3) occurred during the winter months among *Anolis carolinensis* which were stimulated by injections of sheep pituitary extract (Parke Davis) and of human pregnancy urine extract (Parke Davis). If a male that was dominating a cage was removed, a second male soon established himself in the rôle of dominance, but only after fighting and defeating other challenging males. Control males did not respond in this way during mid-winter. Because of this observed dominance the present study was carried out on normal males of *Anolis carolinensis* to discover whether true social hierarchies occur among captive *Anolis* or can be established by experimental means.

Before describing the material and the very simple technique used, it would be well to call attention to certain details in the social behavior of lizards in general and *Anolis* in particular.

Noble and Bradley (6) list 23 species of lizards of which the males are known to fight. And since a complete bibliography of the subject is included in their paper it is only necessary to refer to it here.

---

*Accepted for publication by Carl Murchison of the Editorial Board and received in the Editorial Office, June 24, 1935.

[1]I wish to thank Professors T. Barbour, A. B. Dawson, G. Pincus, and W. M. Wheeler of Harvard University for their very kind aid and interest in connection with this problem.*

It is thus obvious that fighting is a common phenomenon among lizards. However, this activity seems to be induced, for the most part, by the urge of males to defend particular and restricted territories against other males (Noble and Bradley, 6; Wiedemann, 9). If a wandering male, *A*, should chance to linger in the chosen territory of male *B*, a fight ensues, the outcome of which will decide which of these two will hold the territory. If a female approaches, her response to the male ordinarily is passive so she is permitted to remain.

A full description of the fighting and mating behavior of *Anolis* has been given by Noble and Bradley (6). It is only necessary to mention here that in both sexes a dewlap or fan beneath the lower jaw may be extended by the forward movement of a pair of slender bones beneath the loose skin. In males this extended dewlap measures about 3 cm. across, but only approximately 0.8 cm. on females. It is red in males, becoming a brilliant scarlet when the direct rays of the sun strike it. The body can be flattened laterally when the animals are enraged. Even females flatten the body in this way, as was observed among castrates. Incidentally, castrated females were observed to fight exactly as the males do. Normal females rarely fight. Males possess a crest along the nape of the neck and back which rises to a height of $3\frac{1}{2}$ mm. during fighting activity.

The stages which ordinarily occur in fighting are: (1) Both males move slowly toward each other while each animal raises its crest and flashes its dewlap (not usually to its fullest extent, at first, however). The body is flattened and the animals move sidewise toward each other. One, usually the resident male, slowly turns green. This stage is apparently an attempt to impress or frighten. (2) If, after repeated circling, strutting, and display of dewlap neither gives way, they come closer, each attempting to get a position higher than the other. Soon they turn their heads toward each other, their mouths open wide, and with jaws less than a centimeter apart each attempts to bite the snout of the other. No actual biting occurs for several seconds, but they thrust and withdraw their jaws repeatedly until one secures the desired hold on the other's snout. This may end the fight. But a persistent challenger will return to snap and bite two or three times. However, the one that secures the first jaw-hold usually wins. (3) The defeated male is chased away and the victor takes a prominent position at the top of the cage, flashes his dewlap to its fullest extent, and struts back and forth with body flattened. Later, he may pursue the defeated male again. (4) The dominant male usually takes no further notice of the other unless the latter attempts to approach. A flash of the dewlap is enough to frighten the victim away.

It should be particularly emphasized that not a single vocal sound is uttered by either animal during the fight. Often, the only sounds to be heard are those produced by the click of teeth when the jaws of the two combatants come together, or the rustling of leaves as the defeated male retreats. This silent mode of fighting seems to be characteristic of the great majority of diurnal lizards. In contrast, the nocturnal lizards, or geckos, do not utilize bright colors or special structures as a means of frightening rivals, but fight desperately, frequently to the death. Geckos have well developed vocal powers and constantly utter chirps or squeaks in a considerable range of pitch, especially during courtship and fighting.

MATERIALS AND TECHNIQUES

Nineteen males of *Anolis carolinensis* were used in the experiments. They ranged in weight from 3.9 grams to 7.5 grams. All but male *A* were sent from Louisiana in February, 1935. *A* had been kept in the laboratory since February, 1934. Figure 1 shows the males used in this study.

The six cages, each two feet high and 18 inches square, were covered with bronze screening and were arranged in two tiers of three each in a large window with a southern exposure. The floor of each cage was covered with sand and gravel. Branches, bits of bark, and leaf debris were scattered in each cage.

The animals were watered by sprinkling the cages twice daily. They were fed daily with adult flesh flies, Tenebrio larvae, and occasionally with spiders.

Three large aquarium tanks filled with water together with the sprinkling of the cages seemed to produce enough humidity to keep the animals in good condition. The temperature ranged from 70 to 80 degrees Fahrenheit.

The experiments extended from March 7 to May 13, 1935.

FIGURE 1
THE NINETEEN MALES USED IN THE EXPERIMENT
Top row: Categories I and II, middle row: Category III, bottom row:
Category IV.

The technique was as follows: With all 19 males in the first cage they were watched until one male had definitely become dominant by defeating all others who challenged him. This male was immediately caught, weighed, and marked by clipping off the distal end of a particular toe. (Suctorial pads on each toe supplement the claws so that the loss of one or two claws does not handicap the animal.) He was then placed in Cage 2, and was registered as male *A*. The second male to become dominant was registered as *B*, the third as *C*, and so with all 19 animals. As soon as two or more were placed in Cage 2, fights occurred, establishing dominance in that cage. The dominant male of Cage 2 was then placed in Cage 3. Thus, it continued until all of the six cages had been occupied by each of the 19 animals in turn (with a few exceptions which will be discussed later).

Cage 3 contained, besides the male *Anolis,* several castrated female *Anolis* and a variable number of *Sceloperus undulatus* of both sexes. Cage 4 contained, in addition to the male *Anolis,* three geckos (*Gymnodactylus kotschyi*), while Cage 5 also contained three geckos (*Hemidactylus turcicus*). These geckos, Sceloperus, and castrated females in no way destroyed the dominance pattern. The customary fights took place in each cage regardless of the other species present. The rivals often crawled over the Sceloperus in Cage 3 while going through their fighting maneuvers. When this happened, no notice was taken of the Sceloperus. In fact, a male in combat never took his eyes off his opponent for an instant. When shifting his position he pushed one leg out, groping for a foothold, then slowly extended the other, feeling his way along.

Each cage was numbered, from one to six. The following, Table 1, shows the rank of each male in each cage together with the date and hour when each male became dominant.

Table 2 shows the weight of each animal when he was marked and registered after becoming dominant for the first time, and the weight of each again, May 8, a few days before the experiment was concluded. In Table 3, nine witnesses are recorded with the date of the particular combat observed, the victorious male, and the cage number.

Because of the difficulty of photographing the males in action without altering the conditions of the experiment, I felt it was im-

TABLE 1

THE DATE OF DOMINANCE OF EACH MALE IN EACH CAGE

	Cage 1				Cage 2	
A	March 7			A	March 20	1:30 p.m.
B	March 9	10:30 a.m.		B	March 21	10:00 a.m.
C	March 16	2:00 p.m.		C	March 22	10:40 a.m.
D	March 19	3:00 p.m.		F	March 30	11:30 a.m.
E	March 20	1:30 p.m.		D	March 31	12:15 p.m.
F	March 22	a.m.		G	March 31	1:30 p.m.
G	March 25	12:00		H	April 3	10:15 a.m.
H	March 27	3:00 p.m.		J	April 3	10:30 a.m.
I	March 27	3:30 p.m.		I	April 3	11:00 a.m.
J	March 28	a.m.		K	April 4	11:00 a.m.
K	March 29	11:00 a.m.		M	April 5	10:40 a.m.
L	March 29	11:30 a.m.		L	April 6	12:00 a.m.
M	March 30	11:00 a.m.		O	April 14	11:30 a.m.
N	April 3	10:20 a.m.		N	April 16	12:30 p.m.
				P	April 18	10:10 a.m.
				Q	April 20	9:00 a.m.
				R	April 25	10:20 a.m.
				S	April 28	12:00 a.m.
				E	April 30	11:00 a.m.

	Cage 3				Cage 4	
				A	March 31	11:00 a.m.
A	March 29	1:30 p.m.		B	April 3	11:25 a.m.
B	March 31	12:10 p.m.		C	April 4	11:00 a.m.
C	April 3	9:15 a.m.		D	April 4	11:30 a.m.
D	April 3	11:10 a.m.		F	April 5	1:00 p.m.
F	April 4	10:30 a.m.		G	April 6	10:00 a.m.
G	April 4	11:30 a.m.		H	April 11	8:40 a.m.
H	April 6	9:45 a.m.		L	April 14	11:45 a.m.
L	April 11	8:30 a.m.		J	April 14	11:55 a.m.
I	April 11	8:55 a.m.		O	April 15	11:15 a.m.
J	April 11	9:00 a.m.		I	April 16	10:00 a.m.
M	April 11	9:40 a.m.		K	April 18	10:25 a.m.
O	April 15	11:00 a.m.		Q	April 20	10:00 a.m.
K	April 18	10:20 a.m.		M	April 25	10:45 a.m.
Q	April 20	9:45 a.m.		P	April 28	12:25 p.m.
R	April 25	10:40 a.m.		E	May 5	8:50 a.m.
P	April 25	3:40 p.m.		R	May 9	10:30 a.m.
E	May 5	11:45 a.m.				

	Cage 5				Cage 6	
A	April 3	11:20 a.m.		A	April 11	12:00
B	April 4	11:00 a.m.		B	April 14	12:15 p.m.
H	April 11	8:55 a.m.		H	April 15	11:00 a.m.
C	April 14	12:15 p.m.		D	April 16	10:15 a.m.
D	April 15	10:30 a.m.		C	April 18	9:15 a.m.
G	April 15	12:00		E	April 18	10:58 a.m.
L	April 16	1:00 p.m.		L	April 18	1:05 p.m.
F	April 18	10:40 a.m.		G	April 18	1:05 p.m.
J	April 18	10:50 a.m.		Q	April 22	10:15 a.m.
Q	April 20	10:45 a.m.		J	April 22	10:15 a.m.
I	April 22	9:50 a.m.		K	April 25	9:25 a.m.
K	April 23	10:20 a.m.		I	April 26	9:55 a.m.
M	April 26	9:40 a.m.		M	May 3	4:00 p.m.
P	May 1	3:40 p.m.		P	May 9	11:45 a.m.
E	May 9	10:33 a.m.		O	May 11	11:30 a.m.
				E	May 13	10:00 a.m.

TABLE 2

EACH MALE IS DESIGNATED BY A LETTER IN COLUMN 1; COLUMN 2 INDICATES
THE WEIGHT OF EACH WHEN HE FIRST BECAME DOMINANT; COLUMN 3
INDICATES THE WEIGHTS WHEN THEY BECAME DOMINANT IN CAGE 6

A	7.5	6.4
B	6.8	7.5
C	5.8	6.0
D	6.3	6.0
E	5.0	5.3
F	5.5	5.5
G	6.1	6.2
H	6.3	6.5
I	5.1	5.6
J	4.5	4.5
K	4.7	4.5
L	4.7	4.9
M	5.0	4.7
N	3.9	2.25
O	5.5	6.1
P	4.7	4.6
Q	5.6	4.3
R	4.4	4.3
S	4.8	4.5

portant to have some of my colleagues at Harvard witness the combats of the males in at least a few cases. I, therefore, wish to thank the witnesses, Miss Alice Beale, Mr. L. A. Hansborough, Miss G. Hermes, Mr. L. H. Kleinholtz, Mr. B. Renshaw, Mr. G. L. Wood-

TABLE 3

WITNESSES

Hansborough, L. A.	Cage 2	April 3
	Animal H	
Kleinholtz, L. H.	Cage 2	April 4
	Animal K	
Woodside, G. L.	Cage 5	April 11
	Animal H	
Kleinholtz, L. H.	Cage 5	April 15
	Animal C	
Renshaw, B.	Cage 5	April 15
	Animal G	
Hoadley, L.	Cage 6	April 22
Hermes, G.	Animal Q	
Beale, A.	Cage 6	April 26
	Animal K	
Renshaw, B.	Cage 6	May 1
	Animal I	

side, and Professor Leigh Hoadley, for their patience and interest in the problem.

Because of low dominance capacity on the part of a few males, especially those designated as *E*, *R*, *S*, and *N*, only 14 males of the 19 performed in all six cages. Fifteen went through the last five cages. After *N* had become dominant and was removed to Cage 2, no further fights were observed to take place to decide the supremacy of the cage among *O*, *P*, *Q*, *R*, and *S*. After waiting ten days these five males were placed in Cage 2. Table 1, as well as Figures 2 to

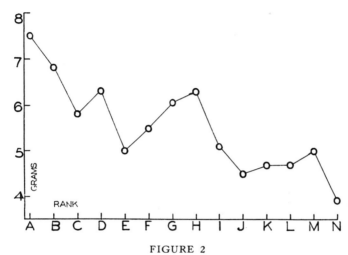

FIGURE 2

THE ORDER IN WHICH THE MALES BECAME DOMINANT IN CAGE 1 PLOTTED AGAINST WEIGHT

Males *O*, *P*, *Q*, *R*, and *S* failed to dominate.

7 show that *P* and *Q* dominated at one time or another in the last five cages. *R* dominated in his turn only in Cages 2, 3, and 4, while *S* only dominated in Cage 2. *E*, as the figures indicate, dropped down the scale from fifth place in Cage 1 to the nineteenth place in Cage 2 and thereafter did not stand higher than fifteenth place in any cage. *O* was dominant in Cages 2, 3, 4, and 6. None of the three males *O*, *R*, and *S* showed dominance in Cage 5 after several days' residence. They were placed, therefore, in Cage 6.

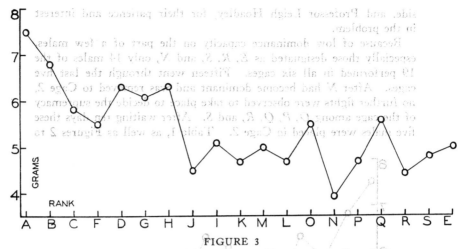

FIGURE 3

THE ORDER IN WHICH THE MALES BECAME DOMINANT IN CAGE 2 PLOTTED
AGAINST WEIGHT

Note the addition to Cage 2 of males *O*, *P*, *Q*, *R*, and *S*.

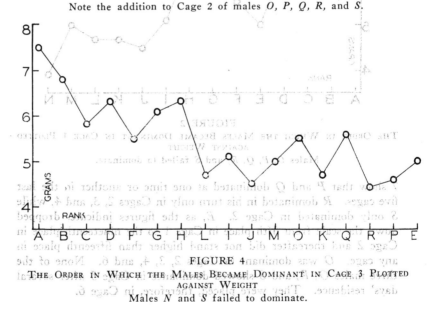

FIGURE 4

THE ORDER IN WHICH THE MALES BECAME DOMINANT IN CAGE 3 PLOTTED
AGAINST WEIGHT

Males *N* and *S* failed to dominate.

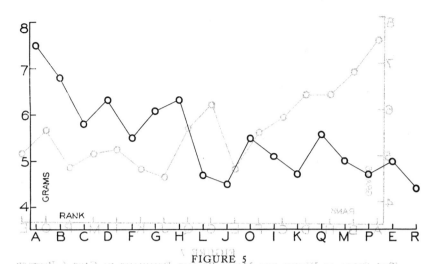

FIGURE 5

THE ORDER IN WHICH THE MALES BECAME DOMINANT IN CAGE 4 PLOTTED
AGAINST WEIGHT

Males *N* and *S* failed to dominate

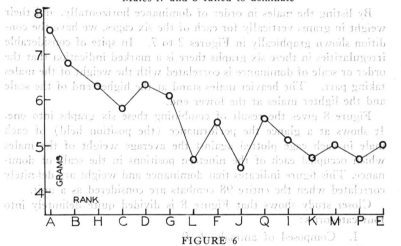

FIGURE 6

THE ORDER IN WHICH THE MALES BECAME DOMINANT IN CAGE 5 PLOTTED
AGAINST WEIGHT

Males *N, O, R,* and *S* failed to dominate.

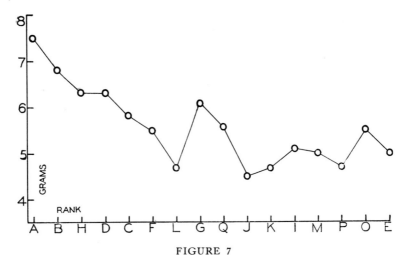

FIGURE 7

THE ORDER IN WHICH THE MALES BECAME DOMINANT IN CAGE 6 PLOTTED
AGAINST WEIGHT

Males *N*, *R*, and *S* failed to dominate.

By listing the males in order of dominance horizontally, and their
weight in grams vertically for each of the six cages, we have the condition shown graphically in Figures 2 to 7. In spite of considerable
irregularities in these six graphs there is a marked indication that the
order or scale of dominance is correlated with the weight of the males
taking part. The heavier males stand at the higher end of the scale
and the lighter males at the lower end.

Figure 8 gives the result of combining these six graphs into one.
It shows at a glance the performance (the position held) of each
male in each cage plotted against the average weight of the males
which occupied each of the nineteen positions in the scale of dominance. This figure indicates that dominance and weight are definitely
correlated when the entire 98 combats are considered as a unit.

Closer study shows that Figure 8 is divided quite definitely into
four categories:

    I.   Composed of animals *A, B*

    II.   Composed of animals *C, D, F, G, H, L*

    III.   Composed of animals *I, J, K, L, M, O, Q*

    IV.   Composed of animals *E, N, P, Q, R, S*

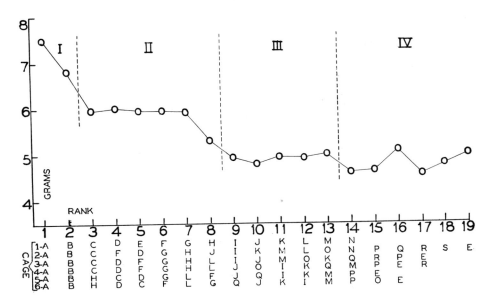

FIGURE 8

THE CLOSE CORRELATION BETWEEN ORDER OF DOMINANCE AND WEIGHT OF
19 MALES OF *Anolis carolinensis*

The 19 points on the graph were found by taking the average of the weights
of the males which are listed in the vertical column beneath each point.
Note that the order of dominance for each of the six cages is given if the
letters are read horizontally. The four categories are labelled and show
males *A, B, C, D, F, G, H,* and *L* ranked in I and II, or highest in dominance, and averaging 6.12 grams; males *I, J, K, M,* and *O* ranked in III, or
intermediate in dominance, and averaging 4.96 grams; males *E, N, P, Q,
R,* and *S* ranked in IV, or lowest in dominance, and averaging 4.73 grams.

Obviously, then, animals in categories I and II are ranked high
in dominance, the animals in IV are low in dominance, while those
in category III stand intermediate. A grading system was adopted
to discover just how each animal stood in rank. If an animal kept
the rank which he secured in the first cage throughout the entire six
cages, his grade is unity. If he held his rank in five cages, but gave
way to another male in Cage 6, his grade is minus one. If he held
his rank unchanged in five cages, but advanced his rank three places
in Cage 6, his grade is plus three. On the basis of this grading

system the males of each category receive the following numerical rating:[2]

		Grams						
I	A	+ + (7.5)			I	— 7 (5.1)		
	B	+ + (6.8)		III	J	+ 4 (4.5)		—11
					K	— 3 (4.7)		
					M	+ 3 (5.0)		
	C	— 3 (5.8)			O	— 8 (5.5)		
	D	— 2 (6.3)						
II	F	+ 2 (5.5)	31					
	G	+ 3 (6.1)			E	—43 (5.0)		
	H	+13 (6.3)			N	—12 (3.9)		
	L	+18 (4.7)			P	+ 1 (4.7)		—36
				IV	Q	18 (5.6)		
					R	— (4.4)		
					S	— (4.8)		

The dominance status of the males in the four categories shown in Figure 8 in relation to weights is thus correlated numerically by utilizing the grading system.

When the positive or negative numerical status of each animal and each category is considered, the true significance of the four categories or parts of Figure 8 is revealed. It is noted that the animals of categories I and II (combined) have a numerical value of -31, and an average weight of 6.12 grams, category III has a num rical value of —11 and an average weight of 4.96 grams, while category IV has a numerical value of —36 and an average weight of 4.73 grams. Figure 9 gives graphically the relationship between the

---

[2]It might be argued that this grading system does not give proper credit to such animals as A, B, R, and S, the males at the top and at the bottom of the scale of dominance. However, if a percentage basis is used instead, the males are found to be arranged in essentially the same order, although A, B, R, and S receive a numerical rating:

		Percentage			Percentage
I	A	100		I	48
	B	95		J	54
			III	K	42
				M	38
				O	17
				E	32
	C	87		N	18
	D	87		P	21
II	F	72	IV	Q	33
	G	68		R	10
	H	78		S	5
	L	58		T	0

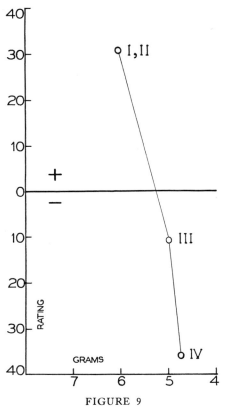

FIGURE 9

The average numerical rating of each category of males (I, II have +31, III has —11, IV has —36) plotted against the average weight of each category (I, II have 6.12 grams, III has 4.96 grams, IV has 4.73 grams).

numerical status of the four categories and the average weight of the males of each category. This method of grading seems reasonable since each male is graded solely on his own ability to maintain or to improve the rank which he secured in the first cage. Under this system no premium is attached to position of high rank nor is penalty attached to position of low rank provided this same rank is maintained in each of the six cages. It is interesting to note that categories

I and II stand almost as high above the line (+31) as category IV stands below the line (—36).

However, if weight and numerical status were true correlates of each other, that is, if the graph, Figure 9, were a straight line (assuming that category III retains its numerical status of —11), then

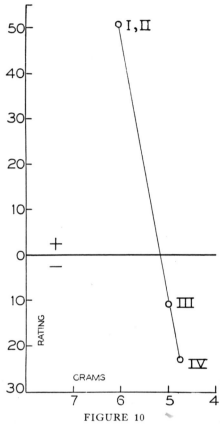

FIGURE 10

BASED ON THE THEORETICAL ASSUMPTION THAT AVERAGE NUMERICAL RATING AND AVERAGE WEIGHT OF THE MALES IN EACH OF THE FOUR CATEGORIES ARE DIRECTLY CORRELATED

To show direct correlation between rating and weight it is necessary to ascribe a rating of +51 to I and II, a rating of —11 to III, and a rating of —23 to IV.

it would be necessary to raise the numerical rating of categories I and II to some point above +36. Also, the rating of category IV should be raised somewhat. The points where rating and weight meet, for categories I and II and IV, are found to be +51 and —23, respectively. Categories I and II (combined) should be +51 while category IV should be —23. Figure 10 shows graphically these theoretical assumptions. In other words, if weight were the sole factor to be considered, the males of categories I and II exerted themselves in combat only to the extent of 60 per cent of their actual fighting capacity. Likewise, the males of category IV performed only to the extent of 63.8 per cent of their capacity.

Description of Representative Combats

It is useless to give details of each of the 98 combats witnessed during the period of the experiments. But several of the encounters should be described since they have considerable bearing on the general problem of diurnal lizard behavior.

Frequently the establishment of the dominance of a particular male over a group occurred during my absence from the laboratory. Among males of the same category in the same cage, especially in categories III and IV, the dominating male, after the initial defeat of other males, only exhibited his dominance for a few seconds during a period of three or four hours. If I missed this brief exhibition of leadership, it was relatively simple to discover the dominant male by introducing a male from another cage. Usually in less than one minute after the newcomer was placed in the cage, if the sun was shining, the dominant male invariably revealed himself by extending his dewlap and challenging the newcomer.

Male N was often thus used to promote fighting. He weighed only 3.9 grams at the start of the experiment and 2.25 grams on May 11 when he died. He was very agile but apparently had met with an accident some time before since his spinal column was bent out of alignment in both vertical and horizontal planes. Autopsy revealed only one testis present although both vasa deferentia were present. Probably because of his physical handicap, for no other male behaved so queerly, he would deliberately advance toward a larger male, flash his dewlap, then run and hide when pursued by the challenged male. Within a few minutes N would show himself again, flash his dewlap, and the performance would be repeated. Five actual instances better illustrate his behavior. (1) May 31, 1935; 1:30 P.M. *N* was placed in Cage 2. Immediately he extended his dewlap while advancing toward five males, *G* and *E* among them, and apparently challenged them since they all turned quickly in his direction and extended their dewlaps as though they accepted the challenge. *N* retreated and hid beneath a branch. The five males thus aroused proceeded to circle and challenge each other. Male *G* soon chased away all but *E* who held his ground until *G* locked jaws with him. After being bitten twice *E* withdrew from the field leaving *G* dominating the cage. This particular episode was witnessed by Mr. L. A. Hansborough. (2) *N* was placed in Cage 5, April 16, 1:00 P.M. He extended his dewlap and *L* immediately chased not only *N* but *O* and *I* as well. All three left the field with *L* dominating the cage. (3) April 3, 3:00 P.M., *N* bit *M* on the neck but was shaken off and chased away. (4) April 4, 12:30 P.M., *N* was seen climbing the screen of the cage toward *M*. When he attained a position higher than *M*, he thrust out his dewlap and compressed his body laterally. *M* responded in a similar way and quickly chased *N* down the screen. (5) On April 11, 8:30 A.M., *N* was placed in cage 3 to incite fighting. He immediately challenged all males present. *L* took up the challenge, chased *N*, and dominated the cage. *L* was then removed from the cage. At 8:55 the same morning male *I* chased *N* and others, thus dominating the cage. Male *I* was quickly removed and at 9:00 A.M. *J* took up *N*'s challenge, chased the others, and became dominant. *J* was removed and at 9:40 *M* took up *N*'s challenge and became dominant in turn in that cage.

A, as the records show, was not beaten by any other male during the course of the six series of combats. During his first few weeks in the laboratory a year previously, February 1934, he permanently injured his mouth while trying to escape. But in spite of this handicap he held his rank as the only unconquered male in the hierarchy of 19 males. Very few males had the courage actually to fight with him. One such combat was particularly interesting. Male *H* became dominant in cage 4 at 8:40 A.M., April 11, and dominated Cage 5 at 8:55, fifteen minutes later. Accordingly he was placed in Cage 6 at 9:00 A.M. *H* immediately extended his dewlap and strutted sidewise with body compressed laterally toward *A* who dominated Cage 6. *A*'s bright green color may have guided *H* toward him, since the dominant male is commonly green while all the rest are brown. It was quite apparent that *A* was surprised by this boldness and it was several seconds before *A* took the fighting pose and advanced toward *H*. *B*, in the brief interval, came from behind a branch and took up the challenge by advancing toward *H* with dewlap flashing. *H* continued to advance toward *A*, ignoring *B* who soon retired from the scene. *B*'s behavior in this episode was the only case observed that resembled in any way the pecker-peckee type of dominance described for the fowl by Schjelderup-Ebbe (7, 8) and by Murchison (4, 5). Both *A* and *H* persisted through the first or bluffing stage of the typical fighting pattern and entered the second or biting stage. *A* opened his mouth very wide and literally thrust his lower jaw into *H*'s mouth. *H* pulled back his head, thrust forward and caught *A* by the angle of upper and lower jaw. A moment later he swung *A* over backward holding him suspended in mid-air. After struggling, *A* broke the hold and dropped to the floor of the cage. He came toward *H* and was again caught and held in the same way. During this struggle *A*'s color had changed to brown. After breaking loose both males kept apart for ten minutes, but each eyed the other constantly. Then *A* challenged *H* again. This time *H* retreated slowly and *A* again established himself as the dominant male of the cage. This encounter, lasting thirty minutes, was the longest and most interesting of all those witnessed during the course of the investigation.

On May 9, at 10:00 A.M., *R* had been in Cage 4 for several days when a small male, *T*, weighing 4.5 grams and which had never fought, together with male *S* was placed in cage 4 with *R*. Within three minutes *R* crawled sidewise toward both *S* and *T* with body compressed and dewlap flashing. Neither of the newcomers responded to the challenge. *T* moved toward *R* and passed him to climb to a branch. *R* followed *T*, threatening all the way. *T* stopped, *R* came alongside and continued to flash his dewlap and to compress his body. Still no answering challenge came from *T* who moved away. *R* then advanced toward S in like manner and received a similar negative response from *S*. *R* was obviously puzzled, but made no attempt to touch the other males although he continued to threaten them for half an hour.

DISCUSSION

From the evidence given in Table 1 it is clear that out of 98 observed encounters, in which dominance was definitely established, 90 were won by the dominating resident male, which had been in the cage at least 24 hours previously, against a strange male newly placed in the cage in spite of the fact that the latter had dominated the cage from which he came. The eight encounters in which the reverse was the case are:

F. April 18, 10:58 A.M., 18 minutes after being admitted, dominated Cage 6.

H, April 11, 8:55 A.M., 15 minutes after being admitted, dominated Cage 5.

K, April 18, 10:25 A.M., 5 minutes after being admitted, dominated Cage 4.

O, April 15, 11:05 A.M., 15 minutes after being admitted, dominated Cage 4.

Q, April 20, 9:45 A.M., 45 minutes after being admitted, dominated Cage 3.

Q, April 20, 10:00 A.M., 15 minutes after being admitted, dominated Cage 4.

Q, April 20, 10:45 A.M., 45 minutes after being admitted, dominated Cage 5.

R, April 25, 10:40 A.M., 20 minutes after being admitted, dominated Cage 3.

Of these eight encounters won by non-resident males only the first four are really significant, since in these cases the defeated resident males were known to fight readily. The three successive cases of dominance on the part of *Q* and the one for *R* are easily explained by the low dominance status of the males which they encountered.

The evidence strongly supports the contention of Noble and Bradley (6) that the male *Anolis* fight to retain a chosen territory against rival males. It further emphasizes the fact that little actual social life exists among adult Anolis of the same sex, particularly among males. In fact, although this purports to be a study of a social hierarchy in *Anolis,* it actually emphasizes the anti-social behavior of adult males. It is this feature of the behavior of the male *Anolis* which prevents anything approaching a straight-line dominance

90

or a "pecking order" from occurring. In fact the "pecking order" so common in fowls presupposes that members of the same species are bound together by a very positive social reflex as Murchison (4, 5) has clearly shown to be the case in the chick.

The cases cited also emphasize an additional fact in connection with behavior of male *Anolis*. He has become so specialized structurally (referring particularly to the dewlap, dorsal crest, and ability to flatten the body laterally) that fighting can only take place after the males involved have extended their dewlap in a challenging manner. If one male ignores the dewlap display of the other, no fighting takes place in the majority of cases. This was specially brought out in the case of *R* on May 9, 10:00 A.M., as mentioned above. However, a male that stands fairly high in the scale of dominance will almost invariably respond to a dewlap of a challenging male by a similar display of his own.

The behavior of caged male *Anolis* is probably a modification of the behavior in the field. Under natural conditions when a strange male approaches a particular territory which is in possession of another, a fight results, the outcome of which will depend upon two factors: first, the weights of the combatants; second, the psychological or emotional elements of the urge to hold territory. The beaten male retreats, leaving the victor in possession of the territory.

In the cage the defeated males cannot leave the territory so the victorious male is kept continually in an aggressive mood because he constantly sees others in his domain. Their proximity serves as an ever-present stimulus to fight to hold territory. Each day, therefore, the dominant male may be expected to express his territorial urge by the customary challenge of flashing dewlap, sidewise advance, and lateral flattening of the body.

Since the dominant male was ordinarily removed and placed in the next cage as soon as he had defeated others, either by mere display or by actual fighting, and since dominance and weight are correlated (Figure 8), it follows that the weight of the average resident male which dominates a particular cage will be slightly greater than the weight of the average newly introduced male. Figures 2 to 7, however, show that there were many individual exceptions to this as evidenced by the zig-zag appearance of these graphs. These exceptions (where a lighter male becomes dominant before another heavier male in the same cage) reveal very markedly the part played by the territorial urge.

As a matter of fact, in all six cages, out of the total of 98 combats, 38 were won by males weighing less than other males in the same cage. When these particular combats took place, had the 38 cases occurred in the same numerical position in each cage, it is very doubtful if the weight-dominance correlation could have been shown to exist. But because they were scattered haphazard throughout the six cages they had little effect on the correlation. Figure 8 proves that weight and dominance are correlated to a considerable degree. This is contrary to the findings of Murchison (4, 5) who records little or no correlation between mass and dominance in his careful and extensive studies of chick hierarchies.

In conclusion it might be suggested that the fighting pattern of males of *Anolis carolinensis* to defend or secure territory is made up of a chain of reflexes which are expressed in a more or less orderly sequence. Those which can be objectively studied and their order of sequence are:

1. Dewlap reflex
2. Dorsal crest reflex
3. Sidewise approach to rival male
4. Reflex causing a flattening of the body laterally
5. The biting reflex
6. Retreat reflex of the defeated male
7. Pursuit reflex of the victor
8. Dewlap reflex of the victor

Both males engaged in a combat will exhibit the first four reflexes listed here and the fifth, also, if both persist beyond the bluffing stage in the encounter. But usually the resident male will be the challenger and therefore the first to express each of the first four reflexes. The sequence of the biting reflex will depend on the skill and strength of the rivals. The first five reflexes will appear in both but will vary in degree in each male. The sixth will appear only in the defeated male while the seventh and eighth will be expressed only by the victor.

If the dewlap reflex of the challenger yields no like response from the second male, the only reflex in the chain which will be expressed by the latter will be sixth or retreat reflex, unless he simply ignores the challenger. The retreat reflex is more apt to be the response on the part of the second male if the challenger has already conquered him and then challenges him again later. However, if the second male be newly introduced into the cage, he rarely ignores the challenge.

The first four reflexes produce what might be called an intimidating mechanism which has apparently evolved to cause the males to appear as large and imposing as possible to each other. As a bluffing device it is certainly successful since many encounters end at this point in the fighting pattern.

Summary

Experiment reveals for the first time in any reptile form a social hierarchy which differs in many respects from that of fowls. Nineteen males of *Anolis carolinensis* ranging in weight from 7.5 grams to 3.9 grams arrange themselves in a series of dominance ranking which is closely correlated with weight. The largest males stand at the higher end of the scale and smallest stand at the lower end of the scale of dominance.

The urge to acquire and to hold a certain restricted territory is very marked. The resident male (that has been in a particular cage for 24 hours or more) wins in 91 per cent of the combats not only because he is heavier than the non-resident (42 per cent of such combats being won by lighter males) but also because he fights harder to defend territory than the non-resident does to acquire it.

The males are grouped in three distinct categories: (1) Those ranking very high in dominance with a numerical rating of $+31$ and an average weight of 6.12 grams. (2) Those ranking intermediate in dominance with a numerical rating of -11 and an average weight of 4.96 grams. (3) Those ranking very low in dominance with a numerical rating of -36 and an average weight of 4.73 grams.

The fighting pattern consists of eight reflexes which are overtly expressed in the following order: (1) dewlap reflex, (2) dorsal crest reflex, (3) sidewise approach to rival male, (4) reflex causing a flattening of the body laterally, (5) the biting reflex, (6) retreat reflex of the defeated male, (7) pursuit reflex of the victor, (8) dewlap of the victor.

REFERENCES

1. EVANS, L. T. The effects of pituitary implants and extracts on the genital system of the lizard. *Science,* 1935, **81**, 468-469.

2. ————. Winter mating and fighting behavior of *Anolis carolinensis* as induced by pituitary injections. *Copeia,* 1935, **1**, 3-6.

3. ————. The effect of Antuitrin S on the male lizard, *Anolis carolinensis. Anat. Rec.,* 1935, **62**, 213-220.

4. MURCHISON, C. The experimental measurement of a social hierarchy in *Gallus domesticus*: I. The direct identification and direct measurement of Social Reflex No. 1 and Social Reflex No. 2. *J. Gen. Psychol.,* 1935, **12**, 3-39.

5. ————. The experimental measurement of a social hierarchy in *Gallus domesticus*: II. The identification and inferential measurement of Social Reflex No. 1 and Social Reflex No. 2 by means of social discrimination. *J. Soc. Psychol.,* 1935, **6**, 3-30.

6. NOBLE, G. K., & BRADLEY, H. T. The mating behavior of lizards; its bearing on the theory of sexual selection. *Ann. N. Y. Acad. Sci.,* 1933, **35**, 25-100.

7. SCHJELDERUP-EBBE, T. Beiträge zur Sozialpsychologie des Haushuhns. *Zsch. f. Psychol.,* 1922, **88**, 225-264.

8. ————. Soziale Verhältnisse bei Vögeln. *Zsch. f. Psychol.,* 1922, **90**, 106-107.

9. WILDEMANN, M. Die Spitzkopfeidechse (*Lacerta oxycephala* Dumeril und Bibron). *Blätt. Aquar.-Terrar.-Kde.,* 1909, **20**, 733-736.

10. ZAPF, J. Beobachtungen an Terrarientieren. *Lacerta,* 1911, 10-12.

Harvard University
Cambridge, Massachusetts

4

Reprinted from *J. Social Psychol.*, **11**, 313–324 (1940)

DOMINANCE-QUALITY AND SOCIAL BEHAVIOR IN INFRA-HUMAN PRIMATES*[1]

Division of Psychology, Brooklyn College

A. H. MASLOW

This paper is an attempt to outline in tentative fashion a hypothesis concerning the correlation between quality or type of dominance to be found in various families of infra-human primates, and the social behavior and group organization to be found in these families.[2] Such a presentation seems to be justifiable in spite of the paucity of necessary data, because of its probable usefulness in the event that further accumulation of data shows it to be valid. This usefulness is of the following nature: (*a*) It may give another centrally organizing principle to the study of animal sociology, and (*b*) it suggests a new approach to the problem of how human social behavior and group organization may be based on individual personality. Any attempt to link individual with group behavior is important enough to warrant even tentative presentation, especially if the hypothesis itself suggests new lines of attack that will determine whether the hypothesis stands or falls.[3]

Accordingly, I shall attempt to demonstrate the following three

*Received at the Editorial Office on September 19, 1938.

[1]This paper, in condensed form, was presented at the 1935 meeting of the A.P.A.

[2]These three main groupings are (*a*) the anthropoid apes which include gorilla, chimpanzee and orang-outan (Family Pongidæ) (*b*) the catarrhine or Old-World monkeys which include the baboons, macaques, langurs, mangabeys, etc. (Family Cercopithecidæ); (*c*) the platyrrhine or New-World monkeys, which include the howlers, woolly, cebus, spiders, etc. (Family Cebidæ). There are other families but the writer has had little experience with them. The writer wishes to emphasize that he feels he is riding rough shod over many real differences when he lumps diverse species into a single group. In this particular connection, however, this crude procedure is somewhat justified by the fact that the search is for similarities and not for differences.

[3]This paper was completed in substantially its present form in 1934-5 but was not published because of the disagreement expressed by Drs. Zuckerman and C. R. Carpenter, two notable authorities in the field. Since then further observations by the writer have convinced him increasingly of the probable validity of the hypothesis. The writer wishes to express his thanks to Drs. Zuckerman and Carpenter for their careful criticism of the paper.

subsidiary hypotheses into which the problem may be divided; (*a*) That different kinds or qualities of dominance behavior are demonstrable in the three main families of infra-human primates that have been studied; (*b*) that different kinds of social behavior and group organization may be found in the two families, for which we have adequate data; (*c*) that these differences are correlated.

Although it has been demonstrated that dominance plays an important role in lower animals, we shall confine our discussion to infra-human primates (15, 16, 17, 6).

With regard to our first point, that there exist considerable variations in quality or kind of dominance, this is very obvious to anyone who has worked with more than one of the three main groups that have been experimentally studied (platyrrhine, catarrhine, and anthropoid). It is particularly well brought out by the technique the writer has developed of studying the animals quantitatively in pairs that have been matched for species, sex, age, size, and weight, in order to bring out the more delicate factors in the determination and expression of dominance that are ordinarily obscured by sexual dimorphism, endocrinological differences, and unlimited space in which to react (10). A pair of rhesus macaques (*catarrhine*), a pair of cebus monkeys (*platyrrhine*), and a pair of chimpanzees (*anthropoid*), in the same experimental situation will all behave in different ways. While in all three pairs it will be possible to determine which animal of the pair is dominant and which subordinate, and while in all three pairs the nature of the total dominance syndrome will be *generally* comparable, still the *specific* nature of the behavior will be different. The dominant animal in all three pairs will be the more aggressive, but the quality of the aggressiveness will differ considerably. The dominant animal in all three pairs will play the masculine rôle and the subordinate animal will play the female rôle, but the quality and quantity of this relationship will differ.

It would be possible to describe in great detail all the differences that exist, but so organized are these differences that it is possible to describe them by their general character. We may say that dominance in the chimpanzee is mostly of a friendly kind, that dominance in the macaque is usually brutal in nature, and that dominance in the cebus is in the first place tenuous and in the second place relatively non-contactual. We have, unfortunately, had too few chimpanzees to work with to be positive about our generalization. Let it be understood that what we say about these animals is based on the useful discussions in the literature, and on direct observation of one pair of young chimpanzees at Vilas Park Zoo, in Madison, Wisconsin; one group of three young chimps at Central Park Zoo; one group of four young adult animals (later separated into two pairs) at Central Park Zoo; and a young female chimp and a female gorilla pair at Bronx Park Zoo. Slight contact with old adults indicates that they may show entirely different dominance-quality from young animals. Older animals apparently become vicious and wild after a certain age.

The dominant chimpanzee (at least in young animals) is a friend and a protector of the subordinate chimpanzee. The subordinate animal shows no fear of his partner, nor does he avoid or run away from him. They form a close contact group and are dependent on each other, frequently becoming so inseparable that isolation for experimental purposes is a disturbing factor, so much will they miss one another. They will share their food, they will play together, and aggression will take the form of rough play or teasing rather than of brutal and wanton attack. Chimpanzees in pairs are rarely scarred, whereas other animals will very often show by their scars the life and death nature of the dominance relationship obtaining in their pairings. In moments of stress or danger, the subordinate chimpanzee will fly to the arms of his dominant partner for protection and comfort, at least in the young animal. In the pair of young chimpanzees that we studied, injury to the one was resented by the other, and they were, in general, inseparable.

The subordinate chimpanzee actually shows anger more often than the dominant one. In our young pair, the dominant male would often tease the subordinate female to the point of complete exasperation. In such cases she would fly into a towering screaming rage, and chase him about the cage in the attempt to vent her annoyance upon him. He would run away laughing. This behavior has never been seen in any other animal although it is seen frequently in the chimp. The subordinate macaque never runs at the overlord, nor does the overlord ever flee.

In a group of three young chimpanzees, the smallest and least dominant formed a contact pair with the second dominant animal. This latter animal was her protector and guardian but preferred to play with the most dominant male who was much more active. In one observation when the smallest animal felt generally fearful, (as evidenced by her continual whimpering and attempts to snuggle close to her protector), she resented fiercely the attempts of the biggest animal to take her protector away in order to play. Finally such was her resentment that she attacked the overlord of the group and pummeled him with her fists. When he laughed and ran away, she turned to her guardian and snuggled close to her again. In a cage of macaques, such a pummeling of the overlord would be a life and death matter. In the chimpanzee it called forth the laughing response.

We have, in adult groups, observed the subordinate animal, for one reason or another, become very angry and go into a fit of screaming and chasing the other animals about the cage. Such a tantrum has never resulted in a fight. The other more dominant get out of her way until the screaming stops.

The behavior in the macaque pair will, ordinarily, be very different from this chimpanzee behavior. Catarrhine dominance is rough, brutal and aggressive; it is of the nature of a powerful persistent, selfish urge that expresses itself in ferocious bullying, fighting, and sexual aggression. The subordinate animal is usually afraid of the dominant animal, and is frequently completely terror-stricken and cowed. These animals never show anything that could possibly be called friendship for their fellows.[4] A weak or sick animal is often killed by his fellows who seize the opportunity to take advantage of his weakness (12). A very subordinate animal is in danger of starving if the supply of food is limited, for the overlord may take it all. The behavior of the subordinate animal almost always shows the effect of this social relationship of brutal dominance; his behavior is always oriented with respect to the dominant overlord. He will almost always avoid the overlord, he will look around fearfully before eating or playing or exploring. His very posture shows fear, for he will slink slowly and carefully about the cage, whereas his dominant partner strides about as if he were alone. Previous papers in this series discuss many examples of such behavior, and the reader is referred to them for more detail. This picture holds true roughly for all the catarrhine monkeys studied by the writer

---

[4]Apparently they *may* show friendship for humans, according to Zuckerman (private communication).

(rhesus macaques, pigtails, baboons of various kinds, java monkeys, etc.) although some more minute differentiations may be made among these various species.

Platyrrhine monkeys may also be tentatively lumped in one group for the sake of this treatment. The writer's experience has been with various kinds of cebus and spider monkeys and with a few woolly monkeys. With these animals may be grouped the howler monkey, for there is nothing in Carpenter's monograph (4) on these animals that contradicts the writer's experience with other platyrrhine animals. These animals are generally low in dominance expression and at times it even appears as if dominance is not expressed in their behavior. Careful long-time study will show, however, that it is present and discernible, even though it exists in a much less marked form than in the other animals we have discussed. For the catarrhine monkey, the dominant animal of a pair may usually be detected a few minutes after they have been together. It takes somewhat longer to diagnose dominance in the chimpanzee, but it takes a very long time indeed for the cebus monkey. When it *has* been established, it is never as clear cut as it is in the catarrhine animal. Where in the latter animal, the subordinate animal may get not a single bit of food out of the 20 bits dropped in (see 10, for description of our food test for dominance), the platyrrhine subordinate may get five or even more out of 20 bits of food, and there will rarely be any squabbling about food, nor any attacks on a subordinate animal by the dominant animal because of food. Their dominance is not often harsh or brutal, and there is little of the persistent sadism so often seen in the rhesus pair. The subordinate cebus will usually show less fear than the rhesus subordinate, and may at times even be aggressive toward the dominant animal. The platyrrhine animal, to a much greater extent than any of the other animals, use vocalizations as an expression of dominance. The prize example is, of course, found in the howlers, but this is also true for the cebus and the woolly monkey. A platyrrhine animal, attacked, is less apt to fight or to struggle, than he is to scream and cry in ever increasing intensity until he is let go.

Probably the most important and most interesting social characteristic of platyrrhine animals is that they pay little attention to each other either in pairs or in large groups. They form very loosely cohering pairs, which may rarely be considered to be units as is the case in the chimpanzee. They rarely show anything that may be called friendship, or protection or dependence on each other, but neither do they bother each other very much. Their rôles in sexual behavior are almost never as fixed as they are in the rhesus, or even in the chimpanzee. A subordinate animal may frequently mount the dominant animal although generally it seems to be true that the dominant cebus also plays the masculine rôle more often than does the subordinate animal.

Our discussion of the group behavior of infra-human primates and its correlation with dominance-quality must confine itself largely to the platyrrhine and catarrhine groups, since very little that is useful for our present purpose has been published for the anthropoid ape. Otherwise excellent field studies have been deficient in descriptions of group behavior and organization in these animals. The writer has had the opportunity to observe ape groups under zoo conditions only, and very few of these. The description of anthropoid sociology is an important task that as yet remains largely undone.

The classical description of social behavior among catarrhine animals is Zuckerman's description of the Hamadryas baboon in the wild and in a large group at the London Zoo (22). These animals, and also the catarrhine groups observed by the writer under zoo conditions, show a definitely organized society with strongly hierarchical social status. The baboon horde is a grouping of a large number of separate families, each of which is composed of an overlord male, as many females as he can secure and hold by his dominance and fighting ability, their offspring, and, that peculiar phenomenon, the bachelor baboon. This latter animal is a male who has not been dominant enough to secure any females for himself. The females of the family, in spite of the fact that they are sexually unavailable to him, continue to exert sufficient sexual attraction upon the bachelor to keep him loosely tied to the family grouping.

The females are definitely "owned" by the overlord and are unavailable to any other male. Zuckerman records only three instances of "unfaithfulness" between a wife and a bachelor baboon in all his observations. As might be expected, there are many ferocious fights for these females, fights in which lives are sometimes lost. There is much homosexuality in these animals.

If we may, for the purposes of this paper, follow the procedure of lumping all catarrhine animals together, it is possible to reason from the dominance-quality we have already discussed for pairs of catarrhine animals to the social organizations of the baboon, which is also a catarrhine animal. We have seen the strong persistent urge to selfish, aggressive dominance that the individual animal displays in the pair, and it is easy to see the connection between such a quality of dominance and the ferocious fighting that the baboon displays in the group. We should expect dominance status to be jealously guarded and affirmed. This we find to be true in the group. We find also, what seems to be a natural concomitant of catarrhine dominance-quality, the strongly hierarchical social organization, the definite family groupings, and the exclusive ownership of the female, since sexual attempts on her are also an attack on the dominance of her owner. It is easy to see also how the phenomenon of the bachelor baboon with its enforced celibacy is a consequence of catarrhine dominance-quality. Homosexuality, as we have shown in a previous paper (11), is also an expression of dominance, and we should as a consequence expect to find more homosexuality where dominance struggle exists. Of course we should also expect to find it as a product of enforced celibacy. Zuckerman's observations have been confirmed by our observations of other catarrhine groups in zoos.

It would seem then that certain social consequences flow at least in part as a natural and to-be-expected accompaniment of a certain dominance-quality. This is not to say that this one factor alone is responsible. This can also be shown to some extent for the platyrrhine.

The writer will rely here for his discussion of social behavior in platyrrhine groups on his own observations of cebus groups under zoo conditions and on Carpenter's brilliant monograph on the social behavior of the howler monkey in the wild (4).

The howlers are organized in homogeneous, closed "clans"[5] in which there is very little consistent and permanent internal grouping of, for instance, the family type (the sub-groupings that are seen are temporary). To be sure, there is some evidence of preference by a female for one male over another when she is sexually receptive, and the more usual sexual behavior reported by Carpenter is of single male-female pairings who remain associated until the male is sexually satiated. This is very different from the situation

---

[5]The word "clan" is apparently not used in the anthropological sense by Carpenter.

to be observed in baboons and macaques, however, for there is little or no fighting between males for the receptive female, and no jealousy to speak of is reported by Carpenter, at least not of the vicious baboon type. While, as has been said, the receptive female shows some preference as between two available males, still, to quote from Carpenter:

> So far as I could ascertain, the female which was receptive but unassociated with a male displayed invitational behavior with reference to the most convenient male. This usually meant the one nearest her. If he did not respond, another one was tried. Once I observed two males respond almost simultaneously to the lingual gestures of a female. . . . The one which arrived first copulated with her and remained a close consort for many hours. The second male followed the consorts , . . , but there was never observed any vigorous competition between the two males for access to the female. . . . . It would seem that the sexual relations of howler males and females should not be classed as polygamous or monogamous, but if a classification is to be made, the relations might be called *communal* (pp. 91-92).
>
> When a female comes into a state of oestrus, and becomes associated with a highly motivated male, primary sexual activity occurs with great frequency. Following an initial period of a few hours, there is a rather rapid decline in the frequency curve to the point at which the first male becomes satiated and discontinues association with the female. The female may then become the consort of another highly motivated male. . . . In this manner two or more males may be satiated by a female during an oestrus period. I have observed five instances of different males copulating with the same oestrus female, and I am convinced that typically more than one male copulated with a female during her period of receptivity (p. 91).

There is no "ownership" of the female among the howlers, as there is among the baboons, for the male-female "exclusive" relationship lasts for no more than a few hours. Neither is there much, if any, of the homosexual behavior which is so common in the baboon.[6] The vicious bullying and ferocity to be observed in catarrhine monkeys is almost totally absent in the howler as well as

---

[6]At least it cannot be as common as in the baboon. A personal communication from Dr. Carpenter says "There is no information as to whether or not howlers are at times homosexual."

in the woolly monkey and the cebus.[7]   The main mechanism for expression of dominance has now become vocalization.   The social hierarchy is very weak, or, to phrase it in another way, has a low gradient.   There is little of either coöperation or competition.   It is impossible to select out "rulers" or over-lords in these animals.

To the writer it seems obvious that this general picture "goes with" the dominance-quality we have already described for the platyrrhine, just as the baboon social picture naturally "goes with" the catarrhine dominance-quality.   While there is little justification or usefulness in speaking of causal priority for either the dominance-quality or the social organization, it is certainly justifiable to point out the correlated status of the two groups of data.   The writer is quite willing to concede the possibly questionable character of his assumption that all catarrhine animals may be lumped in one group, and that all platyrrhine animals may be lumped in another group. In favor of such an assumption, however, is the fact that the writer's experience, while confirming the existence of species differences within the catarrhine and platyrrhine group, indicates that these differences are less important than the wide and important differences between the two main groups.   In any case, this is all subject to experimental and observational verification or disproof, and the hypothesis presented here will have amply justified its existence if it stimulates the search for these very important facts.

Zuckerman's noteworthy attempt (22) to base animal sociology on the sexual physiology of the species as well as on dominance and a few other factors, is particularly interesting in view of the foregoing discussion.   We have already disagreed with this interpretation of primate sociology in a previous paper (11), but the impressive mass of data that he accumulates in support of his thesis cannot be brushed aside lightly, especially in view of the fact that recent work in sub-primates indicates a close correlation between dominance behavior and sexual physiology.   It may be that different explanations are necessary for primates, chickens, and lizards, or what is even more likely, the moral may be the old one that simple explanations in terms of one causal factor are not often likely to be valid.   We should then more cautiously affirm that dominance-quality is at least one factor that correlates with social behavior.

---

[7]The howler may be quite ferocious in his attacks on the isolated male who attempts to join the clan group.   The play-fighting among the juvenile animals may also become rough enough to be called real fighting.

The implications of such a hypothesis (that some organization and group behavior are at least partially correlated with deep characteristics of the "individual" personality), are far-reaching and important. Since dominance as a function of the individual is susceptible of careful scientific study, with probable eventual experimental control and prediction, we may yet be able to lay bare on a scientific basis the psychological and physiological basis of sociological science. There is no reason apparent at the moment why the techniques and hypotheses that have come from the study of infra-humans may not be partially applicable to the similar scientific study of the human.[8]

It is at least conceivable that differences in human cultures may be significantly correlated with some isolable variation of one or a few fundamental dimensions of personality. The question of genic versus social determination of these personality factors is not necessarily involved in such a general consideration. It is possible, for instance, to make a comparison, crude though it may be, between baboon dominance-quality and the similar behavior of the Mundugumor of New Guinea. A study of the social forms that are correlated with baboon dominance-quality in humans does not require a preliminary determination of the relative influence of genic and social factors in the etiology of this dominance-quality.

This study need not be confined to differences *between* cultures. A study now in progress on individual differences in dominance-quantity and quality in a very restricted cultural sub-group indicates also that such individual differences may be very potent determiners of different qualities of social behavior within the one cultural group, even to the extent of some evidence of correlation between widely deviant dominance-quantity and widely deviant social and sexual behavior.

Another cultural problem that might conceivably be helped to solution by the use of our concepts is the problem of differential resistance to acculturation by various peoples. It would seem likely that the quality and quantity of dominance which the normal personality showed in that cultural group would help us to understand why he should resist the foreign influences to which a nearby culture succumbs so readily.

[8]Such studies have been made by the writer since the above was written. See *Psychol. Rev.*, 1937, **44**, 404-429; *J. Soc. Psychol.*, 1939, **10**, 3-39; and forthcoming papers.

REFERENCES

1. ALVERDES, F. Social Life in the Animal World. London: Kegan Paul, 1927.

2. BRIGHAM, H. C. Gorillas in a native habitat. Carnegie Inst. Washington: 1932. No. 426.

3. CARPENTER, C. R. Psychobiological studies of social behavior in Aves: I. The effect of complete and incomplete gonadectomy on the primary sexual activity of the male pigeon. *J. Comp. Psychol.*, 1933, **16**, 25-57.

4. ————. A field study of the behavior and social relationships of Howling Monkeys. *Comp. Psychol. Monog.*, 1934, **10**, No. 2.

5. EVANS, L. T. Winter mating and fighting behavior of *Anolis carolinensis* as induced by pituitary injections. *Copeia,* 1935, **1**, 3-6.

6. ————. A study of social hierarchy in the lizard, *Anolis carolinensis.* *J. Genet. Psychol.,* 1936, **48**, 88-111.

7. ————. Behavior of castrated lizards. *J. Genet. Psychol.,* 1936, **48**, 217-221.

8. MASLOW, A. H. The social behavior of monkeys and apes. *Internat. J. Individ. Psychol.,* 1935, **1**, 47-59.

9. ————. The role of dominance in the social and sexual behavior of infra-human primates: I. Observations at Vilas Park Zoo. *J. Genet. Psychol.,* 1936, **48**, 261-277.

10. MASLOW, A. H., & FLAUZBAUM, S. II. An experimental determination of the behavior syndrome of dominance. *J. Genet. Psychol.,* 1936, **48**, 278-309.

11. MASLOW, A. H. III. A theory of sexual behavior of infra-human primates. *J. Genet. Psychol.,* 1936, **48**, 310-338.

12. ————. IV. The determination of hierarchy in pairs and in a group. *J. Genet. Psychol.,* 1936, **49**, 161-198.

13. MILLER, G. S., JR. Some elements of sexual behavior in primates, and their possible influence on the beginnings of human social development. *J. Mammal.,* 1928, **9**, 273-293.

14. ————. The primate basis of human sexual behavior. *Quar. Rev. Biol.,* 1931, **6**, 379-410.

15. MURCHISON, C. The experimental measure of a social hierarchy in *Gallus domesticus:* I. The direct identification and direct measurement of Social Reflex No. 1 and Social Reflex No. 2. *J. Gen. Psychol.,* 1935, **12**, 3-39.

16. ————. II. The identification and inferential measurement of Social Reflex No. 1 and Social Reflex No. 2 by means of social discrimination. *J. Soc. Psychol.,* 1935, **6**, 3-30.

17. ————. III. The direct and inferential measurement of Social Reflex No. 3. *J. Genet. Psychol.,* 1935, **46**, 76-102.

18. MURCHISON, C., POMERAT, C. M., & ZARROW, M. X. V. The post-mortem measurement of anatomical features. *J. Soc. Psychol.,* 1935, **6**, 172-181.

19. NISSEN, H. W. A field study of the chimpanzee. *Comp. Psychol. Monog.,* 1931, **8**, No. 1.

20. SCHJELDERUP-EBBE, T. Social behavior of birds. In *Handbook of Social Psychology* (Ed. Carl Murchison). Worcester, Mass.: Clark Univ. Press, 1935. (Pp. 947-973.)

21. YERKES, R. M., & YERKES, A. The Great Apes. New Haven: Yale Univ. Press, 1929.

22. ZUCKERMAN, S. The Social Life of Monkeys and Apes. New York: Harcourt, Brace, 1932.

Department of Psychology
Brooklyn College
Bedford Avenue and Avenue H
Brooklyn, New York

5

Reprinted from *Biological Symposia*, **8**, 139–162 (1942)

SOCIAL DOMINANCE AND SUBORDINATION AMONG VERTEBRATES[1]

W. C. ALLEE

UNIVERSITY OF CHICAGO

MOST students of group organization in animals have been attracted to the subject primarily by the opportunity to study general and comparative sociology. This is true in our laboratory as well as elsewhere so far as we have found by searching through the literature.

We have not, in these studies, been attempting an oblique attack upon group organizations of men; but all of us who have worked with these problems have found our attention and curiosity caught by certain similarities between group organization in other animals and some of the simpler phases of human society. Perhaps it will be such comparisons that will most interest the reader. I think it is wise for them to be made, providing only that due restraint is exercised. In order to understand the integrations of human society we need to know how much is uniquely human and how much is, on the other hand, the specialized human development of more generalized primate behavior pattern, which derives from mammalian and lower vertebrate patterns while these in turn are related to certain types of invertebrate social organization.

To be more specific, the social integrations of hens, other flocking birds, dogs, cattle, mice, fish, frogs, lizards, turtles, and many other animals have much in common with the organization of certain human groups. Without taking the comparisons too seriously (if primary sex relations are considered the comparison is almost invariably strengthened) these similarities may help explain human social organizations.

[1] This paper is based on a lecture, illustrated by motion pictures, which was given as a part of this symposium. A shortened version has appeared in Science (Vol. 95: 289–293) under the title "Group Organization Among Vertebrates."

I would not say that the human social variant is justified in existence either because it resembles or departs from the more generalized type.

The modern period in the study of social organization of groups of animals in which the individuals were in some way distinguished from each other was initiated for most of us by Thorleif Schjelderup-Ebbe in 1922.[2] An extensive descriptive literature has accumulated in this field; I have summarized much of this work in preceding books (Allee, 1931, 1938). Most of these studies have dealt with loosely caged flocks of various species of birds; the social order of the common domestic fowl has attracted particular attention. Reports of somewhat similar group organizations in nature are also beginning to appear; some of these will be summarized within the next few pages.

At Chicago we have been interested in such social phenomena since they represent one phase of the broad field of animal aggregations. We have studied the organization pattern in several different species of birds and in certain mammals, especially mice. The following discussion will be based mainly on the work at our own laboratory; not because it is most important, but because of our familiarity not alone with the results obtained but also with the observational and experimental errors and personal biases which may affect conclusions. I shall refer frequently to work done elsewhere in an effort to present the more important developments in this field. While the majority of the pertinent ideas will be reviewed, there will be no attempt to present a through-going summary of all the available literature on this subject.

One of the social hierarchies which we have observed is summarized in Table I.

Such a social order among birds is based on what has come to be called peck-right. The higher ranking individuals are able to peck those of lower rank without them-

[2] Schjelderup-Ebbe in his 1935 review cites an earlier report in Revue Naturae, Bergen, Norway, 1913, which I have not seen.

selves being pecked in return.   This right is usually won
at the first, or one of the early pair contacts between each
two adult members of the flock, either as a result of an
active fight or by passive submission of one of the two in-

TABLE I

THE SOCIAL ORDER IN A FLOCK OF WHITE LEGHORN HENS.   THOSE SHOWN IN
BOLDFACED TYPE WERE LATER INJECTED FOR 51 DAYS WITH
THE MALE HORMONE, TESTOSTERONE PROPIONATE

Indivdual	Number pecked	Indivdual pecked								
RG	7	**BY**	**RY**	BG	RR	GG	....	YY	RW	....
RW	7	**BY**	**RY**	BG	RR	GG	BR	YY	....	....
YY	6	**BY**	**RY**	BG	RR	GG	BR	....	....	....
BR	6	**BY**	**RY**	BG	RR	GG	....	....	....	RG
GG	4	**BY**	**RY**	BG	RR	....	....	....	....	....
RR	3	**BY**	**RY**	BG	....	....	....	....	....	....
BG	2	**BY**	**RY**	....	....	....	....	....	....	....
RY	1	**BY**	....	....	....	....	....	....	....	....
BY	0	....	....	....	....	....	....	....	....	....

dividuals in a contact pair.   Some of these reactions are
suggested by the photographs in Fig. 1.

The flock may or may not be organized into a simple
straight-line hierarchy.   One of the most frequent irregularities comes when a hen with medium or even low social
status has the peck-right over some individual that out-
ranks her in general social position.   Thus in the flock
shown in Table I, BR was pecked by RW and YY and yet
had the peck-right over RG which was otherwise the *alpha*
bird in the flock.   Another fairly common complication
occurs when triangle situations arise in which *a* pecks *b*,
b pecks *c*, and *c* pecks *a*, and then all three birds have the
peck-right over those lower in the social scale.   There is
not space here to suggest all such variations or to discuss
what is known about their causation.

The organization in a flock of hens represents a type of
social pattern in which dominance, once won, is relatively
permanent; we have observed one flock in which the social
status of each hen is known to have remained fixed for
over a year.   At least one other type of social structure
also exists among birds.   With pigeons, doves, canaries

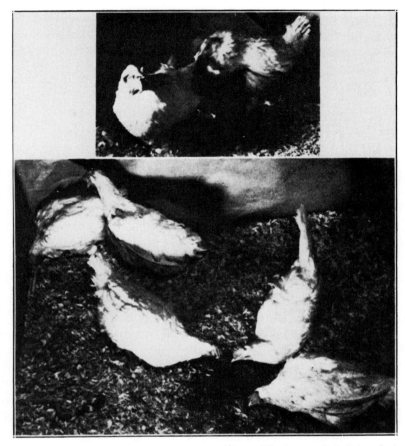

FIG. 1.   The peck order among hens.   Above: The hen to the left has just submitted passively in her first contact with the hen to the right.   As the submissive hen turned away, the other pulled a feather from her back. Below: The *omega* hen at the left is being pecked by the *alpha* bird of the flock.   During the absence of the *alpha* hen, the three intermediates rushed to the food.   These form a triangle of dominance among themselves, which happens to run in clockwise direction in this snap-shot.   The hen to the lower right has just been pecked by her nearest neighbor.   (Photographs by Buchsbaum.   Exposure 0.004 second.)

and shell parrakeets, for example, although the flock organization is no less real, the outcome of any given pair contact is less predictable.   The social structure is based on a sort of differential peck exchange in which domi-

nance is indicated by the ratio of victories to defeats between the members of each contact pair.   Such flocks are
organized on what may be called peck-dominance rather
than on the more absolute peck-right relations that exist
among hens and certain other birds.   As was discovered
several years ago (Masure and Allee, 1934) and confirmed
more than once (Shoemaker, 1939; Diebschlag, 1941), the
peck-dominance type of social order is related in part to
territoriality in that certain birds are dominant in their
own territories and subservient in those of other birds,
hence victories and defeats are determined in part, by
the territory in which the contact takes place.

All group organizations among birds are apparently
based on the ability of birds to recognize and remember
their flock-mates as individuals.   When territory enters
as a factor, recognition of the individual's territory
also becomes a part of the group reaction system.   Shoemaker (1939) has shown that the space available for a
flock of canaries is a matter of importance.   When they
are confined in a small space, the social order becomes
relatively simple and definite; it is little complicated by
territoriality.   Given more space, individual territories
tend to become established in which the territory holder
is usually supreme even though it ranks low when in the
neutral ground about bath bowls, feeding trays and in the
other areas of the canary public service system.   Even a
socially low-ranking male normally dominates other males
in some restricted space about his nest.

In connection with the intermingling of territoriality
and social dominance, I have had access to many unpublished observations made locally in the bird sanctuary in
Jackson Park by Mr. Dale Jenkins and Mrs. Barbara Hale
Brainerd which show that a family of blue geese, as a
family, dominated pairs and single individuals of ducks
and geese during the winter months, and defended the territory about themselves wherever they might be, on land
or in the water.   The defended territory was not precisely
located, but moved with the dominant family.

When there is a recognition both of individuals and of territory it is impossible as yet completely to separate the two. In such cases the question concerns the extent to which the individual plus his territory is the unit in the flock organization, as contrasted with either the individual or the territory as the basic unit.

As was just suggested, there is a possibility that the peck-dominance type of group organization in birds is always associated with territory and is entirely an expression of more aggresive behavior of the birds of a flock in their own recognized territory as contrasted with reduced aggressiveness in the territory of others (Shoemaker, 1939; Diebschlag, 1941). Such differences in the performance of the individuals in the flock would provide the give-and-take type of activity which characterized peck-dominance. If the human observer did not recognize the existence of boundaries while the bird did, then the anomalous situation might arise in which an unaggressive individual might, by human records, stand highest in ratio of pecks administered to those received and so be judged to dominate the flock. Under the suggested conditions, this could happen in the following manner: The bird which was least aggressive would probably remain mostly within its own territory where it would have a psychological advantage over its neighbors who were aggressive enough to invade its territory. Actually, relative social rank in such a system should be measured by the amount of space, and perhaps by the importance of the space in which the different birds are dominant rather than merely by a tabulation of miscellaneous contacts which are won or lost. This could be checked by observations on neutral grounds, if any, and, with caged birds, by staged pair contacts under conditions equally unknown to both participants.

The assumption that peck-dominance is merely the result of differential behavior related to territory is partially supported by the observations of Bennett (1939, 1940) on ring doves. She found evidence of territory recognition by the birds in her homosexual, caged flocks, which were organized on the basis of peck-dominance. She also found that administering the male hormone, testosterone propionate, to either males or females, increased aggressiveness and dominance of the treated birds, a matter which will be discussed in more detail later. Such an increase in male hormone tended to change the peck-dominance behavior pattern into the more definite peck-right type of relationship. Bennett (1940, p. 162) concluded: "The evidence that territorialism fully explained increase in flock stability is inconclusive." She states that, with one exception, ring doves injected with male hormone became more mobile and less confined to a particular location as the injections continued and the birds became more aggressive. Does this mean that the male hormone tended to cause the birds to hold more territory as their aggressiveness increased?

It will be well, as yet, to reserve final judgment concerning the extent to which territory recognition is a factor in the social dominance system of flocking birds.

Early observations upon group organization in birds were limited to penned or caged flocks where individuals may be readily marked and where their detailed behavior is readily followed. Despite the large number of kinds of birds (Schjelderup-Ebbe, 1935) which have been found to have organized flocks, it was natural to question whether this whole matter might not be an artifact characteristic of kept birds. Even the observations of peck-order in the semi-feral conditions that obtain in bird sanctuaries with such birds as the jackdaws of Lorenz (1935) or the geese my students have observed does not prove that peck-orders are characteristic of birds in nature. Recently, however, we have been getting direct information about wild birds on this point. Odum (1941) found a characteristic organization in winter flocks of the Blackcapped Chickadee as did Colquhoun (1942) for the Blue Tit. J. W. Scott (1941a, b) has reported social dominance among males of the sage grouse at their natural breeding grounds and Dr. J. T. Emlen[3] has been able to distinguish high-, mid-, and low-ranking birds in wild coveys of California valley quail.

Such observations leave no room for doubt concerning the occurrence of dominance-subordination organization among some bird flocks in nature. The question remains as to how nearly universal such behavior patterns may be among flocks of wild birds: critical observations on this point are highly important.

We have been considering what may be called social dominance in the strict sense as contrasted with sexual dominance. The two may or may not be easily separated. Thus Noble (1939a) found that the sword-tail fish (*Xiphophorus helleri*) has a straight-line hierarchy ''as rigid as that of the fowl.'' ''In this species,'' Noble said, ''it is possible to distinguish the sex drive from the dominance drive merely by cooling the water, for the sex drive disappears first.''

There is much interesting and even dramatic information concerning inter-sexual dominance, and like our knowledge of many phases of social behavior, it is difficult to make a terse and accurate summary. The usual statement is that the more showy, or the larger, or the stronger sex is dominant. This means that males usually dominate females to a greater or less degree. Schjelderup-Ebbe (1935, p. 960) takes the extreme position that ''In animals female despotism constantly brings about degeneration through the hindering of pairing, thus operating *against* the increase of the species.'' This man, who is often a keen and accurate observer, even concludes that ''The female has no primary aptitude as protector [sic!] and if she becomes the despot, she misuses her power in the most violent way''!

It is true that domestic cocks dominate hens, that during breeding activity male pigeons spend much time chas-

[3] Personal communication.

ing their retreating mates and that male shell parrakeets, which out of the mating season are dominated by the females, turn on them in striking fashion when breeding begins. A similar shift in dominance has been reported for half-wild mallard ducks. Shoemaker (1939) found just the opposite change with his canaries since with them, as the breeding activities increase, females become increasingly dominant. At the height of the breeding cycle females that at other times were low in the canary social system not only dominated their mates, but also some of the other males with which they were associated. Yet there was no lack of successful reproduction in Shoemaker's flock of canaries.

Examples of what Schjeldrup-Ebbe (1935) calls chivalry have been seen in birds, and behavior which can be so interpreted has been reported more recently in caged, mated chimpanzees (Yerkes, 1940). With the latter, during oestrus, the female takes, or is permitted to take, liberties with food priorities for which she would be punished at other phases of the sexual cycle by the usually more dominant male.

SURVIVAL VALUES ASSOCIATED WITH DOMINANCE

What are the long-run biological effects of the dominance-subordination type of group organization at the level of which we are speaking? And what are the known factors that make for dominance? Our observations on flocks of chickens and other birds and on groups of other animals throw some light on these questions.

As to the first: The results of the fact that the highest ranking hens lead the freest lives and low ranking ones are harassed can be shown, among other ways, by relative ease of access to food, or to habitat niches or by studies on egg production. Let us glance at egg production in order to get a concrete illustration. Hens from the lower half of the peck-order lay fewer eggs than do their domineering sisters (Sanctuary, 1932). This is half of the story of the relation of peck-order to reproduction.

The other half is concerned with the effect of social status on mating behavior. Guhl (1941) has issued only a preliminary statement concerning his work on this subject.   He has observed several cocks placed successively or in groups of four with different flocks of hens when all were well accustomed to each other, and has found practically no correlation between frequency of copulation and the social status of the hen.   Mr. Nicholas Collias has unpublished observations like those of Guhl. Skard (1936) had previously found some evidence of such a correlation. Guhl also determined the peck-order in different groups of cocks; again there was no significant relationship between social standing of the cock in relation to his fellows and the frequency of his mating when introduced alone into a flock of these hens.

When, however, the four cocks of a given flock were all placed together in an uncrowded pen of hens, there developed a type of psychological castration of the low-ranking males which, in some individuals, became practically complete.   The details of the whole complicated story have much interest.   I shall give one example from the unpublished records.

The cock Y stood second to R in the peck-order of the males.   Y was sexually aggressive and successful in a rough, forceful way when he was alone with the hens. When the four cocks and seven hens were placed together, the *alpha* cock R would charge at Y and drive him to the roosts whenever Y approached the hens.   Meantime the hens would all scatter and fly to any available perch.   Y soon learned to spend less and less time on the floor and the hens learned to run when he came down to feed.   Y lost weight during this period and his food hunger increased.   For practical purposes Y was no longer sexually effective.

Interestingly enough, these birds show what we can call favoritism as well as antagonism.   Thus the dominant R did not similarly persecute cock G but even allowed him to interfere with R's own courting, and that without punishment.

When Y was placed with another flock of hens he was uninhibited sexually. Placing cocks with strange flocks is known to increase their sexual activity. However later, when Y was again placed alone with the hens that he had been conditioned not to tread, not only did he attempt to copulate less frequently than before he was psychologically castrated, but it was also found that the hens had been conditioned against allowing him to tread them.

Such a system apparently holds in nature for certain species. J. W. Scott (1941a, b) has been studying social dominance in relation to mating in the sage grouse of Wyoming. He has published abstracts only. According to his brief accounts Scott has found an extraordinary system of polygamy. The males have a dominance-subordination organization which is largely determined before the hens appear and is based on bluffing and strutting display, and particularly on the results of fighting. Scott studied a strutting ground which was about one-half of a mile long and about three-hundred yards wide. During the breeding season, this contained over eight hundred birds; by actual count, there were more than four hundred cocks. Five mating spots, each no larger than an ordinary room, were included in this area. "Each spot was occupied by a group of hens, a master cock, his chief rival or sub-cock, and several guard cocks that helped to keep away intruders." The remaining cocks were distributed widely, singly or in pairs. One hundred and seventy-four matings were observed of which the relative dominance position of the cock was known in one hundred and fifty-four cases. Of this number, seventy-four per cent were by master cocks which made about one per cent of the male population; thirteen per cent of the observed matings were made by sub-cocks; these are the chief rivals of the master cocks and like them include about one per cent of the available males. Three percent of the copulations were by guard cocks and the remainder were divided among all the others. Fewer than three per cent of the cocks made eighty-seven per cent of these matings. Copulation occurred only on invitation of the sage hen. Most vigorous, most untiring, most active and usually the largest males attain rank of master cocks but only after strenuous competition.

Conditions such as I have been reviewing indicate that social position in the flock may affect the opportunity of a given male or female for leaving numerous offspring. Those high in their respective peck-orders have the better opportunity for becoming parents of the next generation, as a result of the hen's greater freedom for egg production and of the cock's freedom for copulation.

So much for individual selection as a result of social status. There is a still more important problem in selection to be outlined. Is an organized group sufficiently different from an unorganized one at this level so that selection of the whole lot can occur?

We are coming to realize that groups of animals and other more-or-less integrated population units can be selected somewhat as though they were individuals. In fact, some of us think that natural selection of such populations, rather than natural selection of individuals, is the important basis of evolution (*cf.* Wright, 1940). There is no experimental evidence with regard to the selection of organized as contrasted with unorganized groups. I believe the problem is open to experimental attack.

FACTORS THAT MAKE FOR DOMINANCE

We shall turn from this phase of the subject to consider some of the factors that make for social dominance. Dominance-subordination patterns of behavior may be based on the recognition of other members of the flock as individuals to which a proper reaction must be made. This is the method which holds in many social groups of men, and in all the flocks of birds which we have studied. Opposed to this is a type of impersonal behavior pattern such as is found in many of the groups of mice which have been studied in our laboratory, especially by Uhrich (1938) and by Allee and Ginsburg (1941, and in press). Impersonal group organization depends upon a kind of unoriented, generalized aggressiveness brought in contact with similarly unoriented lack of aggressiveness. This is a type of statistical relationship in which a very bellicose animal dominates its own or foreign groups as a result of his high degree of bellicosity. A dominance-subordination system of behavior with such a basis does not represent the same social mechanism as does one which is based on individual recognition, although the two systems may be related.

We shall consider first the factors which are known to make for dominance in an organization in which the flock mates recognize each other as individuals. For the sake of simplicity we shall limit ourselves to homosexual groups of one species. The following list is based mainly on observations of dominance among birds (Allee, Collias and Lutherman, 1939).

1. The stronger bird usually wins if there is a fight.

2. One bird acts as if intimidated by the appearance of a strange and apparently unfrightened individual and so gives way without fighting.

3. Both birds act as if frightened; one recovers more rapidly and so wins the contact reaction.

4. An individual which is ordinarily aggressive and victorious may be temporarily out of condition (tired, or ill, or moulting severely) and so lose when it might, from other relations, be expected to win. This is a common cause of the irregularities which exist in the social order of many flocks.

5. Mature hens usually dominate immature chickens.

6. The location of the meeting is important; many animals fight best when defending their home territory.

7. Even in strange territory, a bird wins more readily if it is in the company of others with which it has already associated.

8. Birds with young fight more fiercely than they would at other times.

9. With some, canaries for example, intensity of fighting varies with different phases of the reproductive cycle.

10. Birds which stand high in the social order of their own flocks are more likely to dominate a stranger than are birds with low social status.

It is apparent that not all of these factors are of equal importance; they are not listed in the order of their rank and some are more complex than others. The last one stated, social status in the home flock, is much more complicated than is age, weight, or general vigor. At Chicago, we have been much interested in an attempt to analyze the physiological and psychological basis of the dominance-subordination pattern in groups of laboratory birds and mammals.

Many who have observed animals only casually know that males are more aggressive than the females of the species. The males fight more readily and with greater ferocity. Riddle years ago (see Whitman, 1919) postulated that aggressiveness in homosexual fighting between homosexual male or female pairs of ring doves was probably a result of varying degrees of maleness. Allee (1936) suggested that it might be possible by the administration of sex hormones to alter experimentally the social status of selected birds. This suggestion has since been abundantly realized by different experimenters with fish, lizards and turtles, various birds and mammals. This is not the place to discuss all that has been learned about the relation of hormone physiology to social dominance patterns, but the type of results obtained by experimental treatment with the male hormone testosterone propionate, will be illustrated by a typical case in hens.

It is necessary to know that hens produce a potent male hormone which can be extracted chemically from their feces. The material so obtained produced comb growth in capons and this reaction is now generally agreed to be produced only by androgenic stimulation.   The injection of a male hormone into a hen's body does not mean the introduction of a totally strange material into the female system.

We have repeatedly injected testosterone propionate into hens that occupied the *omega* or near *omega* position in the social order of the flock.   The uniform result has been that the hens so treated became more aggressive and moved up in the otherwise stable peck-order of their respective flocks.

Let us consider the case of BY.   As was shown in Table I, this hen was the lowest ranking individual in her flock.   She was given a daily injection of 1.25 mg. of male hormone for over seven weeks.   RY, next to the *omega*

FIG. 2.   BY before (left and after 51 daily injections with testosterone propionate.   (Photograph by Toda.)

hen, received 1.00 mg and BG, which was third from the bottom, received still less.   She became ill and her treatment was discontinued.   The other individuals in the flock were injected daily with neutral oil.

In the fourth week of treatment, BY began to win reversals over hens with higher social status.   Such winning requires hard fighting and persistence on the part of the revolting, low-ranking bird.   She did not overcome the highest ranking hens until somewhat later.   Before the seventh week of daily injections had passed, BY stood highest in her flock in which she had formerly been lowest.

RY alone continued to dominate her and, RY, it will be remembered was also receiving almost as much testosterone as was BY herself. In the meantime BY's whole personal appearance had changed. A part of the transformation is suggested by the before-and-after taking picture given as Fig. 2. In addition to the great increase in size of her comb and other "head furnishings," BY's whole stance had changed, she was more erect and cocky. She was heard to crow repeatedly and three times she was seen to give a somewhat abbreviated male courtship pattern of behavior. She had meantime ceased laying eggs and did not begin again for over three months after cessation of male-hormone treatment, by that time her comb had regressed to that normal for her sex, and her general behavior was again that of one of the meeker hens.

Still, however, she kept her high social rank in her home flock. BY was not inherently an aggressive hen. We could measure this by staged contacts with hens strange to her but whose aggressive fighting ability was known to us. After the effect of the testosterone treatment had worn off, she resumed her habit of losing such contests. Even so, let me repeat, she continued to hold her position as top-ranking bird in her own group. This is apparently a result of memory both on her part and on the part of her flock mates. There is much social inertia in these flocks of hens, and, despite marked changes in personal appearance, in behavior, and in physiology, high social rank once won, tends to be held for a long time.

One other observation about BY helped to stimulate us to make the training tests which are reported in the following section. BB and GY were added to BY's flock soon after treatment with male hormone was discontinued. BY dominated both although they were well-seasoned, mature fighters. This continued for over three months. BY was then being fought in the staged pair contacts I have just mentioned. She lost three such contests in a row and immediately thereafter BB won a revolt over BY in their home pen although the latter seemed to be in good health at the time. Did a succession of defeats produce a psychological reaction which favored further defeats? We turned to mice to secure light from another direction on this problem.

THE SOCIAL ORDER IN MICE

Small lots of male mice caged together develop a social order which, as suggested before, is based on relative aggressiveness and general fighting ability rather than on the obvious recognition of individuals as such. The order is not as stable as that found among hens, but it is sufficiently stable to allow experimentation when this is carefully controlled.

In the work with BY and other hens and in early work with mice, there had been suggestions that a succession of victories tended to condition the individual to be victorious in the next contest, while a series of defeats had the opposite effect (cf. Uhrich, 1940). The fact that mice appear to react primarily to aggressive behavior rather than to a remembered individual made them good subjects for a critical test of this suggestion.

The ease of experimentation was made still greater by the discovery by J. P. Scott (1940) of hereditary strains of highly inbred mice which differed decidedly in fighting tendencies and abilities.[4] Roughly speaking, Mr. Ginsburg and I had available a belligerent strain which was black in color: a strain of pacific mice bearing white coats, and an intermediate agouti strain which generally lost to the blacks and usually won from the whites (See Figs. 3 and 4.) In our hands these brown agoutis fought powerfully once they had started; they were, however, slow to start—perhaps a British type.

These differences in fighting prowess made it possible to expose a high-ranking mouse from the passive white

[4] I am indebted to the Jackson Memorial Laboratory for the gift of these mice.

strain to repeated defeats from the belligerent blacks, and then test the effects of such experiences by again staging intra-strain combats among the white mice. Or, on the other hand, an attempt could be made to "build up" a low-ranking brown or black mouse by repeated contacts with the submissive whites.

FIG. 3. Black defeats brown. (Photograph by Buchsbaum.)

After some 60 fights among themselves, W 1 emerged as the dominant mouse of a group of five white males, and held that status during the next 140 fights. Since this order seemed to be stable, the time was ripe for experimentation. Accordingly W 1 was matched with B 2, the *alpha* mouse of the aggressive blacks, twice a day for eight days. B 2 attacked aggressively even when W 1 was entirely passive. When again matched with his fellow whites, W 1 submitted to every opponent including even the very passive *omega* white mouse. After some 180 fights among themselves, W 1 regained aggressiveness and again became dominant over the white mice. He remained passive when matched against even the least aggressive of the belligerent blacks.

Our experience with other mice indicates that if W 1 had met more active resistance from its own group, it would probably have been slower to reassume aggressiveness. In fact, when W 2 was similarly conditioned downward and then returned to face the other white mice, it was attacked by the dominant W 1 and showed a submissive attitude toward all, until after a series of mild encounters with the *omega* mouse of these pacific whites it again became somewhat aggressive.

It is much easier to cause an intermediate mouse to lose social status by repeated defeats than it is to do the same with a dominant individual. Such an intermediate animal has already been partly conditioned toward submission as a result of losses to the more dominant members of its own group. When intermediate mice were conditioned downward and then kept from meeting dominant mice in their own group, they recovered social confidence just as a dominant individual does under similar conditions.

The dominant white mouse, W 1, was given a longer and more severe experience with repeated defeats. As a result he became so passive that he showed no resistance whatever, and throughout the ensuing two months, he was submissive to all the members of the group which he had formerly dominated. For a time he gave the submissive reaction whenever another mouse came near him; later, only when he was actively threatened. He continued, however, to give up immediately in the face of any show of aggressiveness in another mouse. He regained aggressiveness after being isolated for four months.

Generalizing from a considerable extent of such experience, we have found that it is relatively easy to condition a mouse downward in its social scale, and that the longer and more severe the conditioning, the more lasting the results. We have found that it is also possible to so train a less aggressive mouse that it will become more dominant. This can be done even with low-ranking mice in the most pacific strain. But while it is possible, it is difficult

125

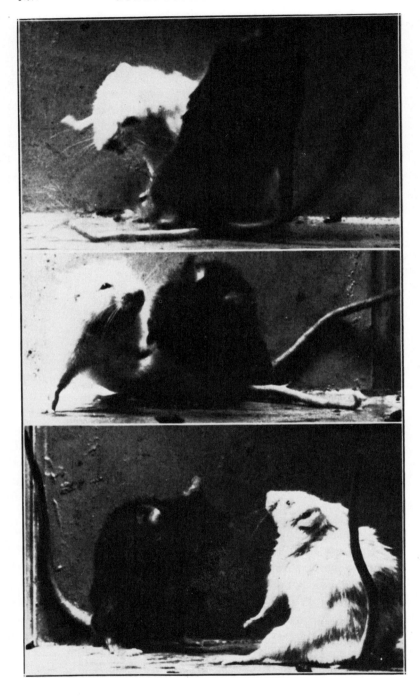

to arrange social contacts such that a mouse at the very bottom of the group organization will show signs of social aggressiveness. The training toward aggressiveness goes exceedingly slowly and must be modified to meet the nuances in the behavior of each individual. As in causing mice to lose social status, it is much easier to train intermediate mice to be aggressive than those which are low in the social scale.

One example must suffice. I choose this particular case since we have a motion picture record of the final battle of such a long conditioning series.

Br 6 was at the bottom of the social order among the brown mice. He was almost completely non-aggressive. Finally Br 6 was mated; and low-ranking, passive white mice were introduced into his home cage. Br 6 had never before made an attack, but now, in the presence of his mate, he. threatened and fought off the mild invaders. Even this show of aggressiveness did not carry over in the absence of a female, and it took six weeks of careful social manipulation combined with a judicious use of isolation, which in these mice helps to build aggressiveness, before Br 6 finally attacked one of the whites when the two were alone together in a neutral cage. After this he was made to encounter several low-ranking whites daily in the fighting cage, and as a result of the total build-up he became definitely aggressive.

The extent of his aggressiveness is indicated by the fact that within an hour after a defeat by B 2, the "fightingest" mouse we had, he vigorously counterattacked and defeated his immediate superior in the social scale among the browns. He also won from other superiors after we had taken the precaution to have these fights staged soon after the latter had been defeated.

FIG. 4. A mouse fight between an aggressive black and a pacific white. Each is typical of his genetic strain. All exposures are 0.001 second. *Upper:* The attack; even at this exposure the lashing tail of the black moved. *Middle:* Black bites white. *Lower:* White "submits" as black poises for a new attack. Note the difference in posture of the front legs. White's right leg is held tensely against his thorax. (Photographs by Buchsbaum.)

Meantime we wanted a hard fight for the motion picture record. B 2, the dominant black, had just suffered two of his rare defeats and was nursing a lacerated shoulder. Even with these handicaps he was an aggressive, hard-fighting mouse. Somewhat optimistically we matched Br 6 against him. It is fortunate that we have this visual record of one of the most decisive inter-strain combats seen in this laboratory. Br 6 lost but only after fighting so hard that he died a few minutes later. There can be little question of the efficacy of the upward conditioning in this case.

This more recent work on hens and on mice has shown that to the ten factors already listed which are known to make for dominance, two more must be added.

11. An increase in the amount of male hormone increases social dominance in either sex.

12. Success tends to make for continued success, and failure tends to produce continued failure.

OTHER CORRELATIONS

The socially dominant animals we have been discussing may or may not be the leaders in their groups. The *alpha* hen in a penned flock does not necessarily lead in foraging expeditions when the hens are turned out into an open lot. In fact, in such a foraging flock leadership changes frequently and the bird at the apex seems always more or less dependent on her followers. With certain other birds, in the flying flocks of which the different individuals can be recognized, the one in front is at times merely the fastest bird in the flock. So far as true leadership is concerned, it is only following along ahead of the main flock. A somewhat similar relationship between leader and followers has been observed among other animals, notably with ants and with men.

In the female herd of cows the dominant animal is the leader. With certain species of deer the female also leads, even when males are present; with other species the male is the leader.   Hence it appears that while group organization with a dominance-subordination pattern occurs among a wide variety of vertebrate animals, the bearing of these patterns on leadership is another matter. Although we now know how to study the problem of leadership in an objective and comprehensive way, actually very little progress has been made in such studies upon non-primate animals.

In general summary we do know, from experimental analysis, that the dominance-subordination social pattern may be influenced by environmental factors and may have its foundations in (a) heredity, as shown by different degrees of aggressiveness in different genetic strains; (b) in the physiological state of the individual, one phase of which is illustrated by studies on the hormonal control of dominance; and (c) on psychological factors associated with experience which with hens and mice may be recent, or remembered from the relatively remote past.

LITERATURE CITED

Allee, W. C.
    1931. ''Animal Aggregations.   A Study in General Sociology.''   University of Chicago Press.   431 pp.
    1936. *Wilson Bull.*, 48: 145–151.
    1938. ''The Social Life of Animals.''   Norton, New York.   293 pp.
Allee, W. C., N. E. Collias and Catharine Z. Lutherman.
    1939. *Physiol. Zool.*, 12: 412–440.
Allee, W. C., and B. Ginsburg.
    1941. *Anat. Rec.*, 80: 50.
Bennett, Mary.
    1939. *Ecology*, 20: 337–357.
    1940. *Ecology*, 21: 148–165.
Colquhoun, M. K.
    1942. *British Birds*, 35: 234–240.
Diebschlag, E.
    1941. *Zeitschr. f. Tierpsychol.*, 4: 173–188.
Guhl, A. M.
    1941. *Anat. Rec.*, 80: 113.
Lorenz, K. Z.
    1935. *J. f. Ornithol.*, 83: 137–213.
Masure, R. H., and W. C. Allee.
    1934. *Auk*, 51: 306–327.

Noble, G. K.
    1939a. *Am. Nat.*, 73: 113–126.
    1939b. *Auk*, 56: 263–373.
Odum, E. P.
    1941. *Auk*, 58: 322–323.
Sanctuary, W. C.
    1932. Master's Thesis, Library of Mass. State Coll., Amherst, Mass.
Schjelderup-Ebbe, T.
    1922. *Zeitschr. f. Psychol.*, 88: 225–264.
    1935. In Murchison's ''Handbook of Social Psychology,'' pp. 947–972.
Scott, J. P.
    1940. *Anat. Rec.*, 78: 103.
Scott, J. W.
    1941a. *Bull. Ecol. Soc. Am.*, 22: 38.
    1941b. *Anat. Rec.*, 80: 51.
Shoemaker, H. H.
    1939. *Auk*, 56: 381–406.
Skard, A. G.
    1936. *Acta Psychol.*, 2: 175–232.
Uhrich, J.
    1938. *J. Comp. Psychol.*, 25: 373–413.
    1940. *Ecology*, 21: 100–101.
Whitman, C. O.
    1919. Posthumous works. H. A. Carr, ed. Carnegie Inst. Wash. Publ.
        257. Vol. III. 161 pp.
Wright, Sewall.
    1940. *Am. Nat.*, 74: 232–248.
Yerkes, R. M.
    1940. *J. Comp. Psychol.*, 30: 147–186.

Reprinted from *J. Abnormal Social Psychol.*, **41**(4), 385, 395–397 (1946)

PROBLEMS IN THE BIOPSYCHOLOGY OF SOCIAL ORGANIZATION

BY T. C. SCHNEIRLA

Fellow of the John Simon Guggenheim Memorial Foundation

THERE is an ancient and understandable tendency to draw moral conclusions from apparent similarities between man and lower animals, for which the social insects have served as convenient material. Solomon's advice to the sluggard, "Go to the ant . . . ," comes readily to mind as an outstanding instance of seemingly infallible repute. Unfortunately, in the light of present knowledge about individual differences among insects in social participation (Combes, 24; Chen, 21), this moral cuts two ways, for the sluggard might well find all his time taken up in contemplating the leisurely ways of those relatively sessile and less productive members which any ant-hill is almost certain to contain. On the other hand, we who are not sluggards may learn much from a more careful comparison of social activities in lower animals and in man, very possibly to the great advantage of better insight into man's social capacities and potentialities.

Scientists exhibit a growing tendency to study comparatively the makeup of what are considered different levels of organization in the inorganic and organic worlds (Redfield, 61). Simple and complex levels are recognized among inorganic phenomena, and in an evolutionary sense certain of the inorganic levels are recognized as prerequisite to the occurrence of organic wholes such as viruses which are regarded as primary. Among biologists and students of behavior there is an increasing alertness for what can be learned from the investigation of one level or type of organization that will assist in the improved understanding of others. An interest in comparative study leads to a closer examination of the various instances of organization describable as levels, for example the individual organism, the animal aggregation, social group and society (Allee, 2). We shall want to inquire how far such studies have advanced beyond the stage of description and naming, exemplified by Alverdes' (5) general survey, and are searching out the essential qualities of different animal organizations. And since there is implicit in the contemporary study of both individual organisms and social groups a conviction that such phenomena must be regarded as unitary wholes in some important sense, we should be concerned about the meaning and validity of this doctrine in the study of different social organizations. In what sense is an individual ant a unitary whole, or an ant colony, a species, a human city, nation, or a "United Nations"?

* * * * * *

Editor's Note: A row of asterisks indicates that material has been omitted from the original article.

"Dominance" Concepts Inadequate For Studying Organization

The role of trophallaxis has not been explored in the general study of vertebrate social organization. Instead, both investigation and theory have featured "dominance" relationships as presumably the most essential factor. As described by Schjelderup-Ebbe (63) in his classical studies with newly assembled groups of barnyard fowl, a social rank or dominance hierarchy is established in time by dint of reciprocal pecking among individuals. After an initial period of fighting and social instability, a ranking is established in which (in the simplest case of a linear peck-order) hen No. 1 pecks all of the others but is not pecked in turn, No. 2 pecks all but No. 1 and is pecked only by No. 1, and so on to the most subordinate member which is pecked by all and may never peck any of the superior members. Dominance orders have been described for a considerable number of bird species and for various other vertebrates including some primates (Allee, 2). Typically, dominance order is considered dependent upon aggressive behavior (Collias, 23); for example Maslow (54) defines the dominant animal as "one whose behavior patterns are carried out without deference to the behavior of his associates."

However, the term does not always have this meaning and in general is not used very consistently. For instance, Noble, Wurm, and Schmidt (57) used the relative height of bill-holding between the members of a heron pair as an indication of dominance, on the assumption that this indirectly resulted from superiority in food-snatching as nestlings. And Maslow, who lays much stress upon dominance in studying primate social relationships, as the above definition suggests, appears to lapse both from consistency and from clarity when he states that "dominance in the chimpanzee is mostly of a friendly kind. . . . The dominant chimpanzee (at least in young animals) is a friend and a protector of the subordinate chimpanzee. . . . They form a close contact group and are dependent on each other. . . ." "Dominance in the macaque is usually brutal in nature, and . . . dominance in the cebus is in the first place tenuous and in the second place relatively non-contactual" (54, pp. 314–315). What these diversified relationships have in common would seem better characterized as "ascendancy," a behavior trait not necessarily dependent upon aggressiveness.

It is highly important to note that describable dominance hierarchies appear under rather special conditions, particularly when groups of birds or primates are confined within a small space, when incentives (i.e., food or drink) are restricted in quantity or in accessibility, or when sexual responsiveness is high. Moreover dominance relationships do not always stand out in grouping lower vertebrates, even in groups of fowl and especially when the groups are large, when social organization displays different characteristics (Fischel, 34). In certain animals studied carefully under field conditions, as for instance the howler monkeys studied by Carpenter (16), an efficient group organization exists without signs of intra-group dominance or aggressive relationships, whereas in others (e.g., baboons) aggression-dominance is prominent (Carpenter, 17, 19).

We must seriously entertain the possibility that dominance theory is an inadequate basis for the study of vertebrate social behavior. "Dominance" appears to be just one characteristic of social behavior, sometimes outstanding in group behavior and sometimes not, according to circumstances. Since the true "dominance" relationship is one of real or abbreviated aggression and withdrawal (Collias, 23), dominance must be considered a factor promoting the isolation and greater psychological "distance" of individuals, and thus more or less counteractive to factors which hold the group together. Perhaps a dominance situation may be viewed as one in which positive unifying factors are relatively weak but are somewhat artificially reinforced by special conditions, such as food scarcity or sexual receptivity (which serve to heighten reactivity to specific stimuli from other individuals), or by physical confinement of the group.

Factors related to social facilitation (Allee, 2, 3), and probably based upon original trophallactic relationships of one kind or another, may well furnish the major unifying basis in groups of lower vertebrates (e.g., schooling fishes), as seems to be the case in the social insects and in mammals. As an alternative to emphasizing dominance in adult groups, it would be well to examine carefully the nature and persistency of early contacts with one or both parents during the period of incubation in relation to "gregarious tendency" in later life. In the domestic fowl, as Brückner's (14) study shows, an intimate trophallactic relationship first exists between hen and chicks, only to be displaced after a few weeks by a condition in which the hen drives off the young. Comparably in mammals, the age of young when weaning occurs, if this change abruptly enforces separation from family conditions, presumably has an influence upon subsequent readiness to group and upon group behavior (e.g., the prominence of interindividual aggressiveness).

In a real sense, aggressive or dominance reactions are an indication of weak social responsiveness either because of individualistic reactions (e.g., sexual reactions) or an incompletely established group organization. Actually, interindividual adjustments in the formation of a new vertebrate group often pass from a stage of overt aggression to one in which tolerance reactions and social facilitation exist (Taylor, 74; Collias, 23). In different animals, the readiness with which early aggressive reactions change into qualitatively different relationships permitting a closer group unity may be opposed to different extents by the individualizing influence of factors such as sexual responsivenes, but otherwise may depend upon species capacity for modifying behavior. Bard (6) and Collias (23) have remarked that as cerebral cortex increases in the animal series, aggressive dominance relationships appear to drop back as prominent characteristics of social behavior. The importance of interindividual grooming, clearly a trophallactic unifying factor of compelling force in primate groups, has been emphasized by Yerkes (86). An infant chimpanzee first raised in isolation from others of its kind (Jacobsen, Jacobsen, and Yoshioka, 43), was at first aggressive when placed with a strange young chimpanzee, but within a few weeks there developed a pacific relationship of mutual dependence between the two.

Concepts such as dominance hierarchy tend toward a particularistic, static type of thinking about social organization, actually distracting attention from the essential problem of group unity. In studying social organization on any level, a theoretical procedure is desirable which centers around the conception of a dynamic integrative process rather than given characteristics such as dominance which may be sometimes absent. We believe that a more thoroughgoing ontogenetic survey of social behavior in the vertebrates will reveal the prerequisite importance of trophallactic processes for group unity wherever it occurs and whatever its strength.

* * * * * * *

Bibliography

2. Allee, W. C. *The social life of animals.* New York: Norton, 1938.

3. Allee, W. C. Social biology of subhuman groups. *Sociometry,* 1945, **8**, 21–29.

5. Alverdes, F. *Social life in the animal world.* New York: Harcourt, Brace, 1927.

6. Bard, P. Neural mechanisms in emotional and sexual behavior. *Psychosomatic Med.,* 1942, **4**, 171–172.

14. Brückner, G. H. Untersuchungen zur Tiersoziologie, insbesondere zur Auflösung der Familie. *Z. Psychol.,* 1933, **128**, 1–110.

16. Carpenter, C. R. A field study of the behavior and social relations of howling monkeys. *Comp. psychol. Monogr.,* 1934, **10**, 1–168.

17. Carpenter, C. R. A field study in Siam of the behavior and social relations of the gibbon *(Hylobates lar). Comp. psychol. Monogr.,* 1940, **16**, 1–202.

19. Carpenter, C. R. Concepts and problems of primate sociometry. *Sociometry,* 1945, **8**, 56–61.

21. Chen, S. C. The leaders and followers among the ants in nest building. *Physiol. Zool.,* 1937, **10**, 437–455.

23. Collias, N. E. Aggressive behavior among vertebrate animals. *Physiol. Zool.,* 1944, **17**, 83–123.

24. Combes, M. Existence probable d'une élite non differenciée d'aspect, constituant les véritables ouvrières chez les Formica. *C. R. Acad. Sci.* (Paris), 1937, **204**, 1674–1675.

34. Fischel, W. Beiträge zur Soziologie des Haushuhns. *Biol. Zent.,* 1927, **47**, 678–695.

43. Jacobsen, C., Jacobsen, M., & Yoshioka, J. G. Development of an infant chimpanzee during her first year. *Comp. psychol. Monogr.,* 1932, **9**, 1–94.

54. Maslow, A. H. Dominance-quality and social behavior in infra-human primates. *J. soc. Psychol.,* 1940, **11**, 313–324.

57. Noble, G. K., Wurm, M., & Schmidt, A. Social behavior of the black-crowned night heron. *Auk,* 1938, **55**, 7–40.

61. Redfield, R. (Ed.) Levels of integration in biological and social sciences. *Biol. Symposia,* 1942, **8**, 1–26.

63. Schjelderup-Ebbe, T. C. Beiträge zur Sozialpsychologie des Haushuhns. *Z. Psychol.,* 1922, **88**, 225–252.

74. Taylor, W. S. The gregariousness of pigeons. *J. comp. Psychol.,* 1932, **13**, 127–131.

86. Yerkes, R. M. Genetic aspects of grooming, a socially important primate behavior pattern. *J. soc. Psychol.,* 1933, **4**, 3–25.

II
Some Examples

Editor's Comments on Papers 7 Through 11

Papers 7 through 11 are included in this collection primarily to give the reader a taste of the descriptive–analytic approach to social hierarchy that characterized the mid-1940s and 1950s. Here, while there was still strong emphasis on the description of social interactions, observations were often moved from the close confinement of the laboratory into a more natural ecological setting, even a barnyard, and attempts were made to correlate rank orders with factors in the animals' biotic environments. Jenkins studied the interaction between territory and social hierarchy in geese in a bird sanctuary; those portions of his paper dealing with social hierarchy are reprinted here (Paper 7). Other examples of the social hierarchy–territory interaction are available in the Benchmark volume *Territory*, edited by A. W. Stokes.

Whereas a territorial organization depends upon the animal's ability merely to distinguish between "insiders" and "outsiders," a hierarchial organization demands not only the ability to distinguish among conspecifics (at least among the immediate neighbors) but also the capability of retaining the distinction over some period of time. Such discrimination and retention capabilities imply a fairly complex nervous system, and so it was generally assumed that stable hierarchies could only be observed in "higher" animals. However, Bovbjerg (Paper 8) was able to demonstrate a rank order in an invertebrate, the crayfish, by eliminating the overriding influence of their natural territories. Thus, while the crayfish utilizes only simple discrimination systems in a natural setting, it is capable of more sophisticated behavior if the occasion warrants.

Guhl's 1953 article (Paper 9) brought the findings of a number of biosociological studies to focus on the chicken, both as a laboratory animal and as a domestic animal. This classical article, long out of print, was a significant step toward a newly emerging emphasis on the behavior of domestic animals and the economic implications thereof. In the same vein, Schein and Fohrman (Paper 10) studied the social organization in a herd of dairy cattle and found a basically stable dominance hierarchy based largely upon seniority in the herd. Subsequent work suggested that social disruptions could have a detrimental effect on milk production; up to a 5 percent decline was reported. McHugh's work with the American buffalo clearly demonstrated that rank order in ungulates is not an artifact of a closely managed barnyard; free-roaming buffalo on open range exhibited rank orders much like those observed in dairy cows. Those portions of McHugh's 1958 article dealing with dominance orders are reprinted here (Paper 11).

7

Reprinted from *Auk,* **61,** 30–38, 46–47 (1944)

TERRITORY AS A RESULT OF DESPOTISM AND SOCIAL ORGANIZATION IN GEESE

BY DALE W. JENKINS

INTRODUCTION

TERRITORIES of birds have been carefully studied and variously defined from the standpoint of the function and result of territory. These studies have been concerned with breeding, nesting, pairing,

Editor's Note: A row of asterisks indicates that material has been omitted from the original article.

adequate food supply, and space between birds. Few detailed or quantitative studies have been made upon one of the main criteria by which we are aware of the presence and extent of territories, namely the contact behavior and relationships with other birds of the same species or other species.

The purpose of the present study was to attempt to find and measure aggressiveness, individual relationships, and social organization in birds under, as nearly as possible, natural social conditions, and to observe territory throughout the year from this viewpoint. The study was made on a Ridgway Fellowship at the University of Chicago, during the academic year, 1939–40. It was undertaken at the suggestion and under the direction of Dr. W. C. Allee. Grateful acknowledgements are made to Dr. Allee, Mrs. M. M. Nice, the late Dr. G. K. Noble, Dr. H. H. Shoemaker, Dr. S. C. Kendeigh and Dr. N. E. Collias for suggestions and helpful criticisms, to Mr. R. E. Smart for permission to use the sanctuary, and to friends who helped make the observations.

The social organization of a few domesticated birds, such as chickens, canaries, and pigeons, and of captive birds, has been carefully studied during recent years, and the types of social relationship are fairly well known.

Two main types of intraspecific social organization have been found in birds. The first, known as 'peck right,' was reported by Schjelderup-Ebbe (1922) in flocks of the common chicken. When schematically developed, this is a straight-line type of hierarchy in which one bird is dominant over all the rest and pecks all birds below it; a second bird immediately below pecks all below it, and so on down the line. Quite a number of species have been found to have this type of social organization. The second type is known as 'peck dominance,' and was reported by Masure and Allee (1934) in the pigeon. This type of social organization presents no absolute peck right over subordinates, but is based on peck dominance after many conflicts. The bird which has won most frequently in a pair-contact cannot always be predicted to win. Peck dominance has also been reported in the Shell Parakeet by Masure and Allee (1934); in Ring Doves by Bennett (1939); and in the Canary by Shoemaker (1939). These two types of peck order are not always distinct and may intergrade.

MATERIALS AND METHODS

The present study was based on three native species of geese and four native species of ducks. These included six Blue Geese, *Chen*

caerulescens, two Lesser Snow Geese, *Chen h. hyperborea,* and twenty-
six Canada Geese, *Branta c. canadensis.* Some of these had been
raised in Minnesota and released in the Jackson Park Bird Sanctuary, Chicago, Illinois, where these observations were made. The
geese raised young, which were included in the observation.

These geese were not tame, and if they were disturbed by intruders,
the Canada and young Blue Geese flew away in characteristic V-forma-
tion and returned only after a considerable lapse of time. The other
geese, having one wing tip clipped, fluttered over the water to the
other side of the lake in the sanctuary.

Various ducks were also on the sanctuary lake, and these included
the Mallard, *Anas p. platyrhynchos;* Lesser Scaup, *Nyroca affinis;* the
Redhead, *Nyroca americana;* and the Wood Duck, *Aix sponsa.* The
Mallards were mostly tame, but the other ducks were wild; many
wild ducks came into the sanctuary and their relationships with the
geese were observed.

Since the Blue and Snow Geese were usually found together and
behaved similarly, they were considered together. The Blue Geese
were divided into two groups—one a family composed of two parents
and two offspring, and the other of a male and a female which were
usually closely associated. The Snow Geese probably were both males
or at least of the same sex. Since geese show no sexual dimorphism
in plumage or size, the sex was not definitely ascertained until their
mating in the spring, when the information given below was finally
completed.

It was necessary to recognize each individual bird in order to establish its social position. Attempts to mark the individuals in various
ways, such as by shooting dye from a distance with a blowgun and
water pistol and by trapping the birds, failed because of their wildness. It was necessary to learn the birds by their plumage, actions,
and other characteristics. The following tabulation shows the symbols used and the status and recognition marks of each goose:

BLUE GEESE

Family
  C—Father of family; all white head; usually main guard.
  A—Mother of family; head white with dark stripes on side.
  S —Immature in family; head partly white; wings unclipped; sex unknown.
  B—Immature in family; head all black, turned white in spring; sex unknown.
Mated Pair
  M—Male and mate of W; head white; dark stripe on back; white-margined
    feathers.
  W—Female and mate of M; head entirely white.

SNOW GEESE
>   L—Left wing tip only dark; probably male; stayed with M and W.
>   R—Right wing tip only dark; usually alone.

CANADA GEESE
>   Certain birds with deformities and size differences were recognized, but since all the birds could not be differentiated, the intraspecific social organization could not be determined.

Observations were made throughout an eight-month period extending from October 18, 1939, to June 15, 1940. The data in this paper are compiled from the observations of about 420 hours in the field. The average peck frequency or rate of conflict was about three pecks per hour. The habits and actions of the geese were observed at all hours of the day and night, but most of the observations were made at feeding time in the late afternoon. The geese were fed grain and bread which was scattered on an island while the geese rested in a lagoon. If the geese were very hungry they would accept food thrown to them from outside the sanctuary fence.

A typical peck was delivered by the dominant bird forcefully striking with its bill at the posterior part of the subordinate bird. The pecked bird quickly moved away, often shaking the rump and tail from side to side, which seemed to denote subordinance, defeat, or loss of food. Actual chasing of a subordinate bird also was counted as a display of dominance.

OBSERVATIONS AND DISCUSSION

The data reported here are observations of despotism and defense behavior of individual, pair, and family; the social inter-actions and resulting territories and social order of geese and ducks. Before the territorial aspects are discussed, the observations on the social organization and despotism will be presented. Observations of the geese and ducks appeared to show not only an intraspecific social organization between members of the same species, but also an interspecific social organization between different species. The intraspecific social organization was found within each of the three species of geese studied, while the interspecific social organization was found between the three species of geese, between the four species of ducks, and between the geese and ducks.

INTRASPECIFIC SOCIAL ORGANIZATION

Fall and Winter Peck Order.—A definite 'peck right' type of social organization, greatly modified by strong family ties and by mated pairs, was observed in the geese. This is shown in the left half of Table 1. No despotism existed in the family C, A, S, B, nor between the pair of mated birds M and W.

TABLE 1

INTRASPECIFIC SOCIAL ORGANIZATION OF BLUE AND SNOW GEESE

Peck order in fall and winter			*Peck order in mating and nesting time*		
Bird	*Number pecked*	*Birds pecked*	Bird	*Number pecked*	*Birds pecked*
C	4	R W L M	C	6	R B S W L M
A	4	R W L M	A	0	
S	4	R W L M	M	4	R B S L
B	4	R W L M	L	4	R B S W
M	2	R L	W	2	R S
L	2	R W	S	0	
W	1	R	B	0	
R	0		R	0	

All members of the Blue Goose family were dominant over the rest of the geese and ducks during the fall and winter and until the time of mating, when the family relations were broken on April 2, 1940.

The pecks occurring in conflicts between individual birds for the five and one-half months during the fall and winter are summarized in Table 2.

TABLE 2

SUMMARY OF INTRASPECIFIC PECKS (FALL AND WINTER ONLY)

C pecks A—0 times	S pecks C—0 times	M pecks L—42 times
S—0	A—0	W—0
B—0	B—0	R—9
M—18	M—24	
L—24	L—32	L pecks W—12 times
W—7	W—10	R—22
R—10	R—15	
		W pecks R—2 times
A pecks C—0 times	B pecks C—0 times	
S—0	A—0	R pecks ——0 times
B—0	S—2?	
M—19	M—4	
L—30	L—10	
W—8	W—1	
R—9	R—7	

This is a modified peck order due to the toleration among members of the family and between the birds of the mated pair. Two pecks were observed in the family by the immatures, B and S; however, this represented attempts by B to get food from the bill of S, rather than true pecks.

The members of the family were always together, and the paired M and W were usually together with L, while R, the Snow Goose at the bottom of the peck order, was nearly always driven away or alone.

The mother, A, had an antipathy against L when attempts were made to invade territory, and gave L thirty severe pecks. The male, M, of the mated pair pecked L forty-two times. Another antipathy occurred in Snow Geese; R was viciously pecked twenty-two times, chased, and kept from eating by L. Geese low in the peck order bothered the ducks more than the superiors did.

Organized Despotism and Facilitation.—Geese were often noticed to defend or take the part of a mate or of another goose. Table 3 shows all possible combinations of the family working together except C, B, S; C, A, B; and A, B, S. M, W, and L made an unstable group. The table may be expanded to read, *e.g.*, C, A, S, and B were observed to dominate 10 Canada Geese, once; 26 Canada Geese, twice; etc.

TABLE 3

COMBINATIONS AND FACILITATION

C, A, S, B pecks (10 Can.) 1; (26 Can.) 2; (4 Can.) 3; (L) 1; (M) 1.
  C, A, S pecks (10 Can.) 1; (2 Can.) 1.
    C, A pecks (2 Can.) 4; (MWL) 1; (L) 3; (M) 1.
    C, B pecks (10 Can.) 1; (10 Mal.) 1.
    C, S pecks (3 Can.) 1; (1 Can.) 2; (MWL) 1.
    A, B pecks (3 Can.) 1; (L) 1.
     B, S pecks (M) 1.

  M, W, L pecks (1 Can.); (S) 1; (R) 3.
   M, W pecks (3 Can.) 2; (1 Can.) 2; (L) 5.
    W, L pecks (R) 1.
    M, L pecks none

The family group showed organized despotism through its strong integration and coöperation, resulting in the dominance of the family. Division of labor was noticeable. The father, C, was usually on guard while the rest of the family was feeding, and defended the family mainly from the larger Canada Geese; the mother, A, often defended the family against the other Blue and Snow Geese. The offspring were kept between the parents. This well-integrated family might be called a family supraorganism, since it performs the activities of a larger, more complex individual, through coördination of its components. This results in the dominance of the family, which is of survival value to its members in that they can feed first and rest in the center of the aggregation, and are not pecked or chased.

This organized despotism appears rather unusual, since Schjelderup-

Ebbe (1935), who has studied peck order extensively in many species of birds, states: "The writer has not been able to prove that organized despotism (by agreement between two or more individuals) exists in birds."

Change of Peck Order in Spring.—Throughout the fall, winter, and early spring, the peck order remained the same. During the coldest weather, while the lake was frozen, the geese and ducks were often huddled in an aggregation at the edge of the ice. During the latter part of March, the male Blue and Canada Geese became more active, and their response to the females became more evident. The males began chasing each other, honking loudly, and, after chasing other geese or ducks, ran back to the female and gave a 'Triumphgeschrei,' in which both male and female stood facing each other and bowed their heads together, honking loudly all the while.

Fortunately, as a result of extended observation just at this time, I saw what was probably the first actual breaking up of the Blue Goose family. This occurred at 11:00 A. M. on April 2, a very warm day. The father, C, kept the mother, A, close and occasionally pecked their offspring, S and B. Finally B was chased away and then S. The offspring were pecked by Canada Geese and by M, W, and L. They attempted to rejoin the mother, A, but were driven away by the father, C. Later that day, at 5:00 P. M., the offspring were again tolerated, and the father defended them against other geese. From April 2 until April 30, the offspring were driven from the family on warm days, and were tolerated on the cold days of Chicago's variable climate. After April 30, the offspring were never tolerated, and C began courting and mating with A.

A new peck order, which lasted from April 2 until June 15, or longer, is given on the right side of Table 1. This table shows the reversals and loss of dominance of the immatures, S and B, when the family broke up. The parents, C and A, maintained top positions, while M, L, and W became dominant over S and B. Table 4 gives a summary of the pecks observed after the break-up of the family.

The immatures, S and B, were pecked by all but R. The immatures themselves pecked no other birds. A vicious fight occurred on April 30 between M and L. This was probably a challenge fight started by L, perhaps over M's mate, W, with which L had attempted to mate. This attempt of L, a Snow Goose, to mate with a female Blue Goose is very interesting. Some systematists regard the Blue Goose as a color phase of the Lesser Snow Goose, and hybrids are known.

TABLE 4

SUMMARY OF INTRASPECIFIC PECKS (APRIL 2, TO JUNE 15)

C pecks A—0 times   M pecks L—15 times   S pecks B—0 times
       M—5   W—0   R—0
       L—43   S—2
       W—3   B—1   B pecks S—0 times
       S—16   R—1   R—0
       B—7
       R—3   L pecks W—5 times   R pecks - —0 .times
                                S—13
A pecks C—0 times   B—0
       M—0   R—2
       L—1
       W—0   W pecks S—3 times
       S—0   B—0
       B—0   R—3
       R—1

INTERSPECIFIC SOCIAL ORGANIZATION

Interspecific social relations of birds have received little attention
from ornithologists, except for chance observations at feeding stations
and especially at nests. A very definite interspecific social organization was found between the groups and three species of geese and
four species of ducks studied. This was a 'peck-right' type of social
organization. The relationships of the birds in the interspecific peck
order are shown in Table 5.

Size is an important factor but is not the only factor causing interspecific dominance. The members of the family of Blue Geese were
dominant over the Canada Geese, which were almost twice the size
of the Blue Geese. The parents, C and A, were still dominant, even
after the break-up of the family.

A summary of the interspecific pecks for the whole period of observation is given in Table 6. Some reversals were observed which

TABLE 5

INTERSPECIFIC SOCIAL ORGANIZATION

Species	*Species and Groups pecked*
Fam. Blue	pecks Scaup, Wood, Redhead, Mallard, Snow G., M, W. Blue, Canada G.
Canada G.	pecks Scaup, Wood, Redhead, Mallard, Snow G., M, W. Blue
M, W. Blue	pecks Scaup, Wood, Mallard, Snow G.
Snow G.	pecks Scaup, Wood, Redhead, Mallard
Mallard	pecks Scaup, Wood, Redhead
Redhead	—
Wood	pecks Scaup
Scaup	—

TABLE 6

Summary of Interspecific Pecks

Species or group	Species pecked	Pecks	Back pecks	Species or group	Species pecked	Pecks	Back pecks
Fam. Bl. G.	Canada G.	259	0	Snow G.	Mallard	28	0
	M, W Blue G.	99	6		Redhead	1	0
	Snow G.	185	13		Wood	3	0
	Mallard	44	0		Scaup	2	0
	Redhead	3	0				
	Wood	4	0	Mallard	Redhead	3	2
	Scaup	6	0		Wood	6	0
					Scaup	1	0
Canada G.	M, W Blue	65	5				
	Snow G.	87	18				
	Mallard	9	0	Redhead	Wood	0	0
	Redhead	3	0		Scaup	0	0
	Wood	6	0				
	Scaup	7	0	Wood	Scaup	8	0
M, W Blue G.	Snow G.	72	18	Scaup	—	0	0
	Mallard	6	0				
	Redhead	0	0				
	Wood	1	0				
	Scaup	2	0				

deserve explanation. The reversals against the family are due to the ostracizing of the immatures, S and B, at the beginning of the mating season.

Although the relationship among the Canada Geese, the Blue Geese, M and W, and the Snow Geese looks like peck dominance, it is because M, W, and L were dominant over a few crippled and small Canada Geese. There was no pecking back and forth, so that no true reversals were noticed in the geese. The relation between the Blue Geese, M and W, and Snow Geese, L and R, is due to L being dominant over W, while M was dominant over both L and R.

The relations of the ducks were not observed enough for any conclusions to be made. A duck with food was often chased by a subordinate. The order of feeding and the formation in moving toward food followed the sequence of the interspecific peck order in that the most dominant group fed first, followed by the next in dominance, down the scale.

*   *   *   *   *   *   *

SUMMARY

1. A definite 'peck-right' type of intraspecific social organization (modified by family ties and mated pairs) was found in Blue, Snow, and Canada Geese, under approximately natural social conditions.

2. Family relations and ties are strong outside of the mating and nesting season. In the breeding season, the adult males become despotic and cause a disruption of the families, resulting in a change of peck order, due to the offspring being driven from the families.

3. There is definite evidence of organized despotism and facilitation, resulting in dominance of well coördinated groups as families.

4. An interspecific type of social organization was found between the species and groups of geese and ducks.

5. Establishment and maintenance of territorial boundaries appears to be the main cause of pecking, which is a display of despotism due to (a) family including young; (b) female mate; (c) food; (d) nest; (e) nesting and sleeping place; (f) position with reference to the rest of the birds; (g) some combination of these.

6. Territory is used here in a broad sense, being applicable in non-mating and non-nesting times, as well as during mating and nesting times. It is any area defended against other organisms, and is usually centered around some bird or birds or other object. It may be fixed or evanescent and variable in size and shape. The types recognized are: (a) family territory; (b) mated-female territory; (c) feeding territory; (d) nesting territory; (e) resting territory; (f) resting and sleeping territory; (g) moving territory.

7. The size and shape of the territories varied directly with the temperature (within limits), and with such activities of the geese as feeding, resting, sleeping, and moving.

BIBLIOGRAPHY

ALLEE, W. C.
1931. Animal Aggregations. (Univ. Chicago Press, Chicago.)
1938. Social Life of Animals. (Norton, New York.)
ALLEE, W. C., COLLIAS, N. E., and LUTHERMAN, C. Z.
1939. Modification of the social order in flocks of hens by the injection of testosterone propionate. Phys. Zool., 12: 412–440.
BENNETT, MARY A.
1939. The Social Hierarchy in Ring Doves. Ecology 20: 337–357.
EMERSON, A. E.
1939. Social organization and the superorganism. Amer. Midland Nat., 21: 182–209.
HEINROTH, O.
1912. Beiträge zur Biologie, namentlich Ethologie und Psychologie der Anatiden. Verhandl. V. Int. Ornithol. Kong., Berlin, 1910: 589–702.
HOWARD, H. E.
1920. Territory in bird life. (Murray, London.)
LORENZ, K.
1931. Beiträge zur Ethologie soxialer Corviden. Jour. für Ornith., 79: 67–127.
1935. Der Kumpan in der Umwelt des Vogels. Jour. für Ornith., 83: 137–213; 289–413.
MASURE, R. H., and ALLEE, W. C.
1934a. The social order in flocks of the common chicken and pigeon. Auk, 51: 306–325.
1934b. Flock organizations of the Shell Parakeet, *Melopsittacus undulatus* Shaw. Ecol., 15: 388–398.
McATEE, W. L.
1924. Do bird families have any permanence? Condor, 26: 192–194.
NICE, MARGARET M.
1937. Studies in the life history of the Song Sparrow. I. Trans. Linn. Soc. New York, 4: 1–247.
NOBLE, G. K.
1939. The role of dominance in the social life of birds. Auk, 56: 263–273.
NOBLE, G. K., WURM, M., and SCHMIDT, A.
1938. Social behavior of the Black-crowned Night Heron. Auk, 55: 7–40.
SCHJELDERUP-EBBE, TH.
1924. Zur Sozialpsychologie der Vögel. Zeit. für Psychol., 95: 38–84.
1935. Social behavior of birds. *In* Murchison's Handbook of Social Psychology: 947–973. (Clark Univ. Press, Worcester, Mass.)
SHOEMAKER, H. H.
1929. Social hierarchy in flocks of the canary. Auk, 56: 381–406.

University of Minnesota
Minneapolis, Minnesota

8

8

Reprinted from *Physiological Zoöl.*, **26**(2), 173–178 (1953)

DOMINANCE ORDER IN THE CRAYFISH ORCONECTES VIRILIS (HAGEN)

RICHARD V. BOVBJERG

Department of Zoölogy, Washington University, St. Louis, Missouri

THE problem of dominance relations among invertebrate species has received scant attention relative to that of vertebrates. This has been repeatedly recognized by Allee and summarized by Allee, Emerson, Park, Park, and Schmidt (1949, p. 411); by Allee and Douglis (1945), and by Pardi (1948). Records of this phenomenon in invertebrates is therefore important if only as an antidote to the conception that dominance relations are peculiar to vertebrates. Such reports are of interest as well to the student of comparative psychology and evolutionary origins of social behavior. If these social relations are not merely laboratory manifestations, they are also of importance in the structure of ecological communities.

Of the few investigations on dominance relations in the invertebrates, the only one of an extended nature is Pardi's excellent demonstration of dominance order in *Polistes* wasps. This is the only such study concerning insects. Two reports of a preliminary nature describe dominance order in crustaceans (Allee and Douglis, 1945; Douglis, 1946). Two-rank hierarchies were observed among the small hermit crabs, *Pagurus longicarpus;* three- and four-rank hierarchies among lobsters, *Homarus americanus;* and interspecific hierarchies among lobsters, blue crabs, *Callinectes sapdius,*

spider crabs, *Libinia emarginata,* and the large hermit crabs, *Pagurus polycarpus.*

Aggressive behavior as such has long been recognized in social insects, beetles, a recently reported instance in dragonflies (Moore, 1952), and in spiders and crustaceans. Aggressive actions in the crayfish, *Cambarus virilis,* were noted by Roberts (1944). Tinbergen (1939), while investigating chromatophoral changes in mating *Sepia officinalis,* observed aggressive behavior between males confined in aquaria. The author has observed such behavior in female squid (*Loligo* sp. of Monterey Bay, California) following egg-laying. Where these animals live in certain contiguity, it would be of interest to investigate their tendencies to form social hierarchies. Some years ago, working in the laboratory of W. C. Allee, preliminary unpublished investigations were made on crayfish dominance relations. Utilizing techniques emerging from these studies, further observations have been made on the crayfish *Orconectes virilis* (Hagen). This paper reports these findings.

Field observations of *O. virilis* in its characteristic habitat—the rock- and gravel-bottomed streams and lakes of the central states—reveal tension contacts between foraging individuals and those backed into crevices under rocks and logs. Evictions frequently follow such contacts. The dominance relations are obscured by the territorial factor, and systematic study is difficult under natural conditions; only pair contacts were observed. It was the purpose of these investigations to take small groups, large enough to contain more than two ranks, to determine whether under laboratory conditions they would form social hierarchies.

PROCEDURE

The specimens were obtained from Williams' Creek, Creve Coeur, Missouri. All were healthy adults with intact appendages. Groups of four of the same sex and size were established in round, stacking bowls 20 cm. in diameter, with a gravel substratum. This monotonous and confined environment with no places of refuge was highly artificial but eliminated the factor of territoriality or possession of a "home." Water was changed and the animals fed on lettuce and meat scraps every third day.

Ten replicate groups, five males and five females, were each observed in daily $\frac{1}{2}$-hour periods. All tension contacts were recorded in abbreviated notation. Individual crayfish were distinguished by markings of varicolored fingernail lacquers.

Tension contacts were defined as any contact between two crayfish in which one was clearly in retreat following a head-on encounter. This excludes random crawling over one another, retreat from a rear contact, or the general melee following some strong stimulus within, or external to, the group. Within this definition, tension contacts could be categorized as one of four types: avoidance, threat, strike, or fight. (The shorthand notations for these being a, t, s, f. For instance C s D is read as a contact in which animal C struck D, with the retreat of animal D.) These levels of ag-

gression warrant a more complete description.

The most dramatic contact is the fight between two crayfish locking chelae and using the walking legs for leverage to force the antagonist onto his back in a position to be consumed.[1] This is rarely carried to such a conclusion unless one of the crayfish has recently molted. The fight may last several minutes and is almost always terminated by disengagement and retreat by one of the animals.

The strike is a unilateral aggression, in which the aggressor approaches with outspread chelae, which are thrust suddenly at another, eliciting a retreat by the latter.

The threat is an approach with outspread chelae in strike position sufficient to cause retreat in the other animal.

The elicitation of avoidance, with no threatening behavior discernible to the observer, is frequently noted, the subservient animal retreating from a moving individual or giving that animal a wide berth in its own movements.

RESULTS

The data recorded from daily observation periods were compiled into a checkerboard type of table (see Table 1). The letters designating individuals head the columns vertically and horizontally, enabling the investigator to compare any two animals in the group. Dominance is read in horizontal columns and subordinance vertically. Total dominant contacts are recorded at the right of each horizontal column, and total subordinance at the base of each vertical col-

---

[1] An objective attitude is imperative in this type of study, and every effort was made to maintain this. The data were not examined during the weeks of observation; therefore, no knowledge of rank was possible. However, in describing aggressive behavior, it is justifiable to employ certain descriptive words without anthropomorphic implication.

umn. The key data from ten such tables are condensed into Table 2.

The tabulated data from the ten groups include a total of 2,238 observed tension contacts, varying from 92 to 381 contacts per group. Over periods of from

criterion for ranking the group. It is weak evidence when such contacts are few, as between third and fourth ranks. Ranking

TABLE 1*

EXPERIMENT I-4, FOUR MALES: SAMPLE TABULATION OF DOMINANCE-SUBORDINANCE RELATIONS BETWEEN FOUR INDIVIDUALS IN FOUR CATEGORIES OF TENSION CONTACTS

		A	B	C	D	Total	
A.......	Avoid	....	14	11	20	45	
	Threat	....	36	24	23	83	(212)
	Strike	....	25	26	30	81	
	Fight	....	1	1	1	3	
B.......	Avoid	1	....	6	12	19	
	Threat	0	....	3	9	12	(42)
	Strike	1	....	5	4	10	
	Fight	0	....	0	1	1	
C.......	Avoid	1	2	....	4	7	
	Threat	0	1	....	2	3	(15)
	Strike	1	0	....	1	2	
	Fight	0	1	....	2	3	
D......	Avoid	0	2	2	....	4	
	Threat	0	0	1	....	1	(8)
	Strike	0	2	1	....	3	
	Fight	0	0	0	....		
Total...	Avoid	2	18	19	36	....	
	Threat	0	37	28	34	....	
	Strike	2	27	32	35	....	
	Fight	0	2	1	4	....	
		4	84	80	109	277	

* Dominance reads horizontally; subordinance vertically.

10 to 30 days, each of these groups established a consistent dominance order. The dominant animal in each case is readily discernible, as is the subdominant. The third and fourth ranks are frequently difficult to distinguish but in all cases are very subordinate to the top two ranks. These data corroborate the suggestions of this noted in lobsters by Douglis (1946).

Comparison of individuals where they meet as pairs furnishes the most reliable

TABLE 2*

DOMINANCE ORDER OF TEN GROUPS OF FOUR CRAYFISH ESTABLISHED OVER A PERIOD OF 20 DAYS

Social Group	Dominance Order	Dominance Contacts	Per Cent Dominant
1, male.........	C	183	98
	A	42	37
	D	15	14
	B	10	11
2, male.........	D	145	100
	A	73	43
	B	18	20
	C†	....	....
3, male.........	B	119	93
	A	50	50
	C	16	21
	D	9	10
4, male.........	A	212	98
	B	42	33
	C	15	16
	D	8	7
5, male.........	D	311	100
	C	57	27
	A	4	2
	B†	....	....
1, female.......	D	35	90
	C	32	67
	B	21	48
	A	4	8
2, female.......	A	235	99
	C	22	18
	B	10	11
	D	7	7
3, female.......	A	170	99
	B	14	15
	C	11	12
	D†	....	....
4, female.......	D	101	98
	C	23	32
	B	15	25
	A	10	10
5, female.......	B	116	95
	D	48	56
	A	4	8
	C	4	5

* Listed is the rank of each animal, its total number of dominant contacts, and the percentage of its contacts which were dominant.
† Individual died in course of observations.

by total dominant contacts corroborates pair-contact data in each case. Percentage dominance calculations agree with total dominance and pair-contact data. Ranking by total subordinate contacts invariably revealed the dominant, but was unreliable for the other three ranks.

As a check on the interpretation of these observations, a maneuver was made which will be termed "social inversion."

SOCIAL INVERSION

The dominant crayfish were removed from their groups and isolated. Observations were continued on the behavior of the remaining three of the original group. After a 5-day period the emergent dominant was removed and reintroduced to the isolated, original dominant, and a second (inverted) group was established. Observations were then made on the behavior of this inverted group, as well as of the remaining two in the original group. The next emergent dominant from the original group was added to the inverted group after a 5-day period. Finally, the fourth-ranking crayfish was transferred, the inverted group completed, and the original grouping terminated.

Seven of the ten groups were inverted in this fashion. The first- and second-ranking individuals, with one exception, re-established their original position through the inversion. The position of the third and fourth ranks was inconclusive, owing to the death of one of these in almost every instance. In each case when the subdominant, with its 5-day reign as top-ranking animal, was reintroduced to the original dominant, with its 5-day period of isolation, a series of vigorous fights soon established the old relationship and frequently was followed by days of a despotic type of aggression on the part of the dominant.

DISCUSSION

Quantitative analysis of the observed tension contacts justifies the postulation of a dominance order in crayfish. A number of qualifications must promptly be appended. The work was done on one species, *O. virilis*, in groups of four animals under laboratory conditions; the individuals were of the same size and sex. None of these conditions obtain in nature. Nevertheless, the establishment of a dominance order implies intrinsic aggressive differences between the individuals in each group. The important implication of sensory perception and recognition of individual crayfish and retention of this recognition as a form of learning would follow. Sensory mechanisms in decapod crustaceans are generally understood and the ability to learn demonstrated (Warden, Jenkins, and Warner, 1934). But the problem of specific cues involved in recognition of a dominant animal remains to be investigated.

The categories of aggressive behavior and response were adequate to distinguish the approximately four thousand tension contacts recorded. The threat constituted 39 per cent of the total tension contacts in the original series, and the strike, 32 per cent; the elicitation of avoidance was manifested in 23 per cent of the total, whereas the fight was observed in only 6 per cent of the contacts. Very little difference was distinguished between the ranks of a group with respect to these types of aggressions.

No pattern of selective behavior toward the lower-ranking individuals was observed. In the basic series the dominant animals had positive contacts with the second-ranking 438 times, with third-ranking 314, and with fourth-ranking 391. The second-ranking individuals, while having only 18 positive contacts with the dominant, had 147 and 168 such contacts with the third- and fourth-ranking, respectively. The third-ranking crayfish had 8 and 40 positive contacts with the first and second ranks, and 65 with the fourth-ranking. The lowest animals had 4, 25, and 38 positive contacts with the three individuals above it in rank.

With exceptions to be noted, a remarkable social stability was observed. In eight of the ten groups the dominant individual became so established in the initial $\frac{1}{2}$ hour. After the first 5 days, all ten of the dominant crayfish remained in this position for the entire observation period. Seven of the ten groups maintained the initial rankings at all four levels for the entire basic series of observations.

In the Crustacea generally, the period following molting is a perilous one, since the new exoskeleton is very soft. The animal is less capable of rapid locomotion, and the chelae are inadequate as defensive appendages. The effect of this on dominance position is apparent to the observer. Invariably, a newly molted crayfish, regardless of rank, became completely subordinate and, if not removed for a day or two, was canibalized. After recovery from this period, the individual moved back into its previous dominance rank. In natural situations refuge for this critical period could be found in bottom crevices.

A number of interesting aspects remain for investigation. These center about the factors influencing dominance order. Intrinsic factors of body size, sex, and molting are suggested by these observations. Extrinsic factors, such as territoriality and population density and size, are undoubtedly of importance in nature.

Previous observations on heterosexual groups indicate a dominance by the male. When established social groups of the two sexes were joined, there followed many instances of low-ranking males fighting and copulating with dominant females.

Investigation of the effects of humeral substances on the dominance order of crayfish could be a fruitful avenue of study. It would be interesting to ascertain the effect of ablation and implantation of sinus gland on aggressive behavior. The action of hormones from this gland on molting seems well established (Brown, 1944, 1948). While the exact nature and operation of sex hormones in crustaceans is not clear, aggressive behavior could be used as a criterion of relative sexuality in endocrine studies.

Size is a very important factor in determining the dominance order. The most casual observation of groups of variable-sized individuals reveals the direct relationships of size to dominance. Under natural conditions, the crayfish population is comprised of both sexes and in sizes of more than one year class.

The problem of natural versus artificial conditions is very important. Certain aspects of this problem were investigated in a small Michigan stream. Four groups were studied as replicates of the laboratory groups. These were contained in wire-screen cages placed in a flowing stream with natural substratum and foods. They did not have unlimited freedom to forage, and the groups of four individuals are, of course, artificially small populations. Under these conditions, dominance order was observed over a period of a few weeks. One difference was noted; there was less over-all activity, and fewer tension contacts were recorded over observation periods similar to those in the laboratory.

Whether crayfish actually maintain territories has not been determined. Field and laboratory observations do definitely reveal alteration of behavior when individuals have crevices in which to reside. Smaller crayfish can often successfully defend such a "home" against larger individuals. The significance of the laboratory-demonstrated dominance order hinges on the extent to which this occurs in nature. In a stream, crayfish are distributed continuously. How large a group can be integrated in such social hierarchies? It seems probable that no dominance order as such exists but rather a random array of essentially initial contacts, in which the outcome is dependent partly on a type of ephemeral territoriality and partly on the intrinsic factors of size, sex, and, as the observations in the laboratory indicate, some degree of innate aggressiveness. Further investigation should involve both territoriality and variable population size.

SUMMARY

1. Under natural conditions, crayfish display aggressive behavior; small groups of the crayfish *O. virilis* (Hagen) were brought into the laboratory, and their social behavior was observed.

2. The aggressions seen in the field were duplicated by crayfish in groups of four confined in small round aquaria. Subservient reactions were found to follow (*a*) active fighting, (*b*) a strike with the chelae, (*c*) a threatening posture, and (*d*) avoidance frequently was elicited by the mere presence of a more dominant animal.

3. Five groups of females and five of males were observed for an extended time in daily observation periods. On the basis of 2,200 tabulated tension contacts, it was clear that each group established a straight-line dominance order.

4. As a check on the initial observation, a technique of "social inversion" was devised in which the crayfish were transferred in order of dominance and at 5-day intervals to a new aquarium. The order in these re-established groups confirmed the initial observations.

5. Stress is placed on the preliminary nature of these observations and the many aspects open to further investigation.

LITERATURE CITED

ALLEE, W. C., and DOUGLIS, M. B. 1945. A dominance order in the hermit crab, *Pagurus longicarpus*, Say. Ecology, 26:411–12.

ALLEE, W. C.; EMERSON, A. E.; PARK, O.; PARK, T.; and SCHMIDT, K. P. 1949. Principles of animal ecology. Philadelphia: W. B. Saunders Co.

BROWN, F. A., JR. 1944. Hormones in the Crustacea. Quart. Rev. Biol., 19:118–43.

———. 1948. Hormones in crustaceans. In PINCUS, G., and VON THIMANN, K. (eds.), The hormones, 1:159–99. New York: Academic Press.

DOUGLIS, M. B. 1946. Some evidence of a dominance subordinance relationship among lobsters, *Homarus americanus*. Anat. Rec., 94:57 (abstr.).

———. 1946. Interspecies relationships between certain Crustacea. *Ibid.*, pp. 57–58 (abstr.).

MOORE, N. W. 1952. On the so called "territories" of dragonflies (Odonata-Anisoptera). Behaviour, 4:85–100.

PARDI, L. 1948. Dominance order in *Polistes* wasps. Physiol. Zoöl., 21:1–13.

ROBERTS, T. W. 1944. Light, eyestalk chemical, and certain other factors as regulators of community activity for the crayfish *Cambarus virilis*, Hagen. Ecol. Mono., 14:360–92.

TINBERGEN, L. 1939. Zur Fortpflanzungsethologie von *Sepia officinalis* L. Arch. Néerl. de zool., 3:323–64.

WARDEN, C. J.; JENKINS, T. N.; and WARNER, L. H. 1934. Introduction to comparative psychology. New York: Ronald Press Co.

Reprinted from *Social Behavior of the Domestic Fowl*, Technical Bulletin No. 73, Kansas Agricultural Experiment Station, Manhattan, Kan., 1953, pp. 3–48

Social Behavior of the Domestic Fowl[1]

A. M. Guhl[2]

INTRODUCTION

The common domestic fowl has received much attention from scientists, especially in the fields of nutrition, physiology, embryology, and genetics. The task of developing this vast and rapidly expanding body of information has been justified from various aspects. The behavior of chickens, however, has not been studied scientifically until rather recently.

In the early domestication of animals, man recognized the importance of behavior when he selected those species which could be used to satisfy his own ends. In the breeding and selection which followed, behavior received less attention as meat quantity and quality, and other products from animals became focal points of interest. Contrariwise, dogs continued to be developed for characteristic types of behavior as human companions in various pursuits.

Modern management and high efficiency diets have brought renewed interest in behavior, especially in the role it takes in obtaining maximum gains. This attention stems from reports that certain aspects of social behavior influence egg production. Present skills and techniques in poultry breeding are so effective that unless some regard is given to behavior, some undesirable types might inadvertently be produced; for example, strains might be highly combative or have such a low sex drive as to influence fertility.

Fowls have characteristic ways of behaving which should be considered when formulating plans for experimentation and for the husbandry of these birds. Flocks of hens, for example, are organized and man's manipulations or interferences with their social system may cause repercussions that will be reflected in the progress of their development or in the egg production of the flock. Activities associated with mating in breeding pens conform to definite patterns which if recognized and considered in husbandry might influence, along with a number of other factors, the fertility of eggs produced. Furthermore, the selection of cocks for breeding purposes might well take into consideration the relative sexuality of the males. These statements give some indication of the goals toward which studies in the social and sexual behavior of chickens are directed.

Reports of studies in the behavior of chickens are in di-

---

1. Contribution number 246 from the Kansas Agricultural Experiment Station in the Department of Zoology and number 289 from the Department of Zoology. The Department of Poultry Husbandry co-operated in several of the experiments conducted by the author.

2. Associate Professor of Zoology, Kansas State College.

verse journals, some of which may not be readily available to poultrymen. It is the object of this bulletin to review much of the information reported to date, to present some of the procedures for studying social behavior, and to point out the basic implications of these findings to the poultry scientists and poultry husbandrymen. In the development of a background some unpublished material will be included.

METHODS OF STUDYING SOCIAL BEHAVIOR

Aggressive Behavior as a Basis for Social Organization: Hens are aggressive and display this trait by fighting and pecking. This behavior is readily observed at the food hopper, nest boxes, water supply, and at dusting areas. These pecks are directed at the individual, usually at the head, and should not be confused with feather picking and cannibalism. When the birds are marked for individual identification, as with colored legbands or colored dyes on the feathers, one may record these attacks in terms of the hen pecking and the one pecked. A tabulation of these observations will show that the pecks between any two birds are usually in one direction only. In small flocks one hen pecks all in her pen without being pecked in return; another hen is pecked by all and pecks none. The other hens in the group may be arranged in an order between these two according to the number of birds each pecks. This ranking of despotism or "bossism" forms a dominance order or peck-order. It is not unusual for this hierarchy to deviate from a straight line; that is, C may peck D, D may peck E, and E may peck C. Such variations are called pecking triangles, and are common in larger flocks.

Table 1 illustrates some variations in dominance relations among hens in small flocks. The letters represent colors used for identification. In the ranking each hen pecks those listed beneath it unless otherwise indicated by arrows. Flock F has a straight-line hierarchy. The other three flocks have deviations or pecking triangles at various levels in the peck-order.

Cocks also have a peck-order and, as they usually do not peck the hens, a mixed flock has two peck orders, one formed by the hens and the other among the cocks. Changes in dominance relationships are infrequent in very small flocks, are more frequent in large flocks, and occur more readily between cocks than between hens.

Dominance orders were discovered by Schjelderup-Ebbe (1913, 1922, 1935). In this country the work was taken up by Sanctuary (1932), Allee (Masure and Allee, 1934; Allee, Collias, and Lutherman, 1939), and Murchison (1935 series).

Determination of Peck-orders: The number of unidirectional dominance relationships in a flock may be determined by the formula $(n^2-n)/2$. In a flock of 10 hens there would be 45 pecking combinations and in one of 100 hens the number would be 4,950. By comparison, even a flock of 25 birds with 300 dominance relationships might be considered to be a large flock when the work required to determine a peck-order is evaluated.

The following procedure has been followed in the determination of a peck-order. Pecks and obvious uncontested threats were recorded in a code based on color identifica-

Table 1.—Dominance Relationships in Small Flocks of Hens.

Flock F	Flock W	Flock G	Flock P
WG	WR	GB ←— GW	PP
RG	WV	↘ ↗	Pb
PB	WW	GP	PW
YB	WB	GG	PY
RB	↗ ↘	↗ ↘	PR
YG	WP ←— Wb	GV ←— Gb	PB
GB			PG
YR	WG	GR	
WR			

tions. For example, YV – BR indicates that the bird marked with yellow and violet pecked one dyed with black and red. This manner of recording permitted rapid notation. These recorded pecks then were tabulated on a chart (c.f. Fig. 14) from which all dominance relationships could be determined. Any changes in pecking relations between any two birds were readily detected when tabulations were made daily. Even after weeks of observation in comparatively large flocks no pecks were observed between certain individuals whereas pecks were frequent between others. When each pair-relationship had to be known, birds of unknown dominance were placed into a cage together, by pairs, until pecking and avoidance were noted.

The determination of a peck-order in a large flock was a task which involved considerable time in observation and tabulation. The following experience illustrated the manner in which indications of a peck-order were obtained. In a flock of 96 White Rock pullets with 17,851 pecks recorded there were a number of unknown dominance relationships among the 4,460 possible pair combinations. Evidences of a peck-order were found in the fact that unidirectional pecking tended to be very consistent (i.e., there were few reversals), and that all relationships were known for two hens. All but three or fewer relationships were known for 14 other hens. Among these 16 birds for which all or nearly all pecking relationships were known were the top-ranking hens, the hen in the low rank of 94, and the others which were at various intermediate ranks.

Measurement of Relative Aggressiveness: In certain experiments (see, e.g., Inheritance of Aggressiveness) it would be to some advantage to obtain a measure of the relative aggressiveness of the members of the flock, or to be reasonably certain that the rank order correlates with variability in aggressiveness. Especially in flocks of 30 or more birds, the peck-order may not necessarily reflect the potentialities of the individuals. During the introductory period of fighting when a group of strange birds is housed, a highly aggressive bird may meet other highly aggressive individuals in quick succession and become fatigued. If it should then be attacked by a less capable one the former may lose the contest. Observations have suggested that defeats resulting from fatigue, or other disadvantages, are not uncommon for several birds during the assemblage of a large group.

Collias (1943, see also Allee, Collias, and Lutherman, 1939) has devised a technique for staging initial encounters between pairs of hens in which some of the factors that make for winning pair-contests were controlled. The procedure was essentially as follows: Two strange hens, one from each of two flocks, were placed into a neutral pen to settle their dominance relationship. A number of hens were tested in this manner with several strange hens. The number of such initial pair-encounters won was used as an indicator of the relative aggressiveness of the hen. Such scores are not to be construed as absolute and therefore cannot be used in making out-group comparisons.

Pair-contests or initial encounters have been conducted by isolating birds in a laying battery. After two weeks of separation the birds were placed into a strange area, or an exhibition cage, by twos for initial encounters. Each bird met each of the others in its flock to settle dominance relationships. A rest period of a few hours or a day was given each bird before it was reintroduced with another flockmate. Relative aggressiveness was estimated by the number of initial encounters won.

The following accounts are given as examples of experimental situations under which initial pair-contests might be used to good advantage. It was necessary to determine the extent to which there may be an agreement between a peck-order as it develops among chicks or one that is formed when a group of strange birds is assembled, and the ranking of the relative aggressiveness of the individuals. The peck-order of 11 pullets, raised together, was compared with the results of two sets of initial pair-contests and with the social order that developed when the same birds were placed into a pen following a period of isolation.

The data analyzed were the number of individuals pecked by each bird in the flock, and the number of initial pair-contests won by each bird. There are 55 pair relationships in a flock of 11 birds. As there were four sets of data, the total dominance relationships obtained were 220. Of these there was no difference in the dominance relations of 108 pairs (49.1 percent), one change in 66 pairs (30.0 percent), two changes in 24 pairs (10.9 percent), and 22 new (unrepeated) relationships (10.0 percent). These results suggest that, even among birds undergoing maturation, and despite the multiplicity of factors involved, there was a definite tendency for the individuals to show fairly consistent pair relationships.

The same data were used to compare social dominance in flocks with levels of aggressiveness as shown by initial pair-contests. Coefficients of correlation were obtained between (1) ontogenetically developed social relationships, (2) the results of two tests for individual aggressiveness as determined by initial pair-contests during isolation, and (3) the social relationships formed under relatively uncontrolled conditions when the group was reassembled and the meeting of the hens to settle dominance relationships followed a random order.

The results are given in Figures 1 to 4 in the form of scatter diagrams. The data plotted are the number of individuals pecked when the pullets were in a pen and/or the number of contests won when matched by pairs in a neutral area. Each figure also gives the coefficient of correlation and the probability of statistical significance in P-values. A fairly high and statistically significant correlation was found between the number of individuals pecked in the social order formed ontogenetically and the number of contests won (Fig. 1) during the first initial pair-contest. Similar results were obtained between the two sets of paired-encounters (Fig. 2), and between the second initial pair-contest and the social relationships formed by a more or less random initial meeting of pairs when the group was reassembled (Fig. 3). A low and non-significant correlation was obtained between the numbers pecked in the two social orders (Fig. 4). Other calculations (not shown in the figures) were made, as between the ontogenetic dominance relationships and the second test, and between the first test and the dominance relations in the reassembled flock. Both of these were of questionable significance (**r** of + 0.58, **P** of 0.05; and **r** of + 0.63, **P** of less than 0.05, respectively).

Recently, Mr. Charles Miller made similar tests with six different groups in which he determined the ontogenetic peck-order and subsequently paired them in initial encounters during isolation. These flocks of cockerels or pullets

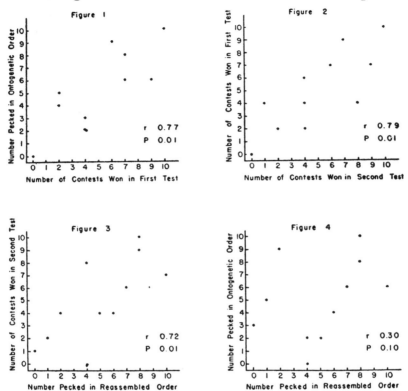

Figs. 1 to 4.—Coefficients of correlation between ranking of 11 pullets based on three methods used for the determination of levels of aggressiveness. Each dot represents one pullet and the number she dominated either in a peck-order or in initial encounters.

contained 6 to 10 individuals. In two of these flocks the coefficient of correlation was less than + 0.50 and lacked statistical significance. The other four groups showed correlations, between number pecked in the ontogenetic order and the number of contests won, varying from + 0.77 to + 0.88 and with significance at the 1 percent level. Three of these flocks were reassembled after isolation. In two flocks the peck-orders formed after reassembly failed to show statistically significant correlations between the ontogenetic orders and the reassembled orders. The other flock showed statistically significant correlations in all comparisons.

From the foregoing tests to measure relative aggressiveness one may conclude that there is a tendency for agreement, and that any one of the methods might be used to evaluate levels of aggressiveness in a flock. The exceptions encountered do show, however, that more than one of these methods should be used when critical results are desired.

These techniques were used also with older birds and with a larger flock. A group of 33 White Leghorn hens was used. These birds were raised together within a large group on a summer range. They were divided into two flocks and paired encounters arranged so that each bird in one flock met each in the other. Peck-orders in these subflocks were determined. All the birds were isolated in laying batteries and all possible paired-encounters were staged. Subsequently all

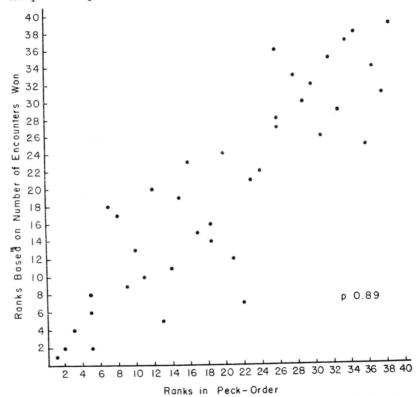

Fig. 5.—Rank correlation between paired-encounters won during isolation and number pecked in the peck-order formed when 39 hens were reassembled by release in the order of rank based on the number of encounters won.

the birds were released simultaneously in a large pen and the reassembled peck-order determined. As there were some deaths and as social ranks in subflocks were not directly comparable with ranks in the combined group, correlations were estimated on the basis of percentage of birds pecked and percentage of encounters won. All the coefficients of correlation between the four sets of data (percent pecked in subflocks; percent of interflock contests won; percent of initial encounters won during isolation; and percent pecked in the reassembled flock) were statistically significant at the 1 percent level and ranged between + 0.77 and + 0.98. Similar tests with other large groups should be made before these results are accepted in a critical manner.

Initial paired encounters were used also in an effort to obtain a peck-order in a flock in which the social ranks would show statistically significant correlations with levels of aggressiveness. Thirty-nine pullets in laying condition were isolated in laying batteries and initial pair-contests were staged in exhibition cages on the floor. At the close of these tests the birds were released into a pen one at a time over a period of 14 days. The order of release was in the order of the ranking based on the number of contests won, beginning with the most successful female. This method gave the advantages to the birds longest on the floor of the pen (see Factors Influencing Attainment of Social Status). The rank correlation between ranks in the fighting-order and the subsequent peck-order was + 0.89. Most of the individuals did not change greatly in their ranking (Fig. 5).

Our experience with staged encounters between isolated birds has suggested certain limitations. Although laying batteries when used for purposes of isolation did conserve space and labor, they presented some disadvantages. The lack of activity made the birds more sluggish and less inclined to engage in combat when they met in a neutral cage. This was particularly evident when a large number of birds were used and isolation extended over a number of weeks. Contests between birds of different flocks, however, did not eliminate certain psychological factors associated with either high or low social position (see also Collias, 1943).

Measurement of Sexual Behavior: Several of the early studies of sexual behavior in chickens were concerned chiefly with the frequency of copulations and focused the attention on the males. Sexual behavior may be considered in the field of social behavior, as at least two individuals are involved. Chickens being polygamous, the social aspects of their reproductive behavior are more complex than in monogamous species. Furthermore, coition in birds is preceded by various behavior patterns, known as displays, which synchronize the sexes. Skard (1937) recognized several types of advances made by cocks to hens which may be placed into a category called courting. Hens may be indifferent to the approaching cock, or they may respond negatively by avoiding the male, or positively by crouching with wings spread. Crouching is followed by male mounting and treading, which may culminate in coition (see Sexual Behavior Patterns).

Sexual behavior can be measured by careful recording of the behavior patterns of males and females in general categories as given above. When observation periods are timed and made repeatedly at the same time of the day, the tabulations can be expressed in frequencies or rates. As the chain of reactions is completed in a few moments, recordings should be made in code (e.g., A-c-T-RY may indicate that the cock approached, the hen crouched, and the treading followed; the RY indicates the colors of the hen involved in the mating. Further elaborations can be made if needed). These data may be tabulated in appropriate charts to indicate rates for each hen and cock.

There are certain limitations to the above procedure. When there is but one observer, the flock size and spatial arrangements are important. Flocks of about 30 hens can be used when one to three cocks are present. Although as many as three males may be followed in a comparatively small flock, recording may be difficult if there is much activity. To the writer's knowledge, no reports have been made in which two or more observers cooperated simultaneously. Such a procedure should be tried when several males are in the pen, and recording could be restricted to certain males for each observer. Differences in accuracy or interpretation of behavior patterns could be compensated partially by daily alternation between the observers of the specific males. As one might expect, skills in recording must be developed, and variations in intensity of actions and reactions might introduce an element of subjectivity. The observer must not distract the birds in any manner if he is within the pen during the period of recording.

SOCIAL ORGANIZATION IN THE FLOCK

Formation of the Social Order: When a number of strange birds are placed together into a pen, fights occur by twos until each bird has engaged all the others. The winner of each initial contest thereafter has the right to peck the loser, and the latter usually avoids the former. Some individuals give way without a fight and others may challenge the winner again before dominance relations are settled. At subsequent meetings one member of each pair pecks or threatens the other, definite dominance-subordination patterns become habitual, and thus the peck-order is established.

The development of a peck-order between birds raised together as chicks has been under preliminary observation. Female chicks in small groups developed a peck-order by the tenth week. Male chicks established a social organization when eight weeks old.

Factors Influencing Attainment of Social Status: A number of factors may be involved in success or failure resulting from fights between two birds of the same sex, breed, or species. These were given by Schjelderup-Ebbe (1935) and Collias (1943), and reviewed by Allee, Collias, and Lutherman (1939). Some of these factors need to be stated here: (1) body weight or strength may be determinant if other factors are equal; (2) one bird acts as if intimidated by an unfrightened opponent and so gives way without a contest; (3) both contestants are frightened and the one which recovers first may win; (4) birds may differ in state of health, fatigue, or severity of molting; (5) age or its usually inseparable factor of skill is an advantage; (6) location of the contest makes a difference, for a bird fights better in its home area; and (7) even in strange surroundings a bird is more successful when in the presence of its penmates.

The Status of Newcomers in the Social Order: Much of the basic information on the social order of chickens was initially reported by Schjelderup-Ebbe. Although his accounts were marked with anthropomorphisms, his conclusions have been confirmed by others. In his single article in the English language (1935) he described the effects produced when strange hens were added to a confined group which had developed a social order.

Typical behavior observed in our flocks when strangers were added was as follows: The resident birds threatened or attacked the new birds. The newcomers fought or, not infrequently, surrendered without a contest and collected in some neutral area such as the roost or retreated to the nest boxes. Being in unfamiliar quarters, the strange hens often darted about while being harassed by the resident birds. If there were several new ones, they collected in a group apart from the home flock. After some days or weeks the newcomers were assimilated into the social order, usually at low levels. In the interim these hens usually failed to approach the food hopper.

Sanctuary (1932) experimented with the shifting of hens from pen to pen. He found that when all the birds placed into an empty pen were strangers to each other, only a few went out of production. When the hens which were moved into a pen of established hens outnumbered the residents, only about one fourth entered a pause. But the introduction of a few birds into an established pen resulted in a cessation of laying by nearly all of the outnumbered newcomers. He concluded that "the nearer the condition approaches that of one bird entering an established pen as a stranger, the larger the percentage of birds so introduced that will be thrown out of production."

Guhl and Allee (1944) shifted birds regularly between flocks of 7 hens each and 21 hens in roomy isolation cages.

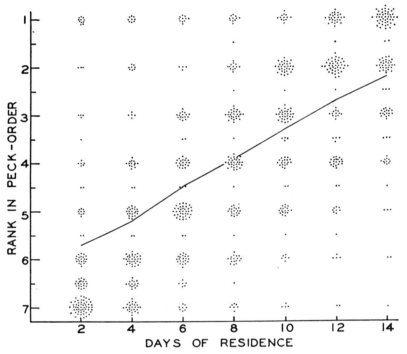

Fig. 6.—Correlation between seniority and social ranks in a flock having its membership changed by a system of rotation between the pen and isolation cages. From Guhl and Allee (1944).

Daily or every other day one isolated bird was placed into each of these pens and the longest resident of each pen was placed in isolation. Typically, the unrecognized newcomer entered the flock at the lowest rank in the peck-order and advanced steadily as new birds were added and resident birds were removed until it attained top status on its last day in the pen. Some individuals (Fig. 6) failed to rise very high in the social order, and a few very aggressive hens managed to gain the alpha position on their first day. The coefficient of correlation between days in the pen and social rank was 0.60 ± 0.02. The correlation between seniority and social rank dropped to 0.24 ± 0.03 when similar shifts or exchanges were made between pens. These statistically significant results show the importance of seniority in the social life of hens. It should be noted that seniority is not an independent factor but includes several of those mentioned earlier in the discussion of the establishment of pair relationships.

The influence of seniority on individual rankings in a large flock may be illustrated by data obtained from a flock of 97 White Rock hens referred to previously. A group of 17 was housed on August 25, 1944. Later, on September 14, 60 more were added, and the rest, or 20, were added the next day. In the final peck-order the first group pecked a mean of 60 hens, those added on September 14 pecked a mean number of 42, and the last group added had an average of 30 hens pecked. An analysis of variance (Snedecor, 1946) showed that there was a statistically significant difference between the three groups in the number of hens they dominated, as groups. An analysis of the statistical significance of the differences between the means showed that they were significant at the 2 percent and 1 percent levels. There was much variation; the range for those penned earliest was 10 to 93 individuals pecked, for the second group it was 9 to 87, and for the latecomers the range was 1 to 59.

It was conceivable that experimental situations might arise in which it may be desirable to have certain hens as members of more than one flock. Douglis (1948) found that part-time members of organized flocks could maintain social positions in as many as five small flocks. The flocks were composed of five to nine White Leghorn hens. Some individuals spent as little as one hour on alternate days in a flock other than the "home" flock. There was a tendency for each part-time member to hold different social positions in each of the flocks into which it was rotated.

The Relation between the Tendency toward Pugnacity and the Tendency toward Flocking: With the information available at present, it would be unwise to attempt a postulation on the evolution of aggressiveness. A more pertinent point here would be to determine how such highly aggressive animals as chickens can become a social species and live in relatively peaceful flocks. Pugnacity would be conducive to the dispersal of individuals or to the disintegration of the group, as is common in many of our song birds during the breeding season.

An antithetical tendency, gregariousness, appears early in the life of chicks. According to Collias (1950d) newly hatched chicks, after experiencing bodily contact, readily come together when separated by short distances. Chicks isolated for 10 days after hatching are less gregarious than

normals. He concluded that the contagiousness of chick behavior facilitates aggregation and socialization. The importance of very early experience in socialization is discussed in a review by Collias (1950b). In our laboratory we
found that aggressiveness in chicks appears gradually and
that peck-orders do not develop until the chicks are 8 to 10
weeks old. Social forces acting centripetally develop first
and forces acting centrifugally toward dispersal develop
later. Both of these opposing tendencies operate within the
flock, the degree of integration depending largely upon the
relative levels of aggression and submission displayed.

When dominance relationships are established the subordinate hens submit to their superiors and thereby are shown
some degree of toleration. The point of interest is that each
bird has learned to react differently to each of its flockmates,
avoiding close contact with some and dominating others.
The cumulative effect in the peck-order promotes some degree of integration in the group and therefore precludes the
recurrence of fighting every time two birds meet in a competitive situation.

The significance of the peck-order may be viewed from
two aspects: (1) the relation of social status to the individual, and (2) the effects of social organization on the flock
as a unit. Both of these merit consideration in certain experiments and in poultry husbandry.

Effects of Social Status on Productivity of the Individual:
The advantages of high social position in a flock of chickens
have been noted by Masure and Allee (1934), Guhl, Collias,
and Allee (1945), and have been discussed for vertebrates in
general by Collias (1944, 1950b). Briefly, it has been observed that birds ranking high in the hierarchy have precedence at the food trough, the nest, the roost, and the dusting
areas, and possess a greater freedom of the pen. Sanctuary
(1932) found that the individuals in the upper half of the
peck-order lay more eggs than do those composing the lower
social level. The most aggressive cocks have a greater freedom to mate and may sire the most chicks (Guhl and Warren, 1946). Contrariwise, the individuals at the lowest positions in the social order may be harassed to the point of
starvation.

Detailed observations on a relatively large flock may prove
to be of somewhat greater interest to poultrymen. Some of
the results obtained on the poultry farm at Kansas State
College may suggest the value of information on large flocks.
A group of 96 White Rock pullets was assembled during late
August and early September when this young stock was
brought in from the summer range. They were cared for
under the standard practices used on the poultry farm. Most

of the pecking relationships were known. Records of pecking were taken from the time the birds were housed until the end of other observations. As the individuals were marked with colored dyes for purposes of identification, it was possible to make a variety of records. The observations reported below were made between November 23, 1944, and January 2, 1945. No adjustments were made in the data for birds which may have been in poor health.

The flock was observed while going to roost at night and when leaving the roost before sunrise. There was a marked movement to the roost when the intensity of light at the windows was about two and one-half foot-candles[3] and all were usually on the roost when the light diminished to somewhat less than one foot-candle. Pecking was frequent during assemblage on the roost but decreased to a vanishing point as darkness set in. In the morning, all the pullets were usually off the roost when the light entering the windows was at about one foot-candle. Pecking and scratching in the litter started at about one or two foot-candles, suggesting that vision and recognition of penmates was impaired at lower intensities.

Records were made of the first 25 and of the last 25 individuals that went to roost, of the last to feed, and also of those located, after dark, at the fringe of the roosting group and more or less isolated. During the morning descent from the roost, records were made of the first individuals to feed, that is, before pecking became frequent. Members of feeding groups were noted also at 10 a.m., 11 a.m., 2 p.m., and 3 p.m. to obtain samples of the frequency of daytime feeding. These data then were related to social positions. Trapnesting provided a record of the first egg laid by each pullet as well as an egg record.

Correlations among certain of the data gave indications of the influence of social position on the activities of the individuals. The coefficients of correlation of some of the more pertinent relationships are given in Table 2. Although they were not significant statistically, there was a positive correlation between the number of females pecked and the first 25 to go to roost, and a negative correlation with the last 25 to go to roost. Also not significant, but interesting by comparison, there was a positive correlation between rank (number pecked) and the frequency of feeding during the daytime, whereas the correlation was negative between rank and the first to feed in the morning. The correlation between the last to feed at night and the first to feed in the morning was statistically significant. These relationships indicate that the low-ranking birds tended to feed after their superiors had gone to roost, and again early in the morning,

---

3. Weston foot-candle meter, model 6J4.

either before many of their dominant penmates descended or while they were unable to recognize them in the dim light. Some individuals were observed bolting food before going to roost as the light faded completely. These inferior pullets also were at a disadvantage at night; they were found among those at the fringe of the group, or isolated, during the cold winter nights, as shown by a negative and statistically-significant correlation between rank and roosting at the fringe of the group.

Table 2.—Correlations Found between Certain Observations Made on a Flock of 96 White Rock Pullets.

Correlates	Means		Coefficient of correlations
Positive correlations			
Number pecked—First to roost	43	11	0.16
Number pecked—Daytime feeding	43	43	0.15
Number pecked—Number of eggs	43	26	0.26[1]
Number of eggs—Daytime feeding	26	43	0.27[1]
Last to feed—First to feed	9	16	0.30[1]
Negative correlations			
Number pecked—Last to roost	43	12	0.17
Number pecked—First to feed	43	16	0.08
Number pecked—Age at sexual maturity	43	222	0.29[1]
Number pecked—At fringe on roost	43	8	0.35[1]
Age at sexual maturity—Daytime feeding	222	43	0.19[2]
At fringe on roost—Daytime feeding	8	43	0.22[1]

1. Statistically significant at 1 percent level.
2. Statistically significant at 5 percent level.
   Others not statistically significant.

Of particular interest was the suggestion that social position, by way of its influence on the frequency of feeding, affects egg production and the rate of attaining sexual maturity. Statistically-significant correlations were obtained between rank and the number of eggs produced, and also between number of eggs produced and the frequency of feeding during the daytime. The egg records used here were for October through December, i.e., after most of the pullets had started to lay. A negative and significant correlation was found between social rank and age at sexual maturity (age at first egg). The relation of social position to egg production was probably an indication of the retardation of development. From these results it should not be concluded that egg production would be always correlated with social position in a well-integrated peck-order among adult birds.

The critical reader rightfully may refrain from making any sweeping generalizations based on the foregoing anal-

ysis of the data taken from a single flock. He would also
reflect on the low correlation coefficients, although some
are statistically significant, and he may wish to examine
some of the data from a study involving a multiplicity of
factors. Three scatter diagrams, Figure 7, are presented to
illustrate the wide variations that occurred. The diagrams
selected show correlations between the number of birds
pecked (ranking) and three sets of data, namely, age at
sexual maturity, number of eggs laid (October through December), and the tendency to roost more or less exposed at
the fringe of the group. Of the data which could be presented
these are probably the most significant biologically as they
are concerned with survival values, and are of interest to
the poultryman, as they deal with production and viability.
The birds at the fringe of the roosting group were probably
the most hungry and therefore the most vulnerable to the
cold in an open-front chicken house.

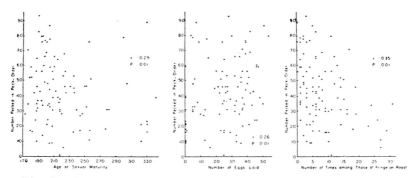

Fig. 7.—Correlations between the number pecked in the peck-order
of 96 pullets and the attainment of sexual maturity, egg production,
and frequency of roosting at the unfavorable fringe of the group.

There are various factors which may influence the behavior of a bird irrespective of its social level. Pecking may
be excessive by a given bird on one of its inferiors, a situation that has been described as an antipathy. The behavior
among others may suggest toleration, as pecking is very infrequent even under competitive situations. The individuality of certain birds may be readily observable in some flocks.
A high-ranking bird may be a "benign" despot, i.e., it pecks
its inferiors infrequently; others are "malign" despots and
attack inferiors with little or no apparent provocation.
Some birds may be described as "fearful," others as "timid,"
whereas still others are "bold," "venturesome," or even suggest "temerity." In rather large flocks there may be evidences of degrees of "frustration" which in extreme cases
show a thwarting of efforts by darting about the pen and

171

taking refuge in nest boxes. Such birds are readily pursued by their penmates, and are usually at the lowest levels in the peck-order. The relation of egg production to social position might be influenced by genetic constitution or certain physiological conditions not necessarily associated with inanition.

Effects of Social Organization on Productivity of the Flock as a Unit: To demonstrate the value of social order to the flock as a whole it was necessary to contrast an organized group with one which lacked organization. Such an attempt was made by Guhl and Allee (1944). Some flocks were kept in a state of social flux by regularly removing a bird in longest residence and replacing it with a stranger. The newcomers were taken from birds in isolation for part of the experiment, and later from flocks also undergoing alternation of membership. By this method the members of the experimental flocks undergoing a steady change of membership were not given the opportunity to formalize their interindividual dominance relationships, i. e., peck-orders formed, but pecking and avoidance tended to remain in the formative stages of habit formation. The time allowed for feeding was restricted to increase competition within the groups. Some of the data obtained included (1) rates of pecking in each flock while feeding, (2) weight of food consumed by each flock, (3) gain or loss in body weight, and (4) the number of eggs laid by each flock.

The results of this 55-week experiment can be given here in general only. The basic data may be found in the original report and are reproduced here, in part, as graphs of weekly rates. In the examination of the figures it should be kept in mind that the controls also had to undergo social organization during the early weeks, and during this time their performance was not significantly different from that of the experimentals. Figure 8 shows the rates of pecking of controls and of unorganized alternates. For the full experiment the controls pecked at the mean rate of 139 and the experimentally unstable flocks at 191 pecks per six hours' observation per week (two half-hour periods daily). The difference is statistically significant beyond the 1 percent level. The mean number of grams of food consumed per bird per week, Figure 9, was significantly greater for the controls (600 grams) than for the alternates (467 grams) for the 55 weeks (at the 1 percent level). With regard to changes in body weight, Figure 10, the controls showed a mean gain of three grams whereas the alternates lost a mean of 33 grams per week (significant beyond the 1 percent level). For a discussion of the egg record the reader is referred to the original publication. It can be seen in Figure 11 that the

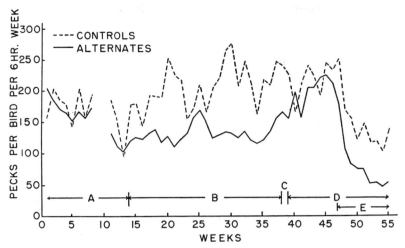

Fig. 8.—The average number of pecks delivered each week based on pecks recorded during observation periods totaling six hours per pen per week. Alternates were rotated daily during period A and every second day in B. No changes were made in the week marked C; during D the rotations were made on alternated days among three experimental flocks rather than between one experimental flock and the birds in isolation, as in A and B. Mash was available all day in E. From Guhl and Allee (1944).

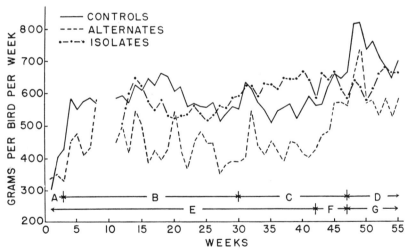

Fig. 9.—Average grams of food consumed per bird per week. During periods A, B, C, and D, controls and alternates were fed twice daily for 30, 60, 90, and 60 minutes, respectively. In D mash was present all day, in addition to the grain fed at the time of observation. Isolates were fed for 120 minutes in E, for 90 minutes in F. During G they received the same feeding treatment as the other birds. From Guhl and Allee (1944).

173

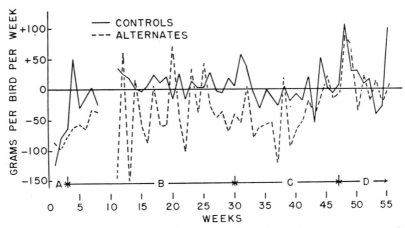

Fig. 10.—Average gain or loss in body weight, in grams per birds per week. A, B, C, and D as in Figure 9. From Guhl and Allee (1941).

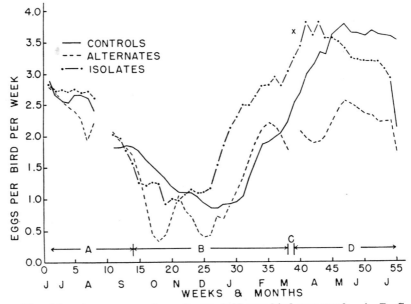

Fig. 11.—Average number of eggs laid per bird per week. A, B, C, and D as in Figure 8. From Guhl and Allee (1944).

laying cycle of the controls lagged behind the laying cycle of the birds alternated. It should be mentioned here that the physical factors for the isolated birds (from which the alternates were shuttled) differed from those of the controls. In the over-all results, the controls laid 2.3 eggs per bird per week, the alternates laid 1.8, and the isolates 2.5. The differences among all these means were statistically significant.

It was concluded that flocks which were well integrated socially pecked less, gained more weight, consumed more food, and laid more eggs than did hens in flocks which were kept in a constant state of reorganization.

Reduction of Social Tension in a Flock and Increased Productivity: Social tension in a flock has been measured by the rates of pecking and threatening. Threats were undelivered pecks or instances in which the dominant bird raised its hackle or in an otherwise threatening manner caused the inferior to which this behavior was directed to avoid the threatening bird. Timed observations were made after grain was fed to induce competition. (An alternative method could be employed by moistening the mash.) These periods of recording were daily and at the same hour of the day. To measure the intensity of pecking, the combs were examined for wounds and scabs. The number of these occurring on the various birds in the flock was compared with those in another flock to indicate differences in social tension.

In the experiment conducted by Guhl and Allee (1944) the relationship of social tension to egg production was illustrated for a complete laying cycle. Figure 8 shows the differences in social interaction between organized and unorganized flocks, and Figure 11 shows the number of eggs laid per bird per week. During the first eight weeks when the controls were undergoing a maturation process in social organization, the weekly rates of pecking and egg laying did not differ statistically from those of the unorganized alternates. During the 11 to 38 weeks the controls pecked at weekly rates which were lower (statistically significant) than those of the birds undergoing rotation. With regard to egg production, there was also no statistically-significant difference during the first eight weeks. However, from the 11 to the 32 week, the controls laid significantly more eggs than did the unorganized flocks under higher social tension. These birds, of the unorganized flocks, when they had their turn in isolation, laid more eggs than did the controls, showing that they were capable of greater production when not under social tension.

When the most aggressive of the organized flocks were used as controls (see D in Figures 8 and 11), the same relationship between social tension and egg production was maintained. Intensity of pecking was also more pronounced among the alternates as revealed by the number of wounds and scabs on the combs of the birds. Counts were made at three different times, in April, May, and July. Among the controls 89 percent had few or no wounds on the comb (42 percent had five or fewer), as compared with 60 percent for the unorganized alternates of which the remaining 40 percent had many scabs or wounds.

This reduction of social tension or the increased toleration which developed as the peck-order became formalized is of utmost interest to poultry management. The implications are that all the birds in the well-organized flock had more opportunity to feed and wasted less energy in pecking and avoiding others. Production, whether of flesh or eggs, might tend to be more uniform among the members of a well-organized flock.

Further experimentation on the influence of social tension on productivity merits consideration. The birds used by Guhl and Allee were culls, and feeding was restricted to relatively short periods of the day. Larger flocks of known productivity could be used with somewhat similar procedure. Tests with young birds might yield some interesting information on the relation of social tension to growth rates of market birds. The relation of flock size to social tension also needs more attention.

Effects of Debeaking on Social Behavior: Hale (1948) working with mature White Leghorn hens performed a series of experiments to determine the effect of debeaking on social behavior. One half or more of the upper beak was removed with an electric debeaker.

In one experiment five unacquainted debeaked hens were placed into one pen and five normal hens, also strangers to each other, were placed into another pen to serve as controls. The debeaked hens formed a social hierarchy in the typical fashion by fighting and bluffing. Following the establishment of a peck-order the debeaked hens pecked one another, as a group, at a considerably higher rate than did the control hens. Pecks delivered by debeaked birds were less effective than those delivered by normals. Over a 10-day period, 58 percent of the pecks delivered by debeaked birds were ignored by their subordinates, whereas less than 1 percent was ignored among the controls. These results show that debeaking does not inhibit pecking nor the formation of a peck-order, but it does demonstrate a decreased efficiency of the altered beak as a means of exercising social control.

Forty-eight paired-encounters were staged between seven debeaked and eight normal hens. Of these, 27 ended in fights and 21 in passive submission, and the debeaked birds won 29 of the 48 pairings. The normal hens won 15 of the 27 fights. It is apparent from these tests, and others not reported here, that debeaking did not alter aggressive behavior although it did decrease the effectiveness of domination by pecking. The use of larger flocks in a similar experiment probably would yield additional information of interest to the poultryman.

SPECIAL ASPECTS OF SOCIAL ORGANIZATION

Recognition of Individuals: The existence of a peck-order is evidence that the birds recognize one another. Without **recognition, pecking would** be promiscuous and unidirectional pecking would not occur. Close observation of hens **feeding from a food trough** suggests that recognition is based **essentially on features of the head,** as a subordinate bird may not avoid its superior until the latter's head is raised and visible. Schjelderup-Ebbe (1922) found when the loppy comb of a hen is turned to the other side and bound, that it would be attacked as a stranger by its penmates, even by its inferiors. Returning the comb to its normal position restored the former social relationships. Accordingly, an experiment was undertaken by the writer to establish the role of the comb in identification of individuals. In one of these tests the comb of a hen at mid-level in the peck-order was dubbed by electric caudery. When returned to her pen of seven hens two days later, along with a nondubbed hen also removed for two days, she was attacked as a stranger by all members of the flock. The nondubbed hen was accepted according to her former relationships. The dubbed bird returned the attack on her inferiors but avoided her superiors. Thus, as with other hens undergoing this treatment, her flockmates failed to recognize her although her behavior toward them as individuals was unaltered. It was also noted that dubbed birds when placed together would form a peck-order, suggesting that other features may be involved in recognition.

Experiments now in progress indicate that recognition of penmates may involve a number of factors. In flocks composed of 8 to 30 White Leghorn pullets, alterations were made which included changes in contour, in color, and in both simultaneously. These modifications were made on various areas of the body. None of these birds was dubbed. The results were variable, but suggest at the present state of the study that changes involving the head and neck were more likely to produce a loss of recognition; however, when the neck only or neck and head were denuded the birds were attacked as strangers.

Chickens, especially cocks, may lose the points on their combs as a result of freezing. These parts are lost gradually, and loss of recognition may not occur. Similarly a hen in a heavy molt may not be treated as a stranger; however, occasionally birds which lose most of the feathers on the neck may be driven about by their penmates. Hens when molting are at a disadvantage in initial encounters (c.f. Collias, 1943). It is interesting to note that changes must be rapid and chiefly in the head region to produce a disruption in the social order.

Memory of Other Individuals: Memory is associated inseparably with recognition and its quality determines the degree to which social behavior patterns are formed and to which such behavior is consistent and endures as social inertia. Separation for two or three weeks results in failure to recognize former penmates (Schjelderup-Ebbe, 1935). It is reasonable to assume that there is a limit (undetermined at present) to the number of individuals one bird can remember. We have evidence of a peck-order in a flock of 96 White Rock pullets which were confined to a pen. As frequent

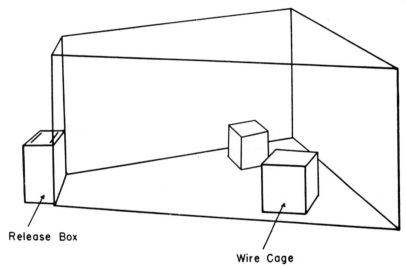

Release Box

Wire Cage

Fig. 12.—Pen used to test for social discrimination. From Guhl (1942).

meetings between any two individuals reinforce memory, the number in the flock and the size of the pen or out-door area influence the extent to which pair reaction patterns are maintained. In penned flocks of about 10 birds the peck-orders may remain unchanged for months. In flocks of about 40 or more penned hens, revolts may occur and some dominance relations be altered. With an expansive out-door range the peck-order of such groups may show somewhat more reversals of dominance.

Are Chickens "Aware" of the Social Order? In raising this question there is no intention to use the term "aware" with the implication of consciousness. Rather, it is a question of whether hens as individuals recognize levels in the peck-order, i.e., whether or not they react differently to penmates at the several levels in the hierarchy. To explore this possibility, Guhl (1942) used a discrimination cage (Fig. 12) similar to one designed by Murchison (1935b), which was covered with cheesecloth. The bird to be tested was placed into a release box at the apex of the triangular discrimination cage. One bird was placed in each of the two cubical wire cages within the larger cage. The test bird was then released. Typically it went to one or the other of the caged birds; there were many non-discriminations. The birds used as "objects" for discrimination were then reversed for a second discrimination, to allow for possible directional effect. Each bird was tested with all possible combinations of penmates, and two flocks were used. The birds were kept in pens of a poultry house where the peck-orders were determined. Both flocks were composed of cocks and hens and both sexes were tested.

Those interested in the analysis are referred to the original report, which showed that neither the cocks nor the hens, as a group, discriminated to the social position of individual cocks or hens. The discriminations were influenced by individuality and by various inter-individual experiences in the home pen. This lack of recognition of social levels partially explains a number of observations made since the experiment. Revolts, when they do occur or when attempted, may involve birds at any two levels, not necessarily contiguous, in the peck-order. The intensity of reactions between any two flock members may vary from pronounced antipathies to extreme toleration, irrespective of the spread between their social ranks.

LEVELS OF AGGRESSIVENESS

Hormones and Aggressiveness: It is apparent that both psychological and physiological factors play a role in the attainment of social position. Cocks are more aggressive than hens and also have larger combs. Capons lose much of their aggressiveness and have very small combs (Domm, 1939). Comb growth occurs when solutions of an androgen are applied directly to the surface of the comb (Koch, 1937). Hens normally secrete an androgenic substance (Juhn, Gustafson, and Gallagher, 1932; see also Witschi and Miller, 1938). These and other considerations suggested that male hormone influences aggressiveness. Allee, Collias, and Lutherman (1939) followed these suggestions and injected hens at low social levels and found that the treated birds rose in the peck-order, even to alpha position. They also reported that hens with large combs won more staged encounters by pairs than did hens with small combs. The size of the comb was the best indicator of the probable winner between birds of the same strain. As compared with normal hens, poulards (ovariectomized hens) with small combs were less aggressive. There is a general agreement that the levels of androgen concentration in the blood contribute to relative aggressiveness in most of the vertebrates studied (Beach, 1948).

The injection of estrogen (estradiol) may produce results which indicate an antithesis when compared with effects of androgenic treatment. Although the effects are less striking than with testosterone propionate, hens treated with estradiol appear to show lowered aggressiveness (Allee and Collias, 1940). Thyroxin had no effect on the social status of treated hens unless the dosage initiated molting and the associated conditions (Allee, Collias, and Beeman, 1940). The injection of adrenalin into normal hens did not alter social position nor aggressiveness in initial paired-encounters (Allee and Collias, 1938).

The relative aggressiveness of a bird, however, should not be attributed solely to the amount of androgenic substance in the blood. Capons penned with hens were able to establish positions in the peck-order which were higher than those of some of the females (Guhl, 1949). Breneman (1951) has considered the possibility that adrenal glands in the chick may secrete an adrogenic substance, but the evidence was not conclusive. Neural mechanisms, although they normally function in aggressiveness in connection with hormones, also might account for individual variations. The extent to which aggressiveness in chickens is inherited still needs to be demonstrated (Guhl and Eaton, 1948). A more adequate discussion of the various aspects of aggressiveness has been presented by Collias (1944).

Inheritance and Aggressiveness: The extent to which certain traits in behavior are either inherited or learned has been in question for some time. Evidences from various sources suggest that certain types of behavior may have a genetic background (Thorpe, 1948). The discussion here will be limited to poultry. Fennell (1945) found that, in fighting, Games were shiftier, faster, and less clumsy than domestic cocks. He also noted differences among varieties of Game cocks in their method of attack. Apparent breed differences in fighting behavior also were reported by Potter (1949). She used hens of seven breeds and found the White Leghorns to be the most dominant and White Cochin Batams the least. Brown Red Game hens tended to initiate the most fights when tested by initial encounters, but the White Leghorns were as successful in winning these contests. Rhode Island Reds appeared to be the most persistent and showed the most endurance. Potter stated that the differences found were not necessarily of the breeds in general, as her small samples were largely from the same strain.

At Kansas State College some attention has been focused on a breeding program based on relative aggressiveness (Guhl and Eaton, 1948). This research was stimulated, in part, by the fact that a commercial poultry breeder inadvertently obtained a highly aggressive strain which was too combative to be maintained. If a relatively unaggressive strain could be developed, certain problems in flock management might be simplified, especially with reference to cocks. Using White Leghorns, taken at random from among five different strains in the initial selection, the general procedure was as follows. Two lines of selection were undertaken by mating the most aggressive males with the most aggressive females, and similarly with the least aggressive individuals of both sexes. The chicks were pedigreed.

We faced several problems; chief among them were the relatively small number of birds in each generation with which we could work, and the reliability of the techniques upon which selection for aggressiveness was based. There was also the need for more experimental evidence that the relative aggressiveness of an individual was maintained within certain limits, unless the birds were ill or otherwise influenced by such conditions as are associated with molting (see Measurement of Aggressiveness).

The original selection and that in the F_1 generation were based on ranks in the peck-order of each flock. Highest ranking males were mated with hens in the top third of the peck-order, and the lowest ranking males were mated to the hens composing the lowest third. As an attempt to improve the measurement of aggressiveness, the F_2 birds were isolated in laying batteries and tested by initial pair-encounters in exhibition cages. The close confinement had debilitating effects, as mentioned previously. Accordingly, the F_3 and F_4 generations were divided into two pens of cocks and two of hens. Initial encounters for each sex then were staged between each hen (or cock) and each of the members of the flock to which it was a stranger. Another change in procedure, suggested by the results given in a preliminary report (Eaton, 1949), was followed in the F_3 and F_4 generations. Instead of selecting chicks from high-ranking groups of hens and from low-ranking groups, a progeny test method was devised. The largest families from high-ranking (based on the number of encounters won) and low-ranking dams were retained for testing by inter-flock pair-encounters. The hens in the F_3 generation were divided into two groups, a "high" line and a "low" line. The males were divided similarly. In this manner the initial encounters were always between individuals from dams ranking high with those of dams having low ranks. The birds of the F_4 generation were divided into two unisexual groups containing approximately equal numbers of individuals from both breeding lines.

The results gave no definite indication that levels of aggressiveness were inherited. Only the F_3 generation suggested a genetic background for aggressiveness. Twenty-five birds in the "high" line had a mean of 66.9 percent of contests won, and 19 birds in the "low" line won an average of 26.6 percent of their encounters. In the F_4 the means were 41.9 and 57.0 percent respectively, with considerable variation in each line. Space does not permit a presentation of the data and their analyses. Furthermore, it was decided to discontinue these two lines for breeding. The above results certainly did not give evidence that aggressiveness was not inherited.

Another experiment is under way which will follow the progeny test method and will utilize some of the more recent information obtained on methods of testing for relative aggressiveness.

SEXUAL BEHAVIOR AND THE SOCIAL ORDER

Sexual Behavior Patterns: In chickens the male typically takes the initiative in sexual behavior. There are a variety of approaches by which a cock may evoke a sexual response in the hen. The most spectacular of these is the wing-flutter, waltz, or dance. This display may be given also to cocks and frequently occurs after the male dismounts subsequent to coition. Our records from small and confined flocks indicate that the wing-flutter is not as effective as the more usual and less marked approach in which the male extends his head over the female with or without raising the hackle. A vigorous approach may take the form of a quick grasp of the hen's comb or hackle, or a wild chase followed by the grasping reaction. These initial sexual reactions of cocks have been described by Skard (1937) and were considered in a category of sexual behavior called "courting" by Guhl, Collias, and Allee (1945).

A hen may be indifferent to courting, or she may respond either negatively or positively. As a negative reaction she may step aside, may walk or run away, or may struggle if captured. These types of "avoidance" of the male by the female may be accompanied by vocalizations varying in intensity from faint screams to loud squawks. A positive reaction to courting takes the form of a crouch, often with head low and wings spread. This behavior has been called a "sex-invitation" (Skard, 1937, Guhl, Collias, and Allee, 1945; Collias, 1950a), but will be referred to here as "crouching." The sexual crouch is a strong stimulus for the cock to mount and tread, particularly when the male approaches from the rear. The male stands on the outstretched wings, grasps the comb or hackle of the hen, and moves the feet up and down in a "treading" manner. Subsequently the male rears up, the hen moves her tail to one side, and each everts the cloaca as the vents meet. The male usually steps off in a forward direction and the hen shakes herself vigorously as she gets to her feet. She may run in an arc and the cock may execute a waltz.

There is a general agreement that sexual behavior is most frequent during the late afternoons (Heuser, 1916; Philips, 1919; Upp, 1928; Skard, 1937; Parker, McKenzie, and Kempster, 1940). A number of observers have noted marked differences in the frequency at which a given male may tread the individual hens in the flock. There may be marked differences in the rates at which several cocks tread the various individual females. This so-called "preferential" mating (Philips, 1919; Warren and Kilpatrick, 1929; Murchison, 1935c; Skard, 1937), has been explained only in part by more recent experiments.

Hormones and Sexual Behavior. Castration was followed by a marked reduction or cessation of pugnacity and of sexual behavior in capons (Goodale, 1913), in turkeys (Scott and Payne, 1934), and in many other birds (Domm, 1939). Bilateral ovariectomy produced similar results; like the capon, these poulards were neutral in behavior (Domm, 1937). In the sinistrally ovariectomized poulard (with left ovary removed and the right rudiment intact) there developed either a testis-like structure or an ovotestis. Such poulards resembled the normal cock in appearance and behavior, depending upon the extent of the hypertrophy in the right rudiment. Intersexes were obtained in Brown Leghorns by the injection of estrogen into the egg before the fourth day of incubation (Domm and Davis, 1941). The males from these eggs developed varying degrees of femininity. There was a close relationship between the degrees of masculinity and femininity to positions in the peck-order, male and/or female plumage, and the display of male sexual behavior. The most feminine ranked lowest in the hierarchy, were most feminine in plumage, and were most neutral in sexual behavior.

An adequate discussion of hormones and behavior is beyond the scope of this review; recent publications by Beach (1948) and Collias (1950a) present comprehensive summaries. A pertinent account may be one concerned with the influence of androgen and estrogen on the behavior of gonadless (i.e., sexually neutral) males and females, and on related experiments which demonstrate the most marked

183

relationships of sex hormones to behavior in chickens. Davis and Domm (1943) injected androgen into capons and poulards, and estrogen into capons, poulards, and cocks. Capons treated with testosterone propionate crowed, trod the hens, and became aggressive. Bilaterally and sinistrally ovariectomized poulards treated with androgen crowed, waltzed, became more aggressive, but never copulated. When bilaterally ovariectomized poulards received estrogen they became more timid and crouched for a normal cock. Capons receiving estrogen executed the wing-flutter, copulated, but did not crow or crouch. Two normal cocks receiving stilbestrol (estrogen) lost their aggressiveness, stopped crowing, and copulated in a listless manner. Copulations by an estrogen-treated capon and an estrogen-treated poulard were observed.

Collias (1950a) injected sex hormones into male chicks. He noted an increase in the frequency of crowing, pecking, attempts to mate, and wing-flutters, with an increase in the dosage of testosterone propionate. This androgen was much more effective in inducing characteristic male behavior in male chicks than was alpha-estradiol benzoate (estrogen). It was found also that testosterone propionate was far more effective in male chicks than in female chicks in inducing typical male behavior.

Beach (1948) developed a schematic representation of the interrelationships between neuromuscular mechanisms in males and females to gonadal hormones. The plan is hypothetical and probably most applicable to certain lower mammals. In the genetic male, both androgen and estrogen increase the sensitivity of the mechanism for the masculine mating pattern. However, there is a marked increase with androgen and only a slight increase with estrogen. The response expected is normal with androgen and sluggish or incomplete with estrogen. The sensitivity of the mechanism for the feminine pattern in the genetic male is markedly increased with estrogen, but the expected response is sluggish or incomplete. With androgen the sensitivity is increased slightly and the response is very weak, incomplete, or absent. In the genetic female the scheme is similar for homologous and heterologous sex hormones and sex behavior.

Heterosexual Dominance and Mating Behavior: Cocks normally do not peck the hens, although they appear to dominate them in a passive manner. A heterosexual flock has two peck-orders, one among the males and the other among the females. Dominance status among cocks is associated with mating rights whereas dominance in hens, at the upper levels, interferes with mating. These observations raise the question as to whether the dominance of males over females exerts any influence over the synchronization of the sexes.

To test this heterosexual relationship it was necessary to devise a peck-order which would include both males and females, and preferably one in which certain cocks would dominate some hens and be dominated by others.

The procedure followed by Guhl (1949) was based on results obtained by Davis and Domm (1941) in which capons that received estrogenic treatment had their sexual behavior restored without becoming aggressive in the social sense. Two flocks containing capons and pullets were used. Heterosexual peck-orders were formed, and after a series of preliminary observations one capon in each flock was injected with estrogen. Both of these capons became active sexually and maintained their intermediate positions in their respective peck-orders. These treated capons were definitely more successful in courting and treading pullets which were their inferiors than they were with females which were superior to them. Pullets which dominated these capons repulsed the males when courted. It was concluded that the passive dominance of males over females facilitated mating.

Unisexual Mating: In flocks of cockerels, whether penned or on a range, unisexual mating is a common observation. Certain males are driven and trod repeatedly to such an extent that far too many of the young males eventually may be killed or mutilated. In a flock of 65 cockerels (Guhl, 1949) it was found that unisexual mating occurred most frequently between males ranking high in the social scale and males of low rank which were pursued and trod. Apparently dominance relations play a role in this aberrant behavior. No indications of receptivity were noted.

Heuser (1916) reported a case in which a hen attempted 35 times to mate with 13 other hens. These pseudo-matings occurred at times of production and non-production, and in the presence or absence of males. Guhl, Collias, and Allee (1945) reported 20 unisexual matings involving three apparently normal hens in the male role with six different hens. The hens trod were socially inferior to those which mounted. Guhl (1948) found that dominance relations facilitated unisexual matings among hens. Of 181 pseudo-matings observed in a flock of 42 White Leghorns, only 8 were on hens superior to those taking the male role. Five hens at various levels of the peck-order trod 27 different hens. None of these females showed any other indications of maleness, such as crowing or waltzing. Such behavior is rare in birds of dimorphic species, but it is not uncommon in such birds as pigeons which lack plumage differentiation in the sexes.

Sexual Satiation and Compensatory Sex Behavior: When a cock was exposed singly and daily for 10 to 15 minutes to a small flock of hens, Collias (1950a) found a decrease in the frequency of wing-flutters, crouchings, and copulations. There was a decline in sex drive with progressive satiation in the flock. Carpenter (1933), working with pigeons, noted

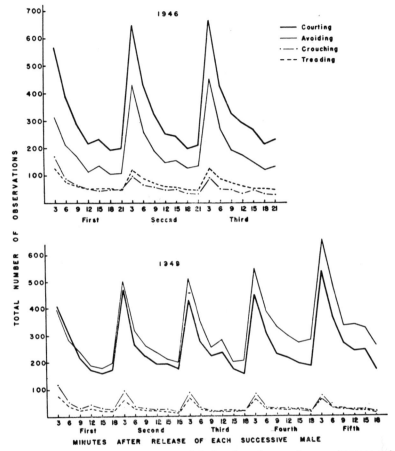

Fig. 13.—Compensatory sexual behavior shown by an increase in courting with a decrease in crouching when sexually active cocks were introduced singly and successively into a pen of hens for short periods. Sexual satiation of each male is shown by its decrease in activity.

that a female mated with a castrated male showed more provocative behavior than with a normal male.

Evidence of the facilitation of treading in chickens by compensatory sexual activity was obtained in flocks in which the male was caged within the pen of the hens, and released singly and successively for short daily periods. The data were tabulated as to the first, second, third, etc., cocks in the order of release. As the males were rotated, these data tend to eliminate individual differences among them and also differences in reactions to them as individuals, by the hens. The results are given in Figure 13 for the 1946 flock, with three males, and for the 1949 flock with five cocks. The data are plotted for each three-minute period after release for a total of 21 minutes in 1946 and 18 minutes in 1949. Total observations are given for courtings, avoidances, crouchings, and treadings occurring over 18 weeks and 10 rounds respectively. The results in a third flock (1947) are not given here, as they are similar to those obtained in 1946.

As the figures illustrate, progressive satiation, for each male, set in quickly during the first nine minutes and then tended to level off more slowly. Avoidance, crouching, and treading followed the same general pattern, indicating that the cocks took the initiative in sexual activity. The recurrence of a higher incidence of crouching with each successive and sexually active male also showed that crouching was essentially a reaction to courting, as was avoidance. Progressive satiation in the hens was made evident by the gradual decline in the maximum crouching with each of the successive males. It should be noted that the peaks and troughs of treading were essentially similar for each male, indicating that mating frequencies, although declining with satiation, tended to maintain a similar pattern for each of the males. Courtings and avoidances tended to increase with each successive cock. Thus, as the hens crouched less often, the males compensated for the decrease in receptivity by an increase in courting. Furthermore, initially and prior to the onset of satiation in the hens, crouching tended to be more frequent than treading, whereas later (especially in 1946 and 1947) there were more treadings than crouchings; this showed that the males released late in the series did not, at times, wait for the hens to crouch before mounting.

In mating flocks under the usual procedures and sex ratios, sex drive would be dissipated more gradually. Nevertheless, satiation and compensatory sex behavior would be expected to occur. In flocks containing relatively inactive cocks the hens may be expected to crouch more often than with males possessing a high level of sex drive. Differences between flocks may occur; for example, the 1949 cocks were avoided more frequently than they courted, whereas the reverse was true in 1946.

With the absence of males, as in laying flocks, often it can be observed that hens crouch readily when threatened by their superiors. The intensification of sex drive becomes a problem in the management of cockerels when in unisexual groups. Inferior males are pursued and driven about until exhausted, and when driven into corners are forcibly mounted and repeatedly trod. We have noted many deaths which could be attributed to this behavior.

Social Dominance among Cocks and Mating Behavior: There have been a number of studies of sexual behavior in chickens, but only a few of these attempted to relate social behavior to sexual activity. Although there were some definite points of agreement, the results cannot be compared strictly, as there were, among other things, variations in experimental procedure. For example, Murchison (1935a) determined social relationships in a "social reflex runway" and not by the now more common tabulation of pecking. In a flock of six cockerels and five pullets, only three of the males trod the pullets. Some of the information given suggested that some males were suppressed by dominance behavior. He stated that the most dominant males worked smoothly and gently, whereas the least dominant seemed to be under a strain at all times.

Heuser (1916) related his observation on mating behavior to "constitutional vigor" and found that "high-vigor" males mated more frequently than did medium or "low-vigor" males. He cited the work of R. H. Wilkins who obtained similar results. The descriptions of behavior offered a strong suggestion that his ranks in "constitutional vigor" may have reflected ranks in the peck-order. The cocks interfered with each other's matings. "High-vigor" males did most of the interfering. A labyrinth which facilitated mating was placed on the floor of the pen. This structure was made of 10-inch boards, placed vertically with the lower edge 8 inches from the floor. There was less fighting among the cocks and less interference with mating.

To determine the influence of social position on mating behavior in cocks, Guhl, Collias, and Allee (1945) tested four males singly and successively with the same group of hens, and after several weeks placed all the males into the pen of the hens for further observations. This procedure was followed with two small flocks. Data were collected to determine the frequencies of courting, crouching, treading, and avoidance of the male by the females. The results are summarized in Table 3 in which the cocks are listed as to their respective dominance orders. The more detailed analyses given in the original publication will not be repeated here. Inter-flock comparisons cannot be made as, among other differences, there were changes in some of the procedures. It will be noted that the males in each flock showed some variation in behavior. These differences, although not statistically significant in all comparisons, showed no direct relationship to the social position of the cocks, nor did the rate of a given male in one category of behavior necessarily reflect his rate in another type of behavior.

When the four males were placed together in the pen of the hens and permitted to remain there, there was considerable interaction between the cocks. A summary of the data

Table 3.—Rates Per Hour of Behavior of Males When Placed Singly and Successively into the Pen of Hens. Summary from Guhl, Collias, and Allee (1945).

Males	Treadings	Courtings	Crouchings	Avoidances
		Flock D		
I	3.05	8.23	0.88	——————[1]
II	4.56	8.15	0.00	——————
III	3.31	8.73	0.28	——————
IV	2.46	7.20	0.52	——————
		Flock F		
V	9.50	24.26	8.19	18.36
VI	8.36	42.13	7.21	29.01
VII	7.05	21.14	6.88	15.24
VIII	9.18	41.31	7.37	49.83

1. Avoidance was not recorded in Flock D.

obtained from the observations under these conditions is given in Table 4, which does not show the gradual changes which occurred over a period of several weeks. The dominant cock (I) in Flock D courted, received crouches from the hens, and trod at the highest rate. Male IV which had gained in

Table 4.—Rates Per Hour of Behavior of Males When All Four Cocks Were Continuously in the Pen of the Hens. Summary from Guhl, Collias, and Allee (1945).

Males	Treadings	Courtings	Crouchings	Avoidances
		Flock D		
I	3.18	8.76	0.76	6.59
II	1.34	7.11	0.00	20.72
IV	2.53	6.38	0.00	6.84
III	0.00	2.50	0.00	23.33
		Flock F		
V	1.60	11.27	0.15	8.07
VI	1.44	9.58	0.83	7.03
VII	0.15	3.15	0.08	3.37
VIII	1.14	10.00	0.04	18.35

social status was tolerated by the dominant I. The other two males were suppressed by cock I, and the hens consistently avoided them. The cocks in Flock F appeared to be less aggressive socially and less active sexually. Suppression of sexual activity was not as marked as in Flock D. In both flocks it was evident that the dominant male had the greatest freedom to mate. The degree to which the sexual activity was suppressed in the inferior males varied greatly. Cock **III** eventually ceased to pay any attention to the females and became psychologically castrated. When tested with hens of another flock, this male courted readily and trod, which he failed to do when alone with the D-hens.

Guhl and Warren (1946) continued this study to determine whether there was any relation between the social position of a cock and the number of offspring sired. The sire of each chick was made evident by the use of males which differed from each other, or from the hens, by some hereditary character. Two flocks were used. One of these (1944 flock) was composed of 30 Rhode Island Red pullets and three cockerels, a Rhode Island Red, a Barred Plymouth Rock, and a White Leghorn. The chicks sired by the Rhode Island Red had red down, those sired by the Barred Plymouth Rock had black down, and those sired by the White Leghorn had white down. The second flock (1945 flock) contained 36 Barred Plymouth Rock pullets, a White Plymouth Rock male known to be homozygous for dominant white factor, a White Plymouth Rock male known to carry recessive white, and a White Wyandotte male known to carry recessive white but to be

Table 5.—Summary of the Data Obtained in a Test to Determine the Relation of the Number of Chicks Sired by Cockerels to Social Dominance. From Guhl and Warren (1946).

Peck-order	Courtings	Total treadings	Completed matings	Eggs fertilized	Viable chicks
1944 flock with 30 Rhode Island Red pullets					
R. I. Reds	219	59	51	236	201
Barred Rock	250	76	35	165	139
White Leghorn	73	8	3	56	54
Total	542	143	89	457	394
1945 flock with 36 Barred Plymouth Rock pullets					
Dominant white	710	175	112	267	221
Recessive white	2184	244	54	129	120
Rose comb	71	8	0	0	0
Total	2965	427	166	396	341

homozygous for dominant rose comb. The chicks sired by the **dominant white** male had white down, those sired by the **recessive white** male had black down, and those sired by the **rose comb** male developed a rose comb. The males for the second flock had the same plumage color and similar size, and thereby improved the experimental situation which prevailed in the first flock. The cockerels were selected on the basis of the results of a fertility test for homozygosity of the character used to trace the sire of the chicks. A further test was made by mixing equal volumes of semen of the three males and by artificial insemination of the mixture into pullets of the same breed to be used later. The results indicated that the sperm of each male was capable of fertilizing eggs when in physiological and numerical competition with those from the other males in its group.

A summary of the observations is given in Table 5. The cockerels were listed in the order of dominance. The top ranking male in each flock did not court or attempt to tread as often as his immediate inferior; nevertheless, the most dominant cockerels were most successful in attaining coition; and they also fertilized a larger number of eggs and sired the most chicks. The lowest ranking males, in both flocks, were the least successful. Most of the interferences with mating were made by the ranking cockerels and only two or none by the males ranking lowest. The statistical analyses given in the original report showed that the results were not as diagrammatic as Table 5 suggests; however, it was evident that the number of offspring sired by each of the cockerels was related to his relative social dominance.

Social Dominance among Hens and Mating Behavior: As has been stated previously, Heuser (1916) grouped his birds according to relative "constitutional vigor" which appeared to be a parallelism of social dominance. The data on the number of times the hens were mated suggested that "low-vigor" females mated somewhat more often than did the "high-vigor" hens. Murchison (1935c), who determined social relations in a "reflex runway," found a positive correlation between ranks in the number of times a hen was mated and her rank in the social order. However, Murchison's data were pooled from observations with males in the pen singly and with all the males in the pen. Schjelderup-Ebbe (1935) observed that high social rank among the females interfered with mating.

The relation between the number of times hens were mated and their social positions was analyzed by Guhl, Collias, and Allee (1945). To avoid the more complicating circumstances which occurred when several males were in the pen, only those data were used which were obtained when the cocks were rotated singly and successively in the pen of the hens. Consistently negative, although not always statistically significant, correlations were found between (1) social position and the frequency of mating, (2) social position and the frequency of being courted, and (3) social position and the frequency of response by the sexual crouch. These results were obtained with several small flocks of White Leghorns.

The crouching response of a hen is a type of submission

and is the best criterion we have to indicate receptiveness in the female. Among hens, those ranking high in the social order are not in the habit of submitting, whereas those with low social status submit freely to their more numerous superiors. Assuming that the relative differences in the habit

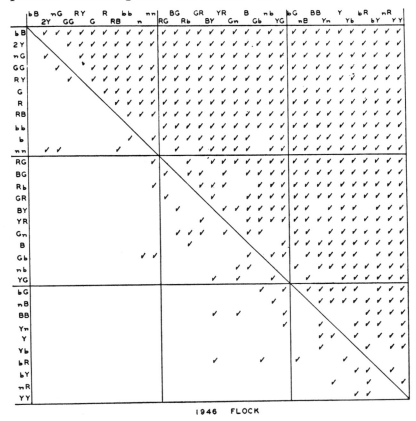

1946 FLOCK

Fig. 14.—Pecking relationships in the 1946 flock. The checks show which birds were pecked by each of the birds listed at the left. The horizontal and vertical lines divide the peck-order into top, middle, and bottom thirds.

of domination and submission at the several levels in the peck-order in hens accounted for the negative correlations between rank and the frequency of crouching, Guhl (1950) devised an experiment in which relative dominance was altered. The procedure was essentially as follows. In a series of three different flocks of White Leghorns, three or more cocks were caged within the pen of about 30 hens. These males were released daily for short (18 or 21 minutes) observation periods, following a system of rotation. At the approximate peak of the laying cycle, each flock was subdivided into three levels of the dominance order and the males were rotated for similar periods among the three pens. During the subflocking phase only one cock was used daily in each pen. The weekly frequency of courting was obtained for each hen and for each of the three levels in the peck-order.

Figure 14 presents the peck-order of the 1946 flock of hens. The top, middle, and bottom thirds of the peck-order are delimited by lines. The check marks indicated that the bird listed on the left has the peck-right over those heading the columns. The number of checks in a horizontal line indicates relative dominance and those in the vertical column show relative subordination and its correlative submissiveness. It is evident from the figure that, in subflocking, the top third, as a group, had approximately two-thirds less pecking or dominating, the bottom third about two-thirds less submission, whereas the amount of dominating and submission was altered almost equally for the middle third of the original flock. The situation was similar for the flocks observed in 1947 and 1949.

There was much variation among individuals in the frequency of crouching. The standard deviations of the means in each of the three levels and in each of the three flocks approximated 50 percent of the means. This analysis shows that the experimentation by groups or levels, rather than by individuals, did obtain about as much control over individuality as might be expected in studies of behavior.

The data are summarized in Figure 15 as graphs which have been smoothed with a 5-point moving average. The original analyses will not be repeated here. The breaks in the curve indicate the point at which the flocks were subdivided. An examination of the figure shows that, whereas the middle and bottom thirds crouched less frequently after subflocking, the top thirds, as groups, tended to crouch more often. Prior to subflocking the top thirds crouched least often. It was concluded that high ranking hens, which were in the habit of dominating, were less inclined to submit to the male than were those at lower levels in the social order. The indications are that in mating flocks, receptivity and presumably mating would be more frequent if flocks are relatively small.

Variations in the Sexuality of Cocks: Interest in a study of relative sexuality in cocks stems from certain assumptions and implications. It seems to be commonly assumed that one male is as good as another for breeding purposes. It is known in rats and some domestic mammals, however, that

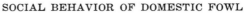
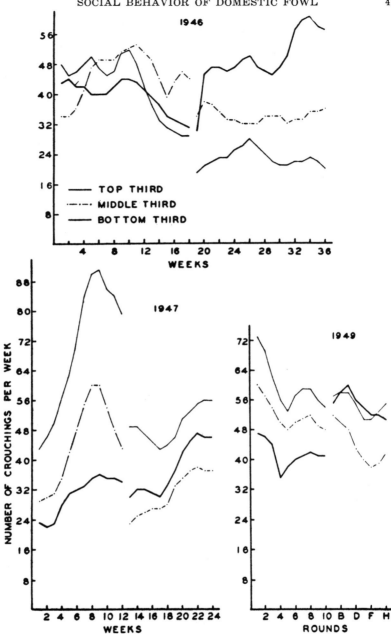

Fig. 15.—The rates at which hens, as groups, in each of three levels of the peck-orders crouched receptively for the males before and after subflocking to alter the degree of domination and submission. From Guhl (1950).

marked variations in sex drive may occur. Breeders of poultry often select males on the basis of close relationship to hens with high egg production, i.e., for genes desired in the females. If males with a low level of sex drive were inadvertently selected, would fertility be decreased in the breeding flock?

Data on the sexual behavior of a total of 14 cocks were presented by Guhl (1951). The males were caged within the pen of the hens and released daily and singly for short periods of observations. Records were made of courtings, crouchings, and treadings. A statistical analysis showed that the cocks, as individuals, tended to be consistent in their respective levels of sexual behavior.

Table 6.—Total Observations of the Sexual Behavior of White Leghorn Cocks. From Guhl (1951).

Males	Courtings	Crouchings	Treadings	Number of ♀♀[1]
1946		18 weeks observation	Jan. 7-May 12	
♂ W	5575	1167	1472	40
♂ R	5496	952	1157	
♂ G	5460	879	1416	
1947		12 weeks observation	Jan. 27-Apr. 18	
♂ R	3873	794	788	38
♂ b	2664	376	410	
♂ G	3664	842	853	
1949		50 days observation	Mar. 16-May 16	
♂ G	3073	386	347	30
♂ B	2323	392	322	
1950		18 days observation	May 26-June 12	
♂ R	276	43	29	31
♂ b	591	180	71	
♂ Y	708	100	69	
1950		18 days observation	June 16-July 7	
♂ RY	570	201	62	30
♂ W	296	42	25	
♂ bY	357	39	36	

[1] Number of hens that survived the experiment. Daily observation period of 21 minutes for each cock in 1946 and 1947, 18 minutes in 1949, and 9 minutes in 1950.

The basic data are reproduced in Table 6 as total observations on each male. The statistical analyses will not be repeated here. In the 1946 flock the hens crouched most frequently for ♂W, and ♂R trod the least often. In the 1947 group ♂b differed from the other two cocks in all three categories of behavior. Of the two males used in 1949, ♂B courted less often than ♂G but did not differ in crouchings received or in treadings observed. The six cocks tested in 1950 showed much variation. These were tested in groups of three with the same hens. Males R, W, and bY were less active sexually than the other cocks. The hens, as a group, crouched most readily for ♂RY and ♂b. Other differences appeared in the statistical analysis of rates in these activities, but are not so obvious in the table which gives totals only. Curves showing sexual satiation in the males show variations that are in agreement with analyses made on the rates of sexual activity and are shown in Figure 16 for the 1950 cocks only.

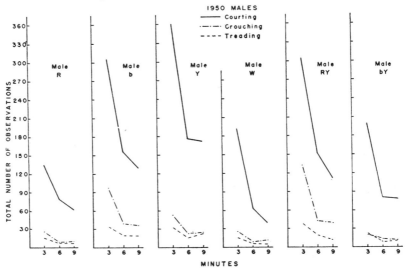

Fig. 16.—Variations in the sexuality of cocks as shown by differences in rates of sexual activity and their decline during daily periods of nine minutes. From Guhl (1951).

The results show that cocks may differ in sexuality and that the level of sexual activity in one category is not necessarily an indication of a male's rate of activity in another category. It was also found that when a sexually inactive cock was socially dominant over the other males in his flock, he interfered with their mating behavior and, as he failed to tread the hens, the frequency at which mating occurred in the flock was reduced considerably. Unfortunately no record was available of the fertility in this group and under these conditions.

SUMMARY

Hens establish dominance relationships among themselves by pecking and fighting. Upon the basis of unidirectional pecking a social organization or peck-order is formed in which individuals may be ranked according to the number of flockmates each bird pecks. The position which a hen obtains is determined when the birds first meet during the assemblage of the flock. There appears to be a marked variation in the relative aggressiveness of individuals when tested under certain conditions. The factors which influence attainment of success in winning initial encounters are more or less controlled in these tests.

Especially during the early weeks after a flock is assembled, the hens which rank high in the peck-order have definite advantages. They have a greater freedom of the pen and more access to the food trough, water supply, nest boxes, and favorable locations on the roost. Hens at the lowest levels in the social order may have little opportunity to feed and may go out of production. As the members of the flock develop appropriate habits of behavior toward each of their penmates, pecking becomes less frequent and less hard. Interactions involving domination and submission become formalized in symbolic threats and head bowing in accordance with the framework of the social pattern. As toleration of presence of others develops, the low-ranking birds are able to improve their situation and productivity in the flock may increase. The degree of toleration developed depends on a number of factors; chief among them are the space available, the size of the flock, and the accessibility of food, water, nest boxes, and the roost. Disruptions in a relatively peaceful flock are most pronounced when strange hens are added to the group.

Cocks in unisexual groups are also organized into a peck-order. Among the males the interactions are more intense and the dominance relations are less stable than among the hens. In the absence of hens the male sex drive becomes intensified, and attempts at unisexual mating may be disastrous to males of low rank. Cocks in a pen with hens usually do not peck the females but dominate them in a passive manner.

In breeding flocks there is considerable variation in the sexual activity among the males and the females. Of the factors involved some appear to be variations in the sex drive of the individuals, and some are associated with rank in the peck-orders of each sex. Socially-dominant cocks have a greater freedom to mate, may be most successful in mating, and sire the most chicks if they are normal in libido. The differential mating of hens, the so-called preferential mating, is partially influenced by previous experience and also by rank in the peck-order among the hens. High-ranking hens are characterized by bossiness and therefore are not inclined to submit readily for the male, whereas low-ranking hens crouch submissively with less provocation.

The above summary is a simplification. Behavior is in-

fluenced by a multiplicity of factors, some of which are unknown. Breed differences are indicated. Conditions in a
given flock may be attributed to variations in the population of which the group is a small sample. The caretaker
and others entering the pen with regularity may influence
the behavior in a flock.

Practical Applications: A number of suggestions are indicated in the body of this review. By way of summary, attention is given to the most critical situations in which
information on social behavior may be most applicable.
Events occurring during and subsequent to the assemblage
of a flock are of particular significance, since at the stage
of peck-order formation the individuals form habits which
may persist for many weeks. Unless the birds being penned
are acquainted they should all, or the majority of them, be
strangers to each other. Assemblage should, if possible, be
completed at the same time, particularly if the flock is small.
During this period of tension, additional food troughs and
sources of water distributed about the pen and the roost,
would alleviate the stressing situations for the most unaggressive birds.

The practice of combining flocks subsequent to culling
presents problems that merit consideration. For best results, the number of birds introduced should be approximately equal to the number in the pen. The greater the
deviation from this ratio, the more serious the social stress
placed on the outnumbered individuals. Additional sources
of food and water would be helpful until it becomes evident
that the birds intermingle quite freely.

Chickens can and do make adjustments to changing situations. However, alteration of habits requires time, and the
effects produced in the interim may be costly to the producer
as well as to the research worker.

Tension between males in a breeding flock may be minimized if the cocks are not strangers to each other when introduced. Relative sex drive and aggressiveness are not
necessarily correlated. A highly aggressive male may prevent other males from mating, and a male with unusually
high libido may cause the hens to avoid him. In either situation the frequency of matings may be low. The characteristics
of the most dominant cock should receive special attention,
as he may sire the most offspring, whereas the least aggressive male may sire the fewest or none. In a breeding program this situation might influence the results obtained.
Further implications are suggested, as the actual sex ratio
and the effective sex ratio are not necessarily equated.

Observations of the activities of the males may be worth
the effort and time. If critical information on performance

is required, techniques for testing aggressiveness and sex drive are suggested. Results of such tests indicate that in a breeding program more attention might be given profitably to certain characteristics of the males.

This bulletin is essentially a progress report. It is anticipated that the review of various studies in the behavior of chickens may be useful to those trained in poultry husbandry and to the poultry scientist. If some of the material appears to be premature it might be concluded that the omission may have impaired the perspective of this field of research. Experimentation might well be extended in breadth as well as depth. It is hoped that this collection of information may point out some of the directions in which further and newer studies might proceed.

REFERENCES

ALLEE, W. C., and N. E. COLLIAS, 1938. Effects of injections of epinephrine on the social order in small flocks of hens. Anat. Rec., Suppl. 72:119.

ALLEE, W. C., N. E. COLLIAS, and CATHARINE Z. LUTHERMAN, 1939. Modification of the social order in flocks of hens by the injection of testosterone propionate. Physiol. Zool. 12:412-440.

ALLEE, W. C., and N. E. COLLIAS, 1940. The influence of estradiol on the social organization of flocks of hens. Endocrinology 27:87-94.

ALLEE, W. C., N. E. COLLIAS, and ELIZABETH BEEMAN, 1940. The effect of thyroxin on the social order in flocks of hens. Endocrinology 27:827-835.

BEACH, FRANK A., 1948. Hormones and Behavior. Paul B. Hoeber, New York.

BRENEMAN, W. R., 1951. A factoral analysis of pituitary-gonad-comb relationships in the chick. Poult. Sci. 30:231-239.

CARPENTER, C. R., 1933. Psychological studies of social behavior in Aves. I. The effects of complete and incomplete gonadectomy on the primary sexual activity of the male pigeon. II. The effects of complete and incomplete gonadectomy on secondary sexual activity with histological studies. Jour. Comp. Psychol. 16:25-98.

COLLIAS, N. E., 1943. Statistical analysis of factors which make for success in initial encounters between hens. Amer. Nat. 77:519-538.

——, 1944. Aggressive behavior among vertebrate animals. Physiol. Zool. 17:83-123.

——, 1950a. Hormones and behavior with special reference to birds and the mechanisms of hormone action. In E. S. Gordon, ed., A Symposium on Steroid Hormones, 277-329. University of Wisconsin Press, Madison.

——, 1950b. Social life and the individual among vertebrate animals. Annals N. Y. Acad. Sci. 51:1074-1092.

——, 1950c. Some basic psychological and neural mechanisms of behavior in chicks. Anat. Rec. 108:64.

——, 1950d. The socialization of chicks. Anat. Rec. 108:65.

DAVIS, D. E., and L. V. DOMM, 1943. The influence of hormones on the sexual behavior of domestic fowl. In Essays in Biology, 171-181, University of California Press, Berkeley.

DOMM, L. V., 1927. New experiments on ovariotomy and the problem of sex inversion in the fowl. Jour. Exp. Zool. 48:31-173.

———, 1939. Modification in sex and secondary sex characters in birds. In E. Allen, C. H. Danforth, and E. Doisy, eds., Sex and Internal Secretion, 227-327. Williams and Wilkins Co., Baltimore.

DOMM, L. V., and D. E. DAVIS, 1941. Sexual behavior of intersexual domestic fowl. Proc. Soc. Exp. Biol. and Med. 48:665-667.

DOUGLIS, MARJORIE B., 1948. Social factors influencing the hierarchies of small flocks of the domestic hen: interactions between resident and part-time members of organized flocks. Physiol. Zool. 21:147-182.

EATON, R., 1949. Breeding for aggressiveness in the fowl. Thesis for M.S. degree, Graduate School, Kansas State College, Manhattan. (Unpublished)

FENNELL, R. A., 1945. The relation between heredity, sexual activity and training to dominance-subordination in game cocks. Amer. Nat. 79:142-151.

GOODALE, H. D., 1913. Castration in relation to the secondary sexual characters in Brown Leghorns. Amer. Nat. 47:159-169.

GUHL, A. M., 1942. Social discrimination in small flocks of the common domestic fowl. Jour. Comp. Psychol. 34:127-148.

———, 1945. Some observations and experiments on social behavior in the domestic fowl. Trans. Kans. Acad. Sci. 47:337-341.

———, 1948. Unisexual mating in a flock of White Leghorn hens. Trans. Kans. Acad. Sci. 51:107-111.

———, 1949. Heterosexual dominance and mating behavior in chickens. Behaviour 2:106-120.

———, 1950. Social dominance and receptivity in the domestic fowl. Physiol. Zool. 23:361-366.

———, 1951. Measurable differences in mating behavior of cocks. Poult. Sci. 30:687-693.

GUHL, A. M., and W. C. ALLEE, 1944. Some measurable effects of social organization in flocks of hens. Physiol. Zool. 17:320-347.

GUHL, A. M., N. E. COLLIAS, and W. C. ALLEE, 1945. Mating behavior and the social hierarchy in small flocks of White Leghorns. Physiol. Zool. 18:365-390.

GUHL, A. M., and R. C. EATON, 1948. Inheritance of aggressiveness in the fowl. Poult. Sci. 27:665.

GUHL, A. M., and D. C. WARREN, 1946. Number of offspring sired by cockerels related to social dominance in chickens. Poult. Sci. 25:460-472.

HALE, E. B., 1948. Observations on the social behavior of hens following debeaking. Poult. Sci. 27:591-592.

HEUSER, G. F., 1916. A study of the mating behavior of the domestic fowl. Thesis, Master of Agriculture degree, Graduate School, Cornell University. (Unpublished)

JUHN, MARY, R. G. GUSTAFSON, and T. F. GALLAGHER, 1932. The factor of age with reference to reactivity to sex hormones in fowl. Jour. Exp. Zool. 64:133-176.

KOCH, F. C., 1937. The male sex hormone. Physiol. Rev. 17:152-238.

MASURE, R. H., and W. C. ALLEE, 1934. The social order in flocks of the common chickens and pigeons. Auk 51:306-325.

MURCHISON, C., 1935a. The experimental measurement of a social hierarchy in **Gallus domesticus.** I. The direct identification and direct measurement of social reflex No. 1 and social reflex No. 2. Jour. Gen. Psychol. 12:3-39.

———, 1935b. II. The identification and inferential measurement of social reflex No. 1 and social reflex No. 2 by means of social discrimination. Jour. Soc. Psychol 6:3-30.

———, 1935c. III. The direct and inferential measurement of social reflex No. 3. Jour. Genet. Psychol. 46:76-102.

PARKER, J. E., E. F. McKENZIE, and H. L. KEMPSTER, 1940. Observations on the sexual behavior of New Hampshire males. Poult. Sci. 19:191-197.

PHILIPS, A. G., 1919. Preferential mating in fowls. Poult. Husbandry Jour. 5:28-32.

POTTER, JANE H., 1949. Dominance relations between different breeds of domestic hens. Physiol. Zool. 22:261-280.

SANCTUARY, W. C., 1932. A study of avian behavior to determine the nature and persistency of the order of dominance in the domestic fowl and to relate these to certain physiological reactions. Thesis for M.S. degree, Massachusetts State College, Amherst. (Unpublished)

SCHJELDERUP-EBBE, T., 1913. Hönsenes stemme. Bidrag til hönsenes psykologi. Naturen 37:262-276.

———, 1922. Beitrage zur social-psychologie des Haushuhns. Zeitschr. f. Psychol. 88:225-252.

———, 1935. Social behavior in birds. Chap. XX in Murchison's Handbook of Social Psychology, 947-972. Clark University Press. Worcester.

SKARD, ASE G., 1937. Studies in the psychology of needs: Observations and experiments of the sexual needs of hens. Acta Psychol. 2:175-232.

SCOTT, H. M., and L. F. PAYNE, 1934. The effect of gonadectomy on the sexual characters of the Bronze Turkey (**M. gallpavo**). Jour. Exp. Zool. 69:123-132.

SNEDICOR, G. W., 1946. Statistical Methods. Iowa State College Press. Ames.

THORPE, W. H., 1948. The modern concept of instinctive behavior. Bull. Animal Behavior No. 7.

UPP, C. W., 1928. Preferential mating in fowls. Poult. Sci. 7:225-232.

WARREN, D. C., and L. KILPATRICK, 1929. Fertilization in the domestic fowl. Poult. Sci. 8:237-256.

WITSCHE, E., and R. A. MILLER, 1938. Ambisexuality in the female starling. Jour. Exp. Zool. 79:475-487.

10

Reprinted from *British J. Animal Behaviour*, **3**(2), 45–55 (1955)

SOCIAL DOMINANCE RELATIONSHIPS IN A HERD OF DAIRY CATTLE

By MARTIN W. SCHEIN
Iberia Livestock Experimental Station, Jeanerette, La.

AND MILTON H. FOHRMAN
Dairy Husbandry Research Branch, U.S. Department of Agriculture,
Agricultural Research Service, Beltsville, Md.

Statement of Problem

Most people realize that if two animals are within each other's range of perception, be it visual, olfactory, tactile or other, then the behaviour of each is somewhat modified with respect to the other. Further, when two or more animals of the same species live in close proximity to each other, then there may exist a pattern of group behaviour fully as complex as the behaviour of the individuals comprising the group.

Dairymen have for years recognised the economic importance of the behaviour of their animals. Such phrases as "fidgety cows," "nervous temperament," "milk from contented cows" and the like, bear testimony to this recognition. However, few dairymen have applied deductions drawn from the actions of the individual cows to the behaviour of the herd as a group; indeed, few investigators have recognized the distinction between a group pattern and a group of individual behaviour patterns. The object of this study, therefore, is to describe a type of social organization existing within a herd of dairy cattle, and to explore the effects of this organization on the behaviour of the individuals.

Little has been reported on the social behaviour of ruminants until fairly recently, when Scott and his co-workers began series of experiments on social organization and leadership in sheep and goats (Scott, 1942, 1946). Alverdes (1935) described the behaviour of herds and packs, with particular emphasis on sex and mating behaviour in many species of free-living mammals but made little mention of dominance relationships. More closely concerned with ruminants, Altmann (1952), noted organizational trends in the social behaviour of free living elk. A certain amount of intensive investigation into the organization and behaviour of deer was initiated by Darling (1937), and interest in this group has continued through the years (Darling, 1952).

An understanding of dominance orders in dairy cattle seems to have been taken for granted by Woodbury (1941) when he discussed how the rank or "hook" order of the herd changed upon dehorning the "boss" cow. Actually, the writers can find no direct reference to dominance orders in cattle prior to Woodbury's article. Dove (1936) discusses the relative merits of a one-horned cow over a two-horned cow, but no mention is made of dominance orders as such. Woodbury's conclusions (*loc. cit.*) were that the "hook" order in horned cows was largely determined by the size, shape and effectiveness of the horns; in dehorned cows, the "bunt" order was determined largely by strength and tenacity in pushing, with tenacity being a somewhat loosely defined variable. During the period from 1944 to 1948, Guhl and his co-workers (Guhl *et. al.*, 1948) started a detailed study of "bunt order" in a large dairy herd with emphasis on the physical and physiological factors affecting intra-herd relationships. Unfortunately this work was never completed, nor was any part of it published. Somewhat more recently, Hancock (1950) recognised the importance of taking "herd laws" in consideration when conducting grazing studies. In the same article, he emphasised the point by delineating the dominance order existing in his herd of ten pairs of monozygotic twins. Other than these few reports, the authors know of no intensive, large scale study of herd behaviour from a "dominance order" point of view.

Background Information

To understand the methods used in this study, it is necessary first to have a clear picture of the management routines and herd composition at this station, and second to be generally familiar with manifestations of "normal" individual behaviour patterns. This study was conducted at the Iberia Livestock Experiment Station, Jeanerette, Louisiana, a field station owned and operated by the Dairy Husbandry Research Branch, Agricultural Research Service, U.S. Department of Agriculture.

Management Routines

For convenience of management, the herd is divided into five major groups, each group representing a particular stage in the development or physiological state of the animal. These groups will best be understood by tracing the development of a typical animal from birth through her first lactation period.

Generally, parturition occurs on pasture; within a few hours after birth, the calf and her mother are placed together in an 11 by 12 foot stall. From this stall, the calf can neither see nor smell any other herd animal except her mother. They remain together for three days, during which time the calf receives the benefit of the colostrum, or "first milk."

Three days post-partum the calf is permanently dehorned (electrically) and isolated in a 3 by 5½ foot stall. She spends 30 days in this stall, from which she can hear others and see the heads of young calves in adjoining stalls; it is possible for two adjacent calves to nose each others' head regions.

After 30 days of age, the calf is transferred to a 5 by 11 foot stall which she will generally share with one or two other calves of about the same age. In this, the "two month stall," the calf has a little more space to romp, and can easily observe other calves up to six months of age. In addition, she can nose, sniff or otherwise investigate her stallmate or even calves in adjacent stalls.

At the age of three months, the calf is introduced into what is termed the "young group," a group of 10-15 heifers between three and six months old. The young group pen inside the barn is only about 11 by 35 feet, but it adjoins a one-quarter acre outdoor lot, with free access from inside to outside. Except for being born outdoors, this is the calf's first introduction to green grass and direct sunlight. At six months of age, she is weaned and transferred to the "middle group," which is comprised of 6 to 12 month old heifers. This group is allowed considerably more space both indoors and outdoors and the animals do a limited amount of grazing on their one acre lot. The size of this group varies from 10 to 19 animals.

When the heifer reaches one year of age, she is transferred to the "front group," which spends all of its time on pasture. This group consists of nulliparous heifers over 12 months of age plus a few dry cows which are due for breeding or calving; the average size of this group is about 35 animals, but this figure may vary from 15 to 50 animals. The front group subsists entirely on pasture herbage which is supplemented with grain-feeding during December, January and February. The heifer is bred for the first time during the first heat period observed after she is 18 months old. Breeding is strictly controlled and copulation is brief: the heifer is stanchioned in a breeding chute and after a short period of vaginal sniffing and licking, the bull will generally mount quickly and complete coitus.

About a month pre-partum, the heifer is transferred to the "wet" or milking group, which consists of all the cows in milk. This is by far the largest single group on the farm, and averages about 70 animals. During the major portion of this investigation, the wet group was made up of Jerseys and Sindhi × Jersey halfbreed cows. The young, middle, and front groups were made up of Jersey and crossbred (Jerseys with varying amounts of Red Sindhi blood) cows and heifers.

As mentioned above, parturition occurs out on pasture, and the cow spends the first three days post-partum in isolation with her calf. After the three-day isolation period, the cow is returned to the wet group and milked twice daily (at 4 a.m. and 2 p.m.) for the remainder of her lactation. A daily record of productivity, in terms of pounds of milk, is kept for each cow as long as she is in the herd. It is perhaps pertinent to note at this point that she generally will not see her calf again until the calf in turn is ready to calve more than two years later.

At the first observed heat period three months post-partum, the cow is again bred and, assuming successful conception, is therefore pregnant throughout most of her lactation. Ideally, the cow should bear a calf each year and lactate for 10½ months out of 12. The cow is dried 305 days post-partum.

When the cow is finally dry, she is transferred to the "bayou group," which generally consists of pregnant dry cows more than 30 days pre-partum. When calving time draws near, she is re-introduced to the wet group. Most animals, after bearing the first calf, spend most of their lives with the wet group and only one or two months of the year with the bayou group. As a consequence, the turnover in the bayou group is rapid and the social pattern often does not get a chance to stabilize for any period of time. Group size varies from six to 20 animals, averaging about 15 individuals.

203

Some cows, however, go dry in less than 305 days (some milk for as few as 100 days) and therefore they spend correspondingly longer periods of time with the bayou group. One cow was used solely for breeding purposes (a damaged udder obviated profitable milking) so that except for about 45 days at calving time, she spent the entire year with the bayou group. Occasionally to facilitate management, the bayou and front groups are combined for several months. At other times, those heifers from the front group which are not due for breeding or calving are put with the bayou group. When this is done, the bayou group contains about 25 to 35 animals, while the front group will consist of about 10-15 animals.

Briefly summarizing, it is important to recognize that the calves used in this study began their social lives (large group conditions) at three months of age; thereafter they were rarely, if ever, by themselves for any appreciable period of time.

Behaviour of the Individual

Before we can attempt to interpret group behaviour patterns, we must have a clear understanding or at least a recognition of the normal individual behaviour patterns. However, since Brownlee (1950, 1954) has already described normal behaviour patterns in cattle, the ensuing discussion will serve to emphasize only those patterns that are pertinent to aggression and social organization.

In cattle, movements that are (subjectively) termed "aggressive" generally follow a definite sequence, each phase of which the author chooses to define as follows: the *approach*, which may be active or passive; the *threat*, or "aggressive intention movement" (Tinbergen, 1952); the *physical contact* (the actual fight) which is followed by victory or defeat.

There is no need to dwell long on the point of *passive approach*, since cattle are extensive wanderers and cross each other's paths any number of times daily. Occasionally, instead of ignoring each other on passing, two cows will engage in a brief manifestation of contesting behaviour. Such contests are usually brief and often go no further than the *threat* stage; however, it is the opinion of the author that these encounters are significant and decisive. (This point will be dealt with more fully during the discussion of "group" behaviour).

Active approach, on the other hand, implies purposeful behaviour and as such requires detailed description and discussion. When a cow is re-introduced to a group after an absence of several weeks or months, she is generally quickly approached and surrounded by many of the group, most of which exhibit investigatory behaviour patterns. A few cows, however, will exhibit "aggressive" behaviour (hard breathing, head lowered, slow deliberate movements, occasional pawing) and move slowly toward the introduced cow. When this aggressive approach occurs, it is fairly safe to predict that *physical contact* will occur between the two animals unless the newcomer flees; the author has termed this action the "active" approach, or "challenge" to distinguish it from the more common and casual "passive" approach. An *active approach* may be made over distances as much as 30 yards, and implies that one animal approaches another for the express purpose of carrying the fight pattern sequence to completion.

The next step of the sequence, the *threat*, can be engaged in only when the contesting animals are within about five feet of each other. In the threat, one of the contestants exhibits aggressive behaviour directed toward the other, i.e., head lowered to about halfway between the normal position and the ground, eyes directed towards the opponent, and the broad forehead on a plane perpendicular to the ground facing the opponent. Despite the fact that all animals at this station are permanently dehorned at birth, all the movements of the threat position would serve to direct the horns towards the opponent, just as if horns were present. The threatened cow has but two alternatives: she may respond by returning the threat in a similar manner (head down, etc.) whereupon direct physical combat ensues; or, more commonly, she will exhibit alarm reactions and retreat. In the alarm reaction, she will lurch out of range of the threatener, with head up, and often will direct all her activities to leaving the immediate vicinity of the aggressive cow. Commonly, a threat is a rather rapid occurrence and is easily overlooked by the casual observer.

Direct physical contact between two animals (aside from accidentally brushing against each other) involves either *butting* or active *fighting*. Frequently, the one action involves or closely follows the other, and thus they are often indistinguishable. However for the purposes of this dissertation, a distinction, perhaps artificial, is made. Butting occurs when one animal uses its forehead to direct a blow at another animal, without any retaliatory action on the part of the struck animal. The entire episode is usually brief, since the struck animal will manifest every desire to escape from the aggressor. The butt pattern clearly follows the *threat*, since threat-position as described above is mostly concerned with readying the forehead (horns) for the blow.

Whereas butts are non-retaliated blows, fights occur when the struck cow strikes back. A fight sometimes occurs following a passive approach, but more often follows the active approach. In both instances, the fight is preceded by the threat, wherein *both* animals exhibit the threatening, aggressive position of head down, snorting, and slow deliberate movements. They may circle each other in this fashion for as few as five seconds or as much as ten minutes, with their foreheads parallel and less than about four feet apart. At some point in the circling, the fight will suddenly erupt. There will be much shifting of position for a better footing and an attempt by one or both to direct its forehead (horns) for a flank rather than a frontal attack. The animal that can successfully manoeuvre itself around the other so as to be able to hit its opponent in the side is invariably the one that will win the encounter. The flanked animal will make an effort to bring its opponent back to the frontal attack, and if this fails she will flee. When flanking attempts are unsuccessful, the opponents will bring all their weight to bear on each other through the forehead. At this turn of events, the lighter animal is clearly at a disadvantage and steadily loses ground, unless she makes up in agility for what she lacks in weight. Upon flight the defeated animal is chased or followed by the victor for a variable period of time.

A *fight* may consist of but one encounter or more commonly a series of encounters. The actual encounters do not persist for more than about one minute at a time and invariably end with one animal taking flight to regain position. The number of severe encounters will vary directly with the evenness of the match. Two well matched cows may struggle over a period of hours and even days. The interval between encounters may vary from but a few seconds to five minutes. During this interval they will often manifest grazing behaviour patterns despite the fact that the contestants are in a highly excited state; this is accompanied by the customary wandering, but all the while the contestants are both close to and wary of each other. This particular type of diversion agrees well with what Tinbergen (1952) has termed a "displacement activity." Unsuccessful efforts have been made by the writer to try to ascertain whether the between-encounter grazing actually involves the ingestion of grass or is merely a nosing of the ground.

An interesting variation of the fighting pattern occurs when two animals are engaged in rather severe combat. Neither Guhl *et al.* (1948) nor Hancock (1950) make reference to this type of action, and for lack of any other name, the author has termed it the *clinch* manifestation of fight behaviour. In this type of activity, two closely matched contestants have been fighting for some time and one or both show signs of fatigue. They are head to head, and as described above, the more aggressive is manoeuvring for a flanking position. Suddenly, the less aggressive will let her opponent slip its head to her flank but at the same time she will turn her body so that she is standing directly alongside and parallel to the more aggressive animal. In the same "slipping" motion, she will lower her head to a position between the hind leg and the udder of the more aggressive cow, with her forehead on a plane parallel to the ground. As long as she is beside her opponent, the other cow cannot inflict a serious blow since she cannot get into position. The more aggressive animal will make every effort to orient herself perpendicular to her opponent so that she can deliver a telling blow, but she is hampered by the nose in her udder (perhaps interfering with the hind leg movement) and the fact that the less aggressive animal will slide and turn with every effort of the aggressor. The *clinch* seems to be ideal for avoiding damaging blows while pausing for a brief rest, but is not manifested very often since most of the fights are quickly decided.

The younger animals also show manifestations of contest behaviour, but the contests are more of the play-fight type without any decisive outcome (see Brownlee, 1954, for a discussion of play in cattle).

Analysis of Social Rank

Experimental Technique

Observations were routinely made of different groups in the herd (young, middle, front, bayou or wet) in such a manner that each group was observed for at least one hour each week. During the observational period, a note was made of every contest that could be clearly discerned as such by the observer. The observer was located so that he could see most, if not all of the animals at one time, which usually involved taking a position on top of a truck in the field. Two persons commonly carried out the observations; one continually observed and called out the contests, while the other observed and recorded the contests. On some occasions the recorder could not keep pace with the contests being called out, and thus many contests went unrecorded; on other occasions, the cows were artificially stimulated to induce contesting during periods of lull. For this reason, it is felt that the actual number of contests recorded *per se* is not of much value as an index of herd excitement or even of an animal's aggressiveness. However, this point will be more fully dealt with in a later section.

It is felt that since the animal is normally accustomed to human proximity the presence of the observer had no effect on the eventual outcome of a contest. This was confirmed by occasionally observing some of the groups from such a position whereby the animals exhibited no awareness as to the presence of the observer; the patterns of behaviour observed under these conditions were indistinguishable from those noted when the observer was in full view of the animals. Similarly, the time of day and weather conditions had no radical effect on the observed contesting behaviour. Observations were usually carried out on pasture while the groups were engaged in normal routine activities, although occasionally, a group would be moved to a new pasture to stimulate contesting behaviour.

Herd Order

Allee *et al.* (1949) state that the "contemporary organization of vertebrate groups. . . . is based on the application of three general principles: the holding of territory; domination-subordination; and leadership-followership." Organization as such implies a bond, in that the movements, behaviour or activities of each individual are not completely independent of the others in the group. Thus, we can easily remove a paper clip from a box of clips (an unorganized group) since each unit is independent of the others. However, if the group were "organized" by fastening the clips together, we could not manipulate one clip without affecting some (in a large group) or all (in a small group) of the other clips. The ensuing discussion is based on a "domination-subordination" type of organization within the herd.

The dominance relationship between two cows can be readily ascertained from direct physical contests; to this end, close to 5,000 individual contests within the herd have been recorded by this observer. These were plotted on a large "master chart" on which was entered every animal in the herd. A section of the master chart, showing the details of the marking system, is presented in Fig. 1. Each dot represents an individual contest: by reading across the chart one can determine the victories won by an individual, while reading down would determine her defeats. For example, the chart shows (Fig. 1) that SX34 won three contests

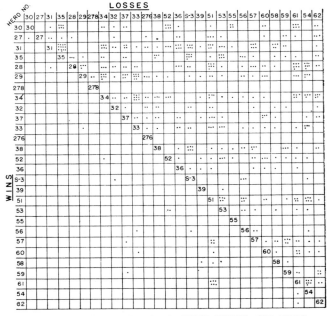

EACH DOT REPRESENTS A SINGLE CONTEST BETWEEN TWO ANIMALS

Fig. 1. A representative section of the Master Chart.

over SX38, while she lost two contests with SX35. As more contests were recorded (and therefore more information made available), the order of the animals was rearranged in an effort to get as few dots as possible to the left of the diagonal. In this manner, the animals would automatically be arranged into what is commonly termed a straight-line peck-order, wherein the alpha, or dominant animal of the group, wins contests from all the others and loses to none; the beta animal wins from all the others except the alpha animal; the gamma animal wins from all except the alpha and beta; and so forth down the line until the omega animal is reached, this individual winning no contests and losing to all the others consistently. The terminology is that used by most workers in the field of social organization and was popularized by Allee (1942) in his work on peck order in domestic fowl. Since the particular terms apparently fit the situation in dairy cattle remarkably well, they will be used freely in the ensuing discussion.

The order is demonstrated by the chart (Fig. 1). If no order existed, then it would not be possible to arrange the chart so that more than about 50 per cent. of the dots would lie to one side of the diagonal. However, of the 4,935 contests recorded, only 248, or about 5 per cent. are to the left of the diagonal. A statistical test is not necessary, in this case, to show that this deviation from 50 per cent. is far from being merely a "chance" arrangement.

It was not possible to stage a contest between every pair of animals of the herd; indeed, this would require over 13,000 contests for a herd of 163 animals. In addition, some of the animals died or were otherwise removed from the herd during the course of the experiment. Furthermore, because of the age spread in the herd many of the older animals had never been together in the same group with the very young animals, and so physical contests were not possible. For these reasons it was necessary to assume that if Cow A dominated Cow B, and Cow B dominated Cow C, then Cow A automatically dominated Cow C (if there were no contests recorded between A and C); this assumption was used as a working rule in organizing the chart.

Observational Errors

Every effort was made to keep the amount of error in the observational system down as much as possible, since the errors themselves were difficult, if not impossible to track down

and correct. It is pertinent at this time to mention the three main sources of error and the methods of minimizing them.

Perhaps the largest source of error stemmed from a subjective misinterpretation of objective fact. Although the outcome of most of the contests was clearly defined, some few would be considered doubtful. To minimize this type of error, it was necessary to set strict limits on what would be recorded as a victory or defeat in a contest, as follows: if one animal showed an aggressive pattern, with lowered head or just a toss of the head while looking at the second animal, and the second animal reacted to the threat by the characteristic alarm and retreat, then the first was considered the victor in a *threat* contest between the two animals. Had the second animal merely walked away without the lurching alarm and/or retreat, then the contest would have been doubtful and not tabulated. Again, if one animal used its forehead to direct a blow at the second, and the second reacted by alarm and retreat, then the first was considered victorious in a *butt* contest between the two animals. The third and last type of clearly defined contest was the actual physical combat (described above) when one animal finally turned and fled. The rule of retreat was strictly adhered to, and in this manner the degree of subjectivity was minimized.

The next largest source of error was in mistaking the identity of a cow. It is possible by diligent study to distinguish between and "know" each animal of the herd, but even so mistakes are easily made when observing from a distance of 100 feet or more. In an effort to reduce this type of error, practically all observations were made with the aid of a trained assistant who knew the herd well, so that some cross-checking was possible. However, when a group was highly excited and unusually active, one observer was so busy recording what the other called out that he barely had time to look up from his pad; in effect, during periods of intense activity, there was no possible check on the veracity of identity, nor is it possible even to estimate the amount of error introduced.

The third source of error, equally as grave as the above two, was in the recording itself. In the press of trying to record events as quickly as they were happening when 30 or 40 animals were stirring about, it was very easy to write 211 instead of 21 or 311, or the like. During the periods of lesser activity or just after an obser-

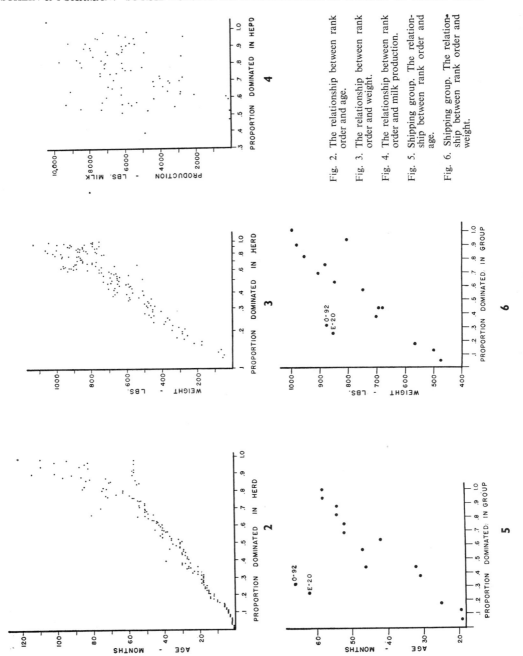

Fig. 2. The relationship between rank order and age.

Fig. 3. The relationship between rank order and weight.

Fig. 4. The relationship between rank order and milk production.

Fig. 5. Shipping group. The relationship between rank order and age.

Fig. 6. Shipping group. The relationship between rank order and weight.

208

vational period, the observer made an effort to check over the field notes to make sure there were no obvious errors, but still it would be impossible to estimate the amount of error from this source.

Factors Determining Rank

In an effort to get a clearer understanding of the complex of factors which determine an individual's social ranking, the herd order was compared with the ages (Fig. 2), weights (Fig. 3) and first lactation milk production records (Fig. 4) of the animals. (In these and subsequent figures, the "rank order" of an individual refers to the proportion of the herd which she dominates). To calculate correlation coefficients it was necessary to first "normalize" the dominance rank order in the herd; this was accomplished by calculating the area under the normal curve at each rank point and then utilizing a standard table of Cumulative Normal Frequency Distributions (Snedecor, 1945).

Correlation coefficients for rank x age (r_a) find rank x weight (r_w) were highly significant, while that for rank x production (r_p) was of a lower order of significance: $r_a = 0.93$, $P < 0.01$; $r_w = 0.87$, $P < 0.01$; $r_p = 0.25$, $P < 0.05$. On the other hand, there seemed to be little, if any, correlation between social position and the failure to complete the full (305-day) lactation period.

The relationship between age, weight and social rank in this herd can be expressed by the following formula (determined by the method of multiple regression):

$R = 0.026 A + 1.532 \log. W - 5.422.$

where R = "normalized" social rank, A = age in months, and W = weight in pounds.

It is impossible to determine from these data whether age and weight are causally or coincidentally related to rank. Age is a good index of "seniority" (length of time in the herd), but it does not necessarily include "aggressiveness" which is probably one of the more important factors in the delineation of social rank. By the same token, weight is used as an index of "strength", but it fails to take into account "agility", which may be equally important.

Strange Cow Experiments

Some additional data on the relationship of age and weight to rank order were obtained by introducing two completely strange cows to various groups of cows from this station's herd. The two strange cows (O-92 and E-20)

were borrowed from the dairy herd at Louisiana State University; prior to this experiment, these cows had never had any contact with animals out of their "home" herd.

After an initial period of isolation during which they could neither see, hear nor smell any other animals, the two strange cows were introduced first to a group of 14 crossbred culls ("shipping group") and later to a group of two Jersey culls from this station's herd. The introduction of the strange cows to the shipping group was not accompanied by any violent fighting, although there was the usual amount of sniffing and threat activity; after eight days of observation the group's dominance order was determined, and the order was plotted against the ages (Fig. 5) and weights (Fig. 6) of the individuals. An examination of these figures indicates that the two introduced Jerseys are very much "out of line" with respect to age and weight, perhaps more so with age. Three weeks afterward the strange cows were introduced to two Jerseys culls from this station's herd. This time, the introduction was marked by violent and prolonged fighting. Observations made one week later indicated that a rather complex dominance order had been established within the newly formed group, as is shown in Fig. 7.

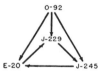

Fig. 7. Dominance relationships between 4 cows. Two cows are from the "home" herd and the other two are "strangers".

The significant lesson learned from this experiment was that the *previously established* dominance relationship between any pair of cows was not disturbed by the presence of the strangers. For example, each of the 14 cull animals in the shipping group was in relatively the same rank position with respect to the others of the "home" herd, despite the introduction of the strange cows. Similarly, J229 had always dominated J245 in the herd, and this relationship continued unchanged (Fig. 7) despite the rather complex order resulting from the introduction of the strange cows. It is interesting to note that O-92 and E-20 ranked fairly low in the shipping group, and yet *all* the animals in the shipping group ranked below J229 and J245 in the herd dominance order.

These results indicate that although age and weight are closely correlated in this herd, they are not necessarily causally related to rank. Certainly, the data are suggestive that in normal herd management, "seniority" as depicted by age or weight, is highly associated with rank position, and that other physical and behavioural factors may play a major role at the same time.

The Individual as a Member of the Group

Oestrus cycles do not materially affect the established social order, since the rank order has already been described (see above) as a relatively stable type of organization. A cow in heat is usually at a somewhat higher level of general excitement than the others of the group and is usually labelled "nervous." Instead of a normal placidity, she will spend a considerable amount of time approaching and sniffing the others (particularly in the vaginal region) without respect to their position in the social order. It is interesting to note that the sex drive of the individual supersedes organizational characteristics. Whereas under normal conditions the subordinate animal will avoid the dominants under penalty of being butted or otherwise punished, this same animal, when in heat, readily approaches her superiors in the order; even after being repeatedly threatened or chased, she will often continue her efforts to mount the aggressor. Consequently, observations made during the heat period of a cow tend to show considerably more contests involving this cow than is usually the case.

It is pertinent to mention that the maternal drive in a cow immediately post-partum seems to supersede the organizational or herd drive. Immediately post-partum, the mother rests and spends a great deal of time licking the calf, and is apparently oblivious to the rest of the group. When, in the course of normal wanderings, the group moves away, the mother stays behind with her calf and *no longer functions* as a member of the group.

Three days post-partum, the mother and calf are separated. The mother is returned to the wet group, but still manifests *maternal* rather than *organizational* drive patterns although she can no longer see nor hear her calf. She will largely ignore the group, preferring to spend her time near the milking parlour and calf barn; she is restless and bellows continually; she becomes visibly excited at the sight of any of the human handlers. However, two or three days after the separation the maternal drive lessens and the cow returns to the group and normal organizational herd patterns.

The effect of the group on feeding behaviour, particularly grazing, has already been noted by other investigators. Hancock (1950) noted that the cows in a group grazed in unison and that the inclination to ruminate was suppressed if most of the other individuals continued to graze. He also noted that if two groups are separated by as little as a single-wire fence, they maintain separate group grazing patterns; that is, one group may be in the midst of a grazing cycle while the other group would be loafing or lying about.

It would be interesting to try to measure the strength of the grazing drive manifested by the group. Normally, grazing can be easily interrupted by the observer simply by caressing one or more of the pets; others will show curiosity and approach in an investigatory attitude. On other occasions, one cow will stop grazing and wander off toward water, and generally all the others follow, indicating that unless the cow is unusually hungry, the grazing drive is not very strong. It has already been noted above that patterns of behaviour strikingly similar to grazing are manifested between encounters in a particularly violent fight, and it was suggested that the grazing as such might well be a displacement activity. If this labelling is correct, then, according to Tinbergen's hypothesis (1951), grazing would be considered a low threshold activity; this particular point is worthy of a considerable amount of further investigation.

Barn feeding behaviour is also affected by organizational patterns, to the extent that under free group feeding conditions the higher order animals apparently get more to eat. For example, when a group of hungry cows is brought into the barn and offered hay, they will line up at the trough with much shuffling and eat as rapidly as possible. The higher order animals will not hesitate to butt or chase their subordinates away from what apparently looks like a desirable pile of hay, while the very lowest order animals spend a good deal of the feeding time searching for an opening whereby they can get to the hay. Although a quantitative measure of the amount of hay consumed in relation to the order of the animal has not as yet been undertaken, there is little doubt that the lower order animal would suffer markedly if she were wholly dependant on trough feeding. (Calhoun, 1950, has indicated a direct relationship between the rank order in wild rats and the amount of food consumed). Apparently, "space" is the environmental factor in short supply (thereby stimulating competition) in barn hay feeding, since no spatial competition is evident in open pasture grazing. (For a more complete discussion on minimum environmental factors, see Orgain & Schein, 1953).

It is interesting to speculate as to when and how the order is established in the normal herd routine. The implication from the above discussion is that seniority plays an important role in the rank order. Therefore, barring sickness, injury or an unusual occurrence detrimental to normal calf development, it would seem that the place of each individual in the herd order is determined by the moment of birth. Furthermore, she will advance in the social scale as the older cows are removed from the group, maintaining largely the same relative position with respect to the others in the group.

Normally, intergroup shiftings are accompanied by a considerable amount of excitement and fighting as the introduced animal apparently strives to reassert her place in the overall social scale. The sole exception to this rule is an introduction into the young group, since the three months old heifer has not had any previous contact with the group, and therefore "is not aware of her place." It is interesting to note that the middle group already has a well defined order readily delineated by the observer, whereas numerous observational periods of the young group have yielded little results. In the three to six month old group, *threats* are rare and contests all seem to be of the play-fight type and therefore inconclusive, while in the middle group, *threats* and decisive contests are as common as play-fighting. Until further evidence is available, the hypothesis is proposed that the mechanism of the order is establishd somewhere between the late young and early middle group ages, with perhaps more emphasis on the latter.

Summary

The object of this study has been to describe a type of social organization existing within a herd of dairy cattle, and to explore the effects of this organization on the behaviour of the individuals. Some of the ramifications of the organizational patterns are explored in an effort to trace the development of the organization.

A rather detailed description of the physical plant and management routines at the station where these studies were conducted is presented in the first section of the dissertation, so that the reader who is unfamiliar with dairy herd practices may better understand the ensuing discussion.

Aggressive behaviour patterns of the individual are also described in detail so as to afford the reader a clearer understanding of the role of the individual in the group. Such behaviour apparently follows a definite sequence, each phase of which is discussed in detail. The various links in the chain of aggression are termed: the *approach*, the *threat*, the *physical contact*, and the ensuing victory or defeat. The *approach* may be active or passive; while passive approach merely describes the chance meeting of two cows engaged in random wandering, active approach implies purposeful behaviour on the part of one or both of the animals.

Direct physical contact between two animals involves either butting or actual fighting. Butting is defined as when one animal uses its forehead to direct a blow at another without any retaliatory action on the part of the struck animal. On the other hand, fights occur when the struck cow strikes back. Commonly, a *fight* consists of a series of encounters; the actual encounters do not last for more than about a minute at a time and invariably end with one animal taking flight. The interval between encounters may vary from a few seconds to five minutes, with "displacement grazing" being the major activity between encounters. Actually, most contesting actions observed did not progress further than the *threat* stage, although these contests are considered just as decisive as if they had progressed through the *fight* phase.

Each group in the herd was observed for at least one hour a week during the course of this study. During the observational period, careful notes were kept of each animal involved in a contest, and the outcome of the contest was recorded. Close to 5,000 individual contests were recorded and were plotted on a large master chart. The pattern of wins and losses clearly indicated that the herd was organized in a straight-line peck-order very much like that previously described by workers in the field of domestic fowl.

The dominance order was compared to the ages, weight and production records of the animals. A highly significant relationship obtained between rank and age and also between rank and weight, but the relationship between rank and milk production was inconclusive. The correlation between rank and the failure to complete a full lactation period was of a low order of significance and therefore not considered further.

In an effort to separate age and weight from seniority, two completely strange cows were introduced into a group of fourteen animals from the experimental herd. After the new group had stabilized, the dominance order was again plotted and it was found that the two introduced cows were very much "out of line" with respect to age and weight. Later, when the two strange cows were introduced to two high ranking animals from the experimental herd, the group of four set up a rather elaborate circular type of dominance order. This particular order did not show any relationship to age or weight, although it was markedly affected by the previously established relationships within the pairs of animals. Hence, it is concluded from these results that although age and weight are closely allied to the social scale pattern in the herd, they are probably not causally related. The data are suggestive that in normal herd management, seniority, as depicted by age or weight, is of primary importance in the determination of rank position. The socal position of each animal is probably determined at three to six months of age, when she first encounters large group living conditions.

Acknowledgments

This work would not have been possible without the assistance and full co-operation of a number of people from entirely different walks of life. The author is particularly indebted to Mr. J. J. Vizinat, Dairy Husbandman, for assistance in many of the field phases of the study, and also to Dr. D. E. Davis, Johns Hopkins University, for overall direction and advice. In addition, the staff of the Iberia Livestock Dairy Station offered many helpful suggestions and criticisms, as did the staff of the Dairy Husbandry Research Branch, Agricultural Research Service, U.S. Department of Agriculture, Beltsville, Maryland.

REFERENCES

Allee, W. C. (1942). Social dominance and subordination among vertebrates. *Biol. Symp.*, **8**, 139-162.

Allee, W. C., Emerson, A. E., Park, O., Park, T., & Schmidt, K. P. (1949). *Principles of animal ecology*. Philadelphia: Saunders.

Altmann, M. (1952). Social behaviour of elk, *Cervus canadensis n.*, in the Jackson Hole area of Wyoming. *Behaviour*, **4**, (2), 116-143.

Alverdes, F. (1935). The behaviour of mammalian herds and packs. In: Murchison, C. *A handbook of social psychology*. Worcester, Mass.: Clark University Press, 185-203.

Brownlee, A. (1950). Studies on the behaviour of domestic cattle in Britain. *Bull. anim. Behav.*, **8**, 11-20.

Brownlee, A. (1954). Play in domestic cattle. *Brit. vet. J.*, **110**, 48-68.

Calhoun, J. B. (1950). The study of wild animals under controlled conditions. *Ann. N.Y. Acad. Sci.*, **51**, 1113-1122.

Darling, F. F. (1937). *A herd of Red Deer*. Oxford University Press.

Darling, F. F. (1952). Social life in ungulates. In: *Structure et physiologie des societes animales*. Paris: Centre National de la Recherche Scientifique. 221-226.

Dove, W. F. (1936). Artificial production of the fabulous unicorn. *Sci. Monthly*, **42**, 431-436.

Guhl, A. M., Atkeson, F. W. & Roark, D. B. (1948). *Social organization in a herd of dairy cows*. Unpubl. MS.

Hancock, J. (1950). Grazing habits of dairy cows in New Zealand. *Emp. J. exp. Agr.*, **18**, (72), 249-263.

Orgain, H. & Schein, M. W. (1953). A preliminary analysis of the physical environment of the Norway rat. *Ecol.*, **34**, (3), 467-473.

Scott, J. P. (1942). Social organization and leadership among sheep. *Anat. Rec.*, **84**, 480-481 (Abst.).

Scott, J. P. (1946). Dominance reactions in a small flock of goats. *Anat. Rec.* **94**, 380-390.

Snedecor, G. W. (1946). *Statistical methods*. Ames: Iowa State College Press.

Tinbergen, N. (1951). *The study of instinct*. Oxford University Press.

Tinbergen, N. (1952). Derived activities; their causation, biological significance, origin and emancipation during evolution. *Quart. Rev. Biol.*, **27**, (1), 1-32.

Woodbury, A. M. (1941). Changing the "hook order" in cows. *Ecol.*, **22**, (4), 410-411.

11

Reprinted from *Zoologica*, **43**, Pt. 1, 1, 17–23 (Mar. 31, 1958)

Social Behavior of the American Buffalo (*Bison bison bison*)

Tom McHugh[1]

Jackson Hole Biological Research Station, Moran, Wyoming

(Plates I-III; Text-figure 1)

CONTENTS

I. INTRODUCTION

THE American buffalo (*Bison bison bison*) not only shaped the life of the Plains Indians but also figured more prominently in American history than any other animal. A vast literature has grown up around the buffalo, but there is still no adequate scientific study of its social behavior. This paper aims to add to the present limited knowledge in that field and to compare its contents with historical literature.

In the gathering of data on which this paper is based, I observed both free-ranging and confined herds of buffalo through the seasons. Mannerisms and basic behavioral patterns of the animals themselves were noted as well as interactions with other buffalo in the coordination, integration and movement of the herds. Interactions between various herd members were also recorded to determine the type of social organization. The herds were further studied to determine their composition. Reproductive behavior was observed during the rut and the subsequent calving season.

Due to the limitations of space, large sections of data had to be condensed into a few sentences. These conclusions, abrupt as they may seem at times, nevertheless rest on a substantial foundation of repeated observations recorded in field notes.

I have used "buffalo" throughout in preference to "bison" because of common usage. The term "herd" refers to all the buffalo in any one geographic area. Each herd is in turn composed of smaller units caller "groups."

* * * * * *

---

[1]Present address: 13001 North Parkway, Cleveland 5, Ohio.

Editor's Note: A row of asterisks indicates that material has been omitted from the original article.

V. THE DOMINANCE HIERARCHY

DISPLAY OF DOMINANCE

The herd in the Jackson Hole Wildlife Park was studied intensively for information on social organization. The outcome of interactions between individuals delineated the social structure of this herd. It was a linear type of dominance hierarchy with dominant individuals exercising a virtual constancy of success in interactions. Subordinates recognized dominants quickly and avoided them.

I distinguished the herd members as individuals by physical differences or painted markings. Their interactions were divided into *passive dominances* and *aggressions*. Out of 1,027 interactions recorded, 72.8% were passive dominances and 27.2% were aggressions.

Passive dominances were the more gentle of the two types of interactions. They were characterized by a lack of any obvious show of force or threat. Typically, the dominant individual walked toward and displaced the subordinate with no aggressive action. The most subtle passive dominance took place when one animal avoided another while moving through the herd.

Aggressions occurred when the dominant individual displaced a subordinate by force or by threat. Threat usually involved an intention movement (Tinbergen, 1951: 79) of an aggressive act. The most gentle form of threat involved a mere look toward the subordinate, causing it to move away. The most common form was horn swinging, where the horns were swung up and down or sideways toward the subordinate with occasional contact. Partial or complete charges were also used to displace subordinates. The dominant animal appeared to be trying to impale or gore the subordinate, but actual contact was uncommon. Occasional violent charges threw the victim for several feet or tossed it as much as twenty feet. One mature bull hooked a cow on his horns and tossed her over his back, where she fell to the ground (in a corral at the Lamar Buffalo Ranch). She was able to walk away, but subsequent autopsy revealed two broken ribs and a punctured, collapsed lung.

Battles were tabulated as aggressions. They consisted of hooking of horns and pushing back and forth that lasted a few seconds or several minutes. The loser usually either moved away or was pushed back a few feet, yet many battles were indecisive. They were started by dominant or subordinate individuals and occurred between younger bulls and between calves, yearlings and two-year-olds. Most battles in the Wildlife Park occurred between one-year-old bulls and two-

year-old cows. This may have been correlated with a later change in rank between these two groups. Battles between cows were rare.

Another indication of dominance was the intention movement of mounting, seeming to occur without any sexual motivation. This was tallied as an aggression if the subordinate moved out when the dominant placed its chin on the rump of the subordinate. Both the intention movement for mounting and genuine mounting also occurred as play, apparently with no regard for dominance. Attempts by a subordinate to mount a dominant resulted in a prompt reversal of the same behavior, playful battles, or other forms of play. This mounting during play produced no withdrawal by the subordinate and was not tallied as an aggression.

SITUATIONS IN WHICH DOMINANCE WAS DISPLAYED

More than 90 per cent. of the interactions in the Wildlife Park were recorded during displacements on the feeding grounds. These buffalo were fed on hay from November to May, approximately. Practically no natural food was available during most of this period. Since the hay was spread in piles over at least 100 feet of ground, there was a surplus of piles at all times except in the first few minutes of distribution. Most buffalo fed from separate piles of hay. They continually shifted every few minutes, even though there was little difference between various piles. A shift of one dominant animal might eventually cause a shift in more than half the herd as the move was passed down the line.

Dominance was also recorded during interactions over special objects as follows: (1) an item of curiosity, (2) a spot for lying, (3) a wallow, (4) a water hole or puddle, (5) a tree for rubbing or horning, (6) choice grass, (7) a salt lick, (8) shade under a small group of trees (dominants clustered in the shade and subordinates were in the sun), (9) sniffing the vulva of a cow, (10) a cow in heat.

Dominance was evident during group movements as well. Most commonly, a dominant animal in the rear or middle of the group pushed subordinates in front or stopped those directly behind. When two dominants stopped on a packed snow trail, subordinate animals wishing to press ahead had to flounder around through deep snow (March, 1951, Wildlife Park). Dominance was also occasionally recorded during grazing.

The preceding paragraphs discuss dominance that was usually traceable to interaction over a certain object, yet much aggressive dominance also occurred apparently independent of any inciting situation. In these cases, one member of a herd suddenly and inexplicably interrupted its feeding to charge another member. This happened most frequently between a cow and juvenile, occasionally between two cows.

STRUCTURE AND DYNAMICS OF THE DOMINANCE HIERARCHY IN THE WILDLIFE PARK

Factors influencing dominance. Table 4 illustrates some factors that determine dominance. Distinct differences in size and weight insured dominance. Seniority in age usually insured dominance, yet one seven-year-old cow was subordinate to two five-year-olds. Within groups of similar size, bulls were dominant over cows; aggressiveness or certain undetermined factors were more important than slight differences in size.

Reversals. Out of 726 interactions in the spring of 1951, there was only one temporary reversal when F4* displaced F3 by a swing of the horns. Occasional reversals were attempted by cows which horned either M1 or M2, yet these bulls did not yield. The two-year-old heifers did this most frequently. It was classed as play. Schein & Fohrman (1955) recorded 248 reversals in 4,935 interactions among 163 dairy cows.

Permanent changes in the dominance hierarchy resulted from the differential growth of bulls and cows. Previous to July and August, 1951, M3 and M4 were slightly smaller than F8, F9 and F10. During those two months, M3

---

*Nomenclature of individuals refers to sex and position in hierarchy. Thus, M2 was the second in position in the hierarchy of males and F4 was fourth among the females.

TABLE 4. DOMINANCE HIERARCHY IN THE WILDLIFE PARK, MARCH 14 THROUGH MAY 4, 1951
(Buffalo listed in order of decreasing dominance; M = male and F = female)

Individual	Total number of interactions	% of total that were victorious	% of total that were aggressions	% of victorious interactions that were aggressions	Age in spring '51	Individual differences within each group
M1	29	100%	28%	28%	6	M1 larger in size and weight than M2.
M2	36	94%	25%	26%	5	M1 stronger than M2 in battles.
F1	112	94%	22%	23%	5	All in this cow group smaller in size and weight
F2	100	75%	10%	11%	5	than M1 and M2.
F3	142	70%	8%	11%	7	No distinct difference in size and weight among
F4	133	76%	23%	29%	5	these cows with the exception of F7, which was
F5	112	38%	18%	7%	5	slightly thinner and shorter in height than all
F6	126	54%	12%	9%	4	others.
F7	84	40%	26%	44%	4	
F8	84	37%	30%	45%	2	Smaller in size and weight than F1-7.
F9	77	40%	51%	68%	2	No noticeable differences in size or weight within
F10	56	18%	32%	60%	2	this group.
						F8 stronger than F9-10 and F9 stronger than F10 in battles. All stronger than following subordinates in battle.
						(The progeny of F1-7 born in Wildlife Park.)
M3	98	32%	43%	68%	1	Smaller in size and weight than F8-10.
M4	78	12%	44%	78%	1	No noticeable differences in size within this group.
						M3 stronger than M4 in battles.
						(The progeny of F1-7 born in Wildlife Park.)
F11	65	5%	45%	67%	1	Smaller in size and weight than M3-4.
F12	54	0	52%	—	1	No noticeable difference in size and weight between these two, although F12 was at least two weeks older than F11.
						F11 stronger than F12 in battles.
						(The progeny of F1-7 born in Wildlife Park.)

TABLE 5. DOMINANCE HIERARCHY IN THE WILDLIFE PARK, FEBRUARY 4 TO 9, 1952

		F12	F11	F10	F9	F8	F7	F6	F5	F3	M4	M3	F4	F2
F1 dominates	13:	F12	F11	F10	F9	F8	F7	F6	F5	F3	M4	M3	F4	F2
F2 dominates	10:	F12	F11	F10	F9	F8	F7	F6	F5	F3			F4	
F4 dominates	10:	F12	F11	F10	F9	F8	F7	F6	F5		M4	M3		
M3 dominates	10:	F12	F11	F10	F9	F8	F7	F6		F3	M4			F2
M4 dominates	9:	F12	F11	F10	F9	F8	F7	F6		F3				F2
F3 dominates	9:	F12	F11	F10	F9	F8	F7	F6	F5				F4	
F5 dominates	9:	F12	F11	F10	F9	F8	F7	F6			M4	M3		
F6 dominates	6:	F12	F11	F10	F9	F8	F7							
F7 dominates	5:	F12	F11	F10	F9	F8								
F8 dominates	4:	F12	F11	F10	F9									
F9 dominates	3:	F12	F11	F10										
F10 dominates	2:	F12	F11											
F11 dominates	1:	F12												
F12 dominates	0.													

and M4 grew to be approximately the same size as these two-year-old heifers. The exact time schedule for the advance in dominance of M3 and M4 is not available, since it occurred during the summer and fall when interactions were uncommon. By September 12, both M3 and M4 had advanced above F8, F9 and F10. By October 9, both had moved above F6 and F7.

By February 4, 1952, M3 and M4 had advanced still farther up the hierarchy, as shown in Table 5. Both M3 and M4 were dominant over all cows except F1, F4 and F5. The irregular gain in dominance of M3 and M4 produced eight triangles (Table 6). By the following summer the advance of M3 and M4 over all cows eliminated these triangles.

The dominance hierarchy was checked during November, 1953, to finish a survey of almost three years. During that period there was no permanent reversal among the two bulls or the top seven cows. There was a dominance hierarchy in the remaining five mature cows, all progeny of the first seven. None had moved ahead of their parents.

Dominance hierarchy among calves. Dominance among calves developed slowly. Although they were alert to advances from adults within three weeks after birth, no dominance hierarchy among calves was detected during the first two months. The first definite signs of dominance were noted at an age of four months. At that time, the only male calf in a group of six was dominant over others and there were interactions among female calves.

I was not able to determine the complete dominance hierarchy among calves until February, 1952. Even during this period of feeding on hay, when there were numerous interactions among other adult herd members, interactions between calves were infrequent. Table 7 shows the hierarchy among calves.

The male calf was dominant over all females and accounted for 41.4% of the interactions in the entire calf group. There was no correlation between the position of dominance of the calf and its seniority in the calf group, small yet noticeable size differences, or the position of dominance of its mother. There was one reversal out of 70 calf interactions when D displaced B with an aggression.

TABLE 6. TRIANGLE SITUATIONS IN THE DOMINANCE HIERARCHY OF TABLE 5.

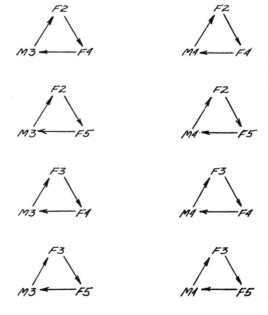

TABLE 7. DOMINANCE AMONG SIX CALVES IN THE WILDLIFE PARK HERD, FEBRUARY 4 TO 9, 1952 (Computed from 70 interactions)

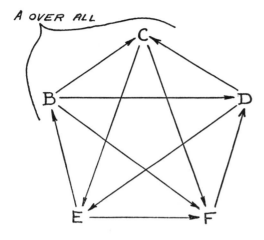

A OVER ALL

Calf	Dominant over	Relative age (date of birth)		Sex	Dominance of mother
A	BCDEF	Sixth	(5/16)	M	Second
B	CDF	Second	(4/29)	F	Fifth
C	EF	Fourth	(5/5)	F	Third
D	CE	First	(4/21)	F	First
E	BF	Fifth	(5/15)	F	Fourth
F	D	Third	(5/1)	F	Sixth

There were no triangles among the four calves in the herd one year previous on March 15, 1951. On this basis, one would expect the five triangles in the 1952 calf group to straighten out in a few weeks.

Factors affecting frequency of interactions. The number of interactions among certain animals or during certain time intervals varied with the factors discussed in the following paragraphs.

Table 8 shows the tendency for dominant cows to have more interactions with the cows immediately below them than with the less dominant cows. "Probability" indicates the probability of getting such a skewed distribution if a column of values was picked at random. It was computed by chi-square. The distribution of interactions for F1 and F2 are statistically significant. While the remaining columns are not, they still show the same distributional trend. Schein & Fohrman (1955) statistically analyzed similar interactions in a herd of dairy cows. They found that "fight contests involved cows closer together on the social rank scale than do threat or butt contests."

The total number of interactions and the per-

centage of aggressive interactions for each individual are listed in Table 4. Bulls M1 and M2 had fewer interactions than any other herd member. With this exception, the more dominant animals tended to have more interactions. This is in line with findings on pigeons (Masure & Allee, 1934: 327), chickens (Allee, 1951: 135) and canaries (Shoemaker, 1939: 404). An analysis of the percentage of aggressions shows just the reverse tendency. The percentages increased distinctly among the less dominant juveniles.

Hunger and palatability of food caused variations in the frequency of interactions. Table 9 shows the pattern for a typical day of feeding on hay. There was a high peak of interactions and aggresssions during the first 20 minutes of feeding. Fewer interactions and no high peak occurred during the afternoon feeding period, when the herd was not as hungry. This positive correlation between hunger and number of interactions was further verified on days when a surplus of hay remained from the previous day's feeding. When fresh hay was put out under such conditions, there were fewer interactions than usual. In addition to hay, the herd was also fed concentrate on some days. The buffalo preferred the concentrate, as evidenced by their quick withdrawal from the hay to feed on it. Even though the concentrate was fed about one hour after the hay, the initial peak of interactions sometimes doubled the initial peak for hay on the same day. The interactions over concentrate were also more aggressive.

Various disturbances also increased the number of interactions in a herd. They increased greatly in any group enclosed in a corral or pen. Interactions were increased by disturbance from the presence of a strange object, such as a car, a recently shed elk antler, a human being or a person concealed under a buffalo hide. One cow in the Wildlife Park Herd increased the fre-

TABLE 8. INTERACTIONS BETWEEN MATURE COWS IN WILDLIFE PARK, MARCH 14 TO MAY 4, 1951

Cow	F1	F2	F3	F4	F5
F2	20 +				
F3	25 +	22 +			
F4	9 −	5 −	17 +		
F5	9 −	14 +	21 +	17 +	
F6	13 −	3 −	15 +	12 +	9 +
F7	5 −	6 −	6 −	8 −	6 −
Mean	13.5	10.0	14.8	12.3	7.5
Probability	.005 *	.005 *	.04	.18	.62

+ = Above the mean for column.
− = Below the mean for column.

TABLE 9. INTERACTIONS PER TEN-MINUTE INTERVAL WHILE WILDLIFE PARK HERD WAS FEEDING ON HAY
(FEEDING TIME: 10 A.M.)

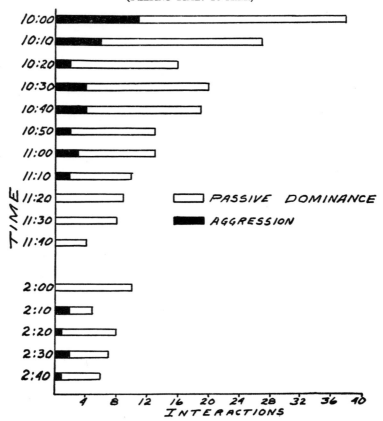

quency of interactions by her numerous aggressions which resulted in a chain reaction of interactions. The herd was noticeably more calm after the cow was removed.

Quality of dominance. Rough or gentle displays of dominance varied with age and sex. Comparatively gentle displacements were made by bulls over cows, by a mother over her calf, among juveniles, among calves, and occasionally among bulls outside the rut. The dominant buffalo used moderate force, and the subordinate moved away slowly for no more than a few feet. The subordinate sometimes returned to feed with the dominant. More forceful displacements were made by cows over yearlings, by a cow to a strange calf, among cows, and among bulls during the rut. The dominant was more aggressive and the subordinate moved faster, farther, and did not return.

There were no obvious behavior characteristics correlated with rank in the hierarchy. Subordinates showed no unusual amount of fear. All buffalo showed a keen awareness of the identity of the animals near them. The approach of a dominant from almost directly behind the subordinate resulted in a passive submission with hardly a glance at the dominant.

Derived dominance. Some buffalo *derived* a higher status in the dominance hierarchy by associating with a more dominant buffalo. Calves thus benefited from the positions of their mothers. The calf was elevated to her status when close to her and if tolerated by her. The mother sometimes pushed close subordinates away from her calf. Most of these instances of derived dominance occurred during hay feeding. Derived dominance was also recorded once during the rut in Hayden Valley: A seven-year-old bull stopped chasing a cow when she started to follow an older bull closely.

DOMINANCE IN WILD HERDS

The extent of development of dominance hierarchies in wild herds could not be determined because enough individuals could not be recognized. Animals in wild groups frequently exhibited passive dominances and aggressions. The frequency of these interactions never exceeded half those recorded for the Wildlife Park Herd during hay feeding. It was usually considerably less. The highest frequency occurred in a cow group of 15 that was feeding in deep snow (Lamar Valley, January 24, 1952). Displacements were frequent as each buffalo moved into an area recently cleared of snow by a subordinate. The more dominant buffalo would be expected to have a particular advantage in a severe winter.

A dominance hierarchy was noted in the above cow group and in a bull group of eight (Madison River, March, 1953) and of five (Hayden Valley, March, 1953). A wild group of 55 that was fed on hay in a Yellowstone Park corral showed numerous passive interactions. But the fact that groups of two to four cows were regularly observed feeding from the same hay pile would seem to indicate a lack of a completely developed hierarchy.

INTERSPECIFIC DOMINANCE

Free-living buffalo sometimes interacted with other species on the same range. These species included elk, antelope, mule deer, bighorn sheep and coyotes. In the Wildlife Park, buffalo fed during the winter in the area with a herd of 29 elk and some pronghorn, mule deer, whitetail deer, moose and horses. In free-living or enclosed herds the number of interspecific interactions greatly exceeded the number of intraspecific ones. The buffalo were dominant over all these species with only occasional reversals among elk and horses.

Elk. In the Wildlife Park, month-old buffalo calves were always successful in observed attempts to displace six-point elk bulls. Interactions involving younger calves were not observed. The buffalo herd completely dominated the hay feeding circuit and could push the elk from any section. They occasionally stampeded back and forth on the hay feeding circuit in order to force the elk into deeper snow. They also chased the elk in a steady, rapid walk during grazing in summer. Buffalo and elk sometimes fed within three feet of each other, but the elk soon became wary of the buffalo and moved farther away. Although elk nervously watched buffalo and usually dodged all charges, the violent charge of a cow buffalo threw one yearling elk three feet into the side of the hay shed.

Five- or six-point elk bulls were dominant over a buffalo cow or yearling in rare cases. The displacement always occurred on the edge of the buffalo herd. Such reversals were considered temporary since the aggression of any buffalo never failed to move the largest elk.

The buffalo herd chased the elk herd on three occasions when a new elk calf was present. The speed of the chase was so fast that the elk calf dropped behind within a few hundred feet. The buffalo then milled about the calf. In the first incident the calf was bruised, cut behind the ear, and bleeding at the mouth, yet still able to walk back to its herd. The buffalo herd bruised the calf again on the next day. It had no broken bones yet was so badly mauled it could not stand. It died in spite of careful nursing by the attendants of the Park. A second calf was rescued just before the buffalo herd reached it. They milled about and smelled the spot where the calf was born. When this calf moved out with the elk herd a few days later, the buffalo followed and slowly chased the elk.

Groups of elk and buffalo in Yellowstone Park moved within 30 feet of each other and even intermingled. The buffalo occasionally charged the elk, but the wary elk withdrew quickly. Buffalo disrupted the elk trapping program since elk would not enter a trapping corral if buffalo were inside or nearby. One incident was observed in the Lamar Valley where the buffalo killed an elk calf they found in hiding. Rush (1942: 225) and Chapman (1937: 148) mention deaths of elk in Yellowstone Park due to buffalo.

Superintendent John Schwartz noted a few cases of dominance of elk over buffalo in the National Bison Range. One elk raised on a bottle showed an attachment for two buffalo bulls at an age of eighteen months despite the presence of other elk. When two and a half years old, he rounded up a group of 15 buffalo cows and bugled. Another bull elk assembled a harem of cow elk during the rut and belligerently chased any buffalo that wandered near. This same elk rammed a tine of his antler through the palate of a yearling buffalo. The buffalo bled profusely and died from this injury within an hour.

Pronghorn. Buffalo in the Wildlife Park occasionally charged pronghorns (antelope) which wandered near the herd. They killed one eight-month-old pronghorn buck. Pronghorns in Wind Cave passed near or through buffalo groups, and the latter occasionally charged them, but the alert and agile pronghorns easily out-maneuvered these attacks. A two-month-old buffalo calf chased one pronghorn buck for almost a hundred feet. The walking of a buffalo group within 150 feet of resting pronghorns caused the pronghorns to rise and watch the group. Bryant (1885: 132) observed in Wyoming that "The deer and antelope are compelled to frequently shelter themselves from the attack of wolves under the strong protection of buffalos and you sometimes see herds of buffalos and antelopes mingle in grazing together."

Mule deer. One charge of a buffalo cow in the Wildlife Park tossed a mule deer ten feet. The attack left the deer lying on the ground, but it arose and rapidly moved away when the cow charged a second time. Another charge by a cow knocked a mule deer to the ground.

Moose. When a moose calf was placed in the Wildlife Park in August, the buffalo sometimes chased it. The buffalo killed the calf when it was seven months old. One rib was broken and the chest was pierced with a hole two inches in diameter (oral comment from James R. Simon).

Horses. Two draft horses were dominant over all cow and yearling buffalo during interactions at a salt lick in the Wildlife Park in October, 1951. The horse made no passes toward the buffalo, which moved slowly away as the horses came in to the salt. In March, 1953, the draft horses had lost some rank with the buffalo and were dominant over only a minority of cows and all yearlings. All buffalo were dominant over one saddle horse.

Head Animal Keeper David Pierson recorded one death of a horse in the Lamar Valley in 23 years of observation. This horse was ranging free in the valley at the time. He also noted three other horses and one mule that were gored by buffalo.

Bighorn sheep. Cy Young observed three cases in the National Bison Range where an older bighorn ram associated with a buffalo bull during the summer.

Dominance hierarchies. The interspecific dominance hierarchy among the big game species in the Wildlife Park was determined for the period February-May, 1951. With the exception of reversals noted below, the order was as follows: adult human beings, buffalo, elk, mule deer, pronghorn, moose and whitetail deer. The exact order of the last two species was not known.

The man who fed these animals maintained his dominant position by the use of a pitchfork. Tame elk, mule deer and pronghorns enjoyed derived dominance over the buffalo when they followed this man closely.

The elk exercised dominance by charges, by swinging antlers or by rapid downward strikes of the forelegs.

The mule deer involved in this hierarchy included two tame bucks and a tame and a wild doe. They used short charges or rapid downward strikes with the forelegs to display dominance. Bucks were dominant over does. The mule deer bucks dropped below the pronghorn buck in dominance when they shed their antlers. The horns of the pronghorn were becoming larger and harder at the same time. The reversal was not absolute; both deer occasionally resisted the horning of the pronghorn buck and drove him away. The pronghorn buck later gained dominance over the tame mule deer doe.

All pronghorns were tame. Pronghorn bucks dominated pronghorn does and other species by horning them. One buck was dominant over a castrated buck of the same age.

*   *   *   *   *   *   *

FIG. 7

SOCIAL BEHAVIOR OF THE AMERICAN BUFFALO (BISON BISON BISON)

Literature Cited

Allee, W. C., 1951. Cooperation among animals with human implications. New York: Henry Schuman. 233 pp.

Bryant, Edwin, 1885. Rocky Mountain adventures. New York: Worthington Co. 452 pp.

Chapman, Wendell and Lucie, 1937. Wilderness wanderers. New York: Charles Scribner's Sons. 318 pp.

Masure, Ralph H. & W. C. Allee, 1934. The social order in flocks of the common chicken and the pigeon. Auk, 51(3):306–327.

Rush, W. M., 1942. Wild animals of the Rockies. New York: Harper and Brothers. 296 pp.

Schein, Martin Warren, & Milton H. Fohrman, 1955. Social dominance relationships in a herd of dairy cattle. Brit. J. Animal Behaviour, 3(2):45–55.

Shoemaker, Hurst H., 1939. Social hierarchy in flocks of the canary. Auk, 56(4):381–406.

Tinbergen, N., 1951. The study of instinct. Oxford: Clarendon Press. 228 pp.

III
Analysis

Editor's Comments on Papers 12 Through 17

From the very outset, there was great interest in elucidating the factors that contributed to an animal's social rank position. Schjelderup-Ebbe (Paper 1) considered "experience" to be very important, but he also touched upon physiological conditions, among other things. In the mid-1930s, Carl Murchison and his colleagues at Clark University published a series of papers in which they applied covariant techniques to analyze social hierarchies in chickens. The ideas presented in these papers were not picked up by other workers in the field, but selected portions of the first and fifth papers in the series are presented here for historical purposes: the first (Paper 12) because it clearly spells out the philosophical basis of the approach used by Murchison and his colleagues and shows some of their data, and the fifth (Paper 13) because it brings together and summarizes the results of their studies.

The other papers in this section all stem from the laboratories of W. C. Allee at the University of Chicago (and later at the University of Florida), where emphasis was more on the biological bases of social interactions. The 1939 paper by Allee, Collias, and Lutherman (Paper 14) set the stage for a number of subsequent studies dealing with the physiological, primarily hormonal, basis of social interactions and dominance. Male sex hormone emerged as a key factor in dominance. (Unfortunately, space limitations preclude reprinting the entire article.) Similarly, the 1942 paper by Ginsburg and Allee (Paper 15), which examines the contributions of genetic endowment and experience in determining the outcomes of agonistic interactions between mice, opened up still another facet of the biological basis of social behavior. This paper led to a number of subsequent studies dealing with the relationship between genetic endowment and dominance interactions. Collias (Paper 16) reexamined the biological factors that contributed to success in interhen social encounters and found that male hormone and thyroxin output were of prime importance; his work serves as an example of a careful analytic approach to social behavior. Finally, the 1957 study by Banks and Allee (Paper 17) explored yet another parameter of social interaction among animals, that of the effects of group size upon the number of interactions.

Taken as a group, the papers in this section illustrate some of the techniques and approaches used to analyze social interactions and to examine the biological basis of social behavior. Papers 12 and 13 seemed to reach a dead end; however, Papers 14 through 17 each opened up new avenues of study and so are still in the forefront today.

12

Reprinted from *J. General Psychol.*, **12**, 3–11, 14–15, 22–37 (1935)

THE EXPERIMENTAL MEASUREMENT OF A SOCIAL HIERARCHY IN *GALLUS DOMESTICUS:* I. THE DIRECT IDENTIFICATION AND DIRECT MEASUREMENT OF SOCIAL REFLEX NO. 1 AND SOCIAL REFLEX NO. 2[1][2]

From the Psychological Laboratories of Clark University

CARL MURCHISON

In the city of Warsaw in 1910 and in the city of Kiev in 1917 appeared separately the two volumes of a remarkable book by E. Spektorsky entitled *The Problems of Social Physics in the Seventeenth Century.* Most of the available copies of this book were destroyed or lost during and following the Great War, and it is to Pitirim Sorokin that we in America owe our knowledge of this book. It is not alone the title of the book that so arrests our attention, but also its thesis that the great scientific and philosophical minds of the seventeenth century were just as much interested in the physics of society and social behavior as they were in the physics of material bodies.

> The social physicists of the seventeenth century tried to do the same as the physicists themselves. In the first place they constructed the conception of a moral or social space in which social, moral, and political movements go on. It was a kind of space analogous to physical space, and superposed upon it. To the position of a material object in physical space, there corresponded in social space the conception of status, as of sex, age, occupation, freedom, religion, citizenship, and so on. In this way they constructed a system of social coordinates which defined the position of man in this moral space as exactly

[1] This is the first of a series of preliminary papers that will appear in *The Journal of General Psychology, The Journal of Genetic Psychology,* and *The Journal of Social Psychology.* The entire preliminary series is being published in this way in order to save time before a second series of more careful repetitions is begun on several new groups of chicks. This will be done in order to make certain of the incidence of these social reflexes in *Gallus domesticus* before the investigation is extended to other social animals.

[2] Reported in part before combined seminars at Yale, Brown, Columbia, and Clark.

Editor's Note: Tables 4 through 9 and 15 through 23 have been omitted owing to limitations of space.

as the system of geometrical coordinates defines the position of a material object in physical space. Time was depicted by a geometrical line; historical processes began to be illustrated by various curves; and an individual's life history by a curve as of a falling body. Straight lines, parabolas, and spiral curves began to be used to describe these processes. (52)

In the light of this tremendous interest in social physics we might expect that Descartes and his followers and admirers might have formulated some brilliant social hypotheses and laws for experimental verification, and that some of these laws might be universally accepted by now. For information on this point we turn to a little book by K. D. Har entitled *Social Laws*. Mr. Har subjects to logical analysis every general formulation that ever so much as claimed to be a social law, and his conclusion is that in all the history of the social sciences Gresham's law and the law of population are the only formulations that even merit consideration as laws, and that these are far from the standards of physical law.

How can such a result as Mr. Har describes be the net result of all the scientific vigor and brilliance that Mr. Spektorsky describes? Are social phenomena immune to law? Are the problems more difficult than are the problems of physics? Does progress in this field depend on some lucky discovery that has not yet been made? In any case one who sets out to do experimental work in the field of social behavior finds the ages pretty much arrayed against him.

If social law is impossible, social experimentation must of necessity be futile and hopeless. One who works in this field must plan his experimentation so that it may furnish the basis for social law, even though three hundred years of such attempts have led to little as yet. It should be possible to reflect about the character of such work. As I have read samples of social physics from Descartes to Herbert Spencer, it has seemed to me that all of them were much too ambitious, that all of them started at the top and forgot all about elements. There is a great deal of talk about societies, religion, human traits, occupations, wealth, climate, evolution, but in no case is there an identification of an element in social behavior that can be measured wherever it may be found. Let us suggest that this may be one of the difficulties that has brought so little reward for so much labor.

What must be the nature of the experimentation that may lead to social law? It must deal with elements that are common to all social behavior, and therefore common to all social animals. Traits of character, intelligence, and accomplishments that are peculiar to restricted classes of social life, while important subject matter to know about, can never be the medium for the discovery of factors describable by social law. After such elements have been identified, they must be investigated by the covariable technique. If there are any factors that are common to all social behavior, and if such factors can be investigated by the technique of the covariable, it is certain that the eventual formulation of social law is highly probable.

What are the pitfalls to be avoided? There must be no uncritical attempt to import into the social sciences the laws of the natural sciences. The laws of motion, conservation, and thermo-dynamics are laws based upon the behavior of non-social matter, and their application to social bodies may leave out of the equation the distinguishing factor. Neither must there be any uncritical attempt to import the concepts of natural science, such as gravitation, magnetic fields, irradiation, refraction, etc. The use of such physical laws, concepts, and principles may lead to little more than the formulation of analogies, and physical scientists have frequently accused social scientists of such futile analogy manufacture.

The concepts of space and time with their subsidiary categories of measurable quantities, measurable qualities, and measurable functions are not subject to such restriction, however, since they are general rational concepts. The social sciences deal with phenomena in the same physical world in which the physical sciences are interested, and so must use concepts that deal with the physical world. The social sciences, therefore, can use time, space, and the necessary subsidiary categories without being accused of plagiarism, and must do so if true social law is ever to be formulated.

Now let us sum up the argument to this point. Social physics in the sense used by seventeenth century scientists is probably impossible, certainly has not led to useful experimental research, and obviously has not resulted in the formulation of widely accepted social laws. Such work has not risen above the level of analogy, and the analogies have usually been inadequate. Social science has made the mistake of trying to do *what* physics has done, instead of trying to do *as* physics has done. In order to do this latter, social science must take the general rational concepts of time, space, and their subsidiary categories, and apply them to the particular types of subject-matter in which the social sciences are interested. If this is done, there is ground for hope that social law may emerge from such operations.

The first major task is to identify simple quanta of social phenomena that by definition are describable in terms of time and space categories. After the identification comes the second major task of developing a method of measuring these quanta in terms of such categories. The third major task involves the application of the technique of covariables. The fourth major task has to do with the formulation of law and theory.

Perhaps the simplest quantum of social behavior that obviously is measurable in terms of time and space categories is the phenomenon of two individuals moving towards each other. This situation is so simple, so easily measured, and so universal, being common to all social behavior in all social animals, that we feel justified in assuming that it is the most elementary form of social behavior. We therefore designate this phenomenon Social Reflex No. 1. This reflex may be in any condition ranging from inhibition to maximum facilitation.

The next most simple and universal social quantum that is easily measurable in terms of such categories is the phenomenon of two individuals fighting each other to a decision. We designate this phenomenon Social Reflex No. 2. This reflex also may be in any condition ranging from maximum inhibition to maximum facilitation.

A third quantum, not quite so easy to measure as the first two, is the sex reflex, and we designate this phenomenon Social Reflex No. 3. This reflex also may be in any condition ranging from a maximum inhibition to a high degree of facilitation.

The above three classes of reflexes may be measured directly or inferentially. If they are measured inferentially, they are measured as discriminations, and discriminations may be measured in terms of motion in a particular direction rather than in some other direction. There are two main classes of social discrimination. The first class involves an inference that another individual is greater than or less than oneself. This class we designate Social Discrimination No. 1.

The second class involves an inference that another individual is greater than or less than some other individual. We designate this class Social Discrimination No. 2. These two classes of discrimination may be applied to any type of social body whatever about which a judgment may be rendered. It matters not whether the judgment involves mass, height, beauty, virtue, strength, repulsiveness, intelligence, honesty, ability, or any other concept in terms of which a judgment may be rendered; these two classes of social discrimination ought to suffice as vehicles in terms of which any social judgment may be measured inferentially. The measurements when made would be made on some social reflex inhibited or facilitated in some particular direction.

Gallus domesticus AS A LABORATORY ANIMAL

The chick is a laboratory animal with a long and dignified history. The modern scientific interest in the behavior of the chick began with Spalding (53) in 1873. The earlier studies were all of a very general nature and were directed at a description of the general behavior of the chick. The important names during this early period before 1900 were Morgan (34-38), Romanes (43), Hunt (18), Kline (25), Mills (33), Preyer (42), and Thorndike (56). From these early studies developed widespread interest in the classical problems of instinct, maturation, and habit, and these particular problems were later investigated more rigorously by Breed (6, 7), Shepard and Breed (51), Bird (3-5), and Moseley (39). Form perception, visual acuity, color vision, eye functions, and other aspects of the general visual problem have been investigated by Hess (15-17), Thorndike (56), Lashley (32), Dunlap and Mowrer (10), Bingham (1, 2), Johnson (19-22), Munn (40), Katz and Révész (23, 24), Stavsky and Pattie (55). Surprisingly enough, straight learning problems have not been investigated frequently with chicks as subjects, but it has been done by Cole (8), Bingham (2), and more recently by Dunlap (9). The fetal behavior of chicks has recently attracted renewed interest, and very ambitious studies have been reported by the Chinese psychologist Kuo (26-31). Some clinical studies on the effects of alcohol have been reported by Fletcher, Cowan, and Arlitt (12). Peculiar patterns of behavior, such as brooding, have been studied by Pearl (41) and by Goodale (13).

Widespread attention during recent years has been paid to a series of very striking and fundamental observations on the presence of pecking hierarchies among domestic hens. This work has been done by Schjelderup-Ebbe (44-50) in various German, French, and Scandinavian universities, and is the work that first interested me in the social behavior of *Gallus domesticus*. I have not attempted to duplicate any of Schjelderup-Ebbe's work, as I have worked primarily with the male. I have been interested, however, in observing many things which he had previously reported, and I have not yet found any contradiction. I consider his early work the product of great insight. He has extended his observations to nearly all types of the better known social birds. It is greatly to be regretted that he has not made more use of experimental techniques and quantitative analyses.

THE SUBJECTS

The subjects for this experiment consisted originally of 18 cross-bred chicks, divided equally as to sex. The chicks were obtained directly from the incubator early in April, 1934. They were a cross of Rhode Island Red with Plymouth Barred Rocks. The purpose of the cross was to make sex identifiable at birth. The chicks were divided into three equal groups, A, B, and C. Group A consisted of 6 males, Group B of 6 females, and Group C of 3 males and 3 females. These three groups were kept separated in three brooders. An infection started in group B, and they were immediately eliminated from the laboratory. Coal gas accidentally destroyed one member of Group A, and chick YY from Group C was added to Group A at the age of 12 weeks. The remainder of Group C was then eliminated. Five pullets of the same age as Group A were then obtained from an outside source and added to the group, which became known as Group D. This group of 6 roosters and 5 pullets has lived as a unit from the 16th week to the present time, the 36th week. They have lived on a wire floor and have been fed a standard all-food dry mash. At all times there has been enough food present to last two days, and plenty of water. It has never been necessary to fight for food or water. The young chicks were cared for in standard brooders, and have lived since in a wire cage situated in the laboratory.

Apparatus

The Social Reflex Runway is described in Figure 1. During the first four weeks I worked with a type just one-fourth as long. The chicks run on a wire floor of 1/4 inch mesh. The release-boxes

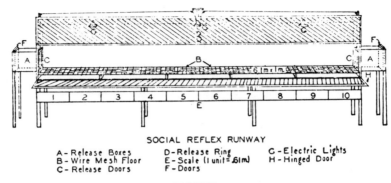

SOCIAL REFLEX RUNWAY

A – Release Boxes	D – Release Ring	G – Electric Lights
B – Wire Mesh Floor	E – Scale (1 unit = .61m)	H – Hinged Door
C – Release Doors	F – Doors	

FIGURE 1

are fitted with glass doors. The entire runway is surrounded with cheesecloth and is brilliantly illuminated on the inside. Two long doors may be lowered on one side for the removal of the chicks. Inside the cheese-cloth and surrounding the runway is a wall of poultry-wire. The release-boxes were altered in size to fit the size of the chicks.

Behavior of Chicks in the Social Reflex Runway

When two chicks are released simultaneously from the release-boxes at opposite ends of the Social Reflex Runway, they rush towards each other and quickly meet at some point in the runway. If only one chick is released, this rush does not take place. The glass doors are important in this connection. If a stopwatch is started with the opening of the release-boxes and stopped at the moment the two chicks come together, and if the point where the two chicks meet is noted on the scale which extends along the side of the runway, the various measurements of Social Reflex No. 1 for each chick may be obtained.

Procedure

Every possible pairing was made of the males in Group A, and each pair was run daily in the Social Reflex Runway, beginning April 3, 1934. After four weeks they were moved to the longer and present runway, and continued daily through the ninth week. The same procedure was being followed with Group C. At the sixteenth week Group D was tested in the runway daily for one week. Following that test, additional tests were made at intervals of four weeks. At each interval starting with the sixteenth week the chicks were tested daily for one week, but gradually the number of such tests was reduced. At the present time it is sufficient to run each pair once at an interval of four weeks. At first the direction of each member of a pair was reversed daily. Eventually it was discovered that direction had no influence on the behavior of a single pair.

Beginning at the sixteenth week and continuing at intervals of every four weeks afterwards, the males were classified in terms of dominance. Each possible pair was observed until a fight took place. At that age these fights happen rather frequently. If a particular pair shows an inclination not to engage in open conflict, they can be encouraged to do so after a few hours of isolation. If the pair meet in the runway after a few hours of isolation, they are especially prone to fight. After a few days of observation at the sixteenth week, one can easily arrange the males in the order of dominance. As time passes, relative dominance becomes more and more obvious. By the thirty-sixth week the order can be determined within an hour.

The Raw Data

The male chicks were weighed each week during the first nine weeks, and at intervals of each fourth week beginning with the sixteenth week. The weight in kilograms for each chick at each weighing period is recorded in Table 1.

At the end of each week the average distance in meters that each male chick ran against each of the other male chicks was recorded, as well as the average time in seconds consumed by the average runs. The complete averages of such data are exhibited in Tables 2-13.

The above raw data were finally elaborated into average velocities per second, momentum per second, and kinetic energy per second, and these elaborated data are exhibited in Tables 14-23.

TABLE 1

AVERAGE WEIGHT IN KILOGRAMS EACH WEEK OF EACH MALE CHICK IN GROUPS
A AND C THAT LIVED THROUGHOUT THE TEST

Week	YY	Blue	White	Red	Green	Yellow
1	.07	.07	.06	.07	.07	.07
2	.13	.12	.11	.12	.12	.13
3	.23	.21	.19	.20	.20	.23
4	.34	.31	.29	.27	.32	.34
5	.50	.47	.44	.40	.46	.51
6	.66	.61	.62	.49	.60	.68
7	.85	.80	.73	.65	.75	.83
8	1.04	.98	.87	.79	.91	.99
9	1.16	1.15	1.06	.94	1.05	1.14
16	2.33	2.28	2.13	1.96	2.16	2.02
20	2.56	2.70	2.49	2.43	2.73	2.49
24	2.80	2.96	2.64	2.91	3.05	2.94
28	3.22	3.38	3.03	3.38	3.51	3.43
32	3.51	3.51	3.40	3.74	3.74	3.74
36	3.71	3.40	3.37	3.63	3.45	3.85

TABLE 2

AVERAGE DAILY DISTANCE IN METERS AND AVERAGE DAILY TIME IN SECONDS
RUN BY CHICK *Blue* EACH WEEK AGAINST THE OTHER
FOUR CHICKS IN GROUP A

Week	White		Red		Green		Yellow	
	d	t	d	t	d	t	d	t
1	.30	1.4	.29	2.3	.11	1.4	.39	2.0
2	.78	1.0	.81	1.0	.79	1.0	.66	1.4
3	.67	1.0	.79	1.2	.82	1.0	.87	1.2
4	.73	1.0	1.26	1.1	.76	1.0	1.00	1.0
5	2.70	3.1	3.83	2.3	2.88	2.0	2.27	2.0
6	2.51	1.3	2.98	2.2	3.05	2.0	2.51	1.4
7	2.37	1.4	2.80	2.0	3.14	2.0	2.74	1.3
8	2.27	3.1	2.19	2.0	2.80	2.0	2.80	2.1
9	2.51	2.0	2.62	2.0	2.95	1.4	2.88	2.4

TABLE 3

AVERAGE DAILY DISTANCE IN METERS AND AVERAGE DAILY TIME IN SECONDS
RUN BY CHICK *White* EACH WEEK AGAINST THE OTHER
FOUR CHICKS IN GROUP A

Week	Blue		Red		Green		Yellow	
	d	t	d	t	d	t	d	t
1	1.22	1.4	.56	1.1	.85	1.3	.85	1.2
2	.83	1.0	.68	1.0	.83	1.0	.67	1.0
3	.85	1.0	1.26	1.3	.73	1.0	.85	1.0
4	.79	1.0	1.39	1.3	.82	1.0	.81	1.0
5	3.39	3.1	3.59	2.2	3.22	2.1	3.29	2.2
6	3.59	1.3	3.66	2.0	3.66	1.4	3.29	1.3
7	3.72	1.4	3.29	1.3	3.47	1.4	3.29	1.4
8	3.83	3.1	3.05	1.4	3.11	2.0	3.11	1.4
9	3.59	2.0	2.78	1.3	3.11	2.0	3.39	2.0

TABLE 10
AVERAGE DAILY DISTANCE IN METERS AND AVERAGE DAILY TIME IN SECONDS RUN BY EACH CHICK DURING THE 24TH WEEK AGAINST THE OTHER FIVE MALE CHICKS IN GROUP D
(Read vertical columns)

	YY d	YY t	Blue d	Blue t	White d	White t	Red d	Red t	Green d	Green t	Yellow d	Yellow t
YY			1.22	4.0	1.22	3.5	2.44	3.2	.0	5.2	.0	2.4
Blue	4.88	4.0			1.22	10.0	2.44	12.0	.0	11.0	.0	4.0
White	4.88	3.5	4.88	10.0			3.66	12.0	.60	6.0	.0	7.1
Red	3.66	3.2	3.66	12.0	2.44	12.0			4.27	8.0	.0	8.0
Green	6.09	5.2	6.09	11.0	5.49	6.0	1.82	8.0			.0	6.4
Yellow	6.09	2.4	6.09	4.0	6.09	7.1	6.09	8.0	6.09	6.4		

TABLE 11
AVERAGE DAILY DISTANCE IN METERS AND AVERAGE DAILY TIME IN SECONDS RUN BY EACH CHICK DURING THE 28TH WEEK AGAINST THE OTHER FIVE MALE CHICKS IN GROUP D
(Read vertical columns)

	YY d	YY t	Blue d	Blue t	White d	White t	Red d	Red t	Green d	Green t	Yellow d	Yellow t
YY			1.22	2.1	.60	3.0	1.82	2.3	1.22	5.0	.0	6.0
Blue	4.88	2.1			1.22	10.0	2.44	4.0	.60	10.0	.0	7.0
White	5.49	3.0	4.88	10.0			4.27	10.0	1.82	7.0	.0	5.0
Red	4.27	2.3	3.66	4.0	1.82	10.0			4.27	8.2	.60	15.0
Green	4.88	5.0	5.49	10.0	4.27	7.0	1.82	8.2			1.22	15.0
Yellow	6.09	6.0	6.09	7.0	6.09	5.0	5.49	15.0	4.88	15.0		

TABLE 12
AVERAGE DAILY DISTANCE IN METERS AND AVERAGE DAILY TIME IN SECONDS RUN BY EACH CHICK DURING THE 32ND WEEK AGAINST THE OTHER FIVE MALE CHICKS IN GROUP D
(Read vertical columns)

	YY d	YY t	Blue d	Blue t	White d	White t	Red d	Red t	Green d	Green t	Yellow d	Yellow t
YY			.61	4.0	.61	3.4	1.22	2.0	1.83	2.1	.0	2.1
Blue	5.48	4.0			1.22	8.0	1.22	6.1	1.22	5.1	.0	2.4
White	5.48	3.4	4.87	8.0			4.87	6.0	5.48	5.0	.0	3.1
Red	4.87	2.0	4.87	6.1	1.22	6.0			4.87	3.1	.0	3.3
Green	4.26	2.1	4.87	5.1	.61	5.0	1.22	3.1			.0	3.4
Yellow	6.10	2.1	6.10	2.4	6.10	3.1	6.10	3.3	6.10	3.4		

TABLE 13
AVERAGE DAILY DISTANCE IN METERS AND AVERAGE DAILY TIME IN SECONDS RUN BY EACH MALE CHICK DURING THE 36TH WEEK AGAINST THE OTHER FIVE MALE CHICKS IN GROUP D
(Read vertical columns)

	YY d	YY t	Blue d	Blue t	Green d	Green t	Yellow d	Yellow t	Red d	Red t	White d	White t
YY			.61	2.0	.61	2.1	.61	3.0	.61	2.2	.61	3.0
Blue	5.48	2.0			1.22	20.0	1.22	20.0	1.22	20.0	.61	20.0
Green	5.48	2.1	4.87	20.0			1.22	10.0	1.22	8.0	.61	18.0
Yellow	5.48	3.0	4.87	20.0	4.87	10.0			.61	7.0	1.22	6.0
Red	5.48	2.2	4.87	20.0	4.87	8.0	5.48	10.0			1.22	20.0
White	5.48	3.0	5.48	20.0	5.48	18.0	4.87	6.0	4.87	20.0		
Average	5.48	2.4	4.14	16.4	3.41	11.6	2.88	9.2	1.71	11.4	.85	13.4

TABLE 14

WITH MASS IN KILOGRAMS, AND VELOCITY IN METERS PER SECOND, THIS TABLE GIVES THE AVERAGE VELOCITY, MOMENTUM, AND KINETIC ENERGY PER SECOND OF EACH CHICK PAIRED AGAINST EACH OTHER CHICK IN GROUP A DURING THE First and Second WEEKS AFTER INCUBATION

The figures for YY are added for convenience of reference

(Read vertical columns)

	v	YY MV	YY KE	Blue MV	Blue KE	Blue v	White MV	White KE	White v	Red MV	Red KE	Red v	Green MV	Green KE	Green V	Yellow MV	Yellow KE
GG	.76	.076	.02														
RR	1.06	.106	.056														
Blue							.072	.03	.85	.066	.019	.59	.084	.037	.58	.058	.016
White				.042	.009	.45				.077	.031	.81	.056	.016	.69	.069	.023
Red				.032	.005	.34	.05	.014	.59				.04	.008	.46	.046	.010
Black				.049	.013	.52	.053	.016	.63	.069	.025	.73	.06	.019	.89	.089	.039
Green				.036	.006	.38	.062	.022	.73	.055	.016	.58				.061	.018
Yellow				.049	.013	.52	.05	.017	.69	.081	.035	.86	.038	.007	.61		
Average	.91	.091	.038	.041	.009	.44	.057	.019	.69	.069	.025	.71	.055	.017	.64	.064	.021

At the sixteenth week the male chicks were classified in the form of a "dominance" hierarchy, and at the end of each fourth week afterwards a similar classification was made. These classifications from the sixteenth to the thirty-sixth weeks are exhibited in Figures 2-5.

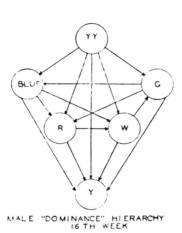

MALE "DOMINANCE" HIERARCHY
16 TH WEEK

FIGURE 2

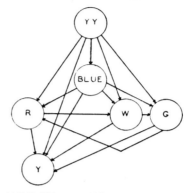

MALE "DOMINANCE" HIERARCHY
20 TH WEEK
24 TH WEEK
28 TH WEEK

FIGURE 3

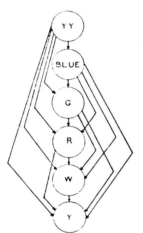

MALE "DOMINANCE" HIERARCHY
32 ND WEEK

FIGURE 4

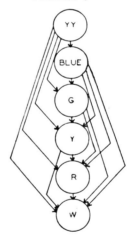

MALE "DOMINANCE" HIERARCHY
36 TH WEEK

FIGURE 5

ANALYSES OF THE DATA

1. *Dominance Hierarchies.* Figures 2-5 demonstrate that YY is the only individual that preserves his rank inviolate through all the 36 weeks. It is significant that polygonal "dominance" is gradually extended into straight-line "dominance." It has been reported by Schjelderup-Ebbe that straight-line dominance is very rare. My data would seem to indicate that straight-line dominance is a function of adjustment over a long period of time within an isolated social group. It could be expected that this length of time would increase with the size of the group, and that it would be greatly interfered with if new members are being added to the group. In a large flock of chicks it might never be observed. This entire matter is to be discussed at some length in a later paper and need not be elaborated here.

2. *Identification of the Two Reflexes.* The measures of Social Reflex No. 2 are plotted as the abscissa in all the graphs where this reflex is analyzed. The unit 0 is a classification and is occupied by that individual that is "dominated" by all the others. The chick that dominates just one other chick possesses just one positive unit of Social Reflex No. 2. The chick that dominates all the other male chicks in the group of six would possess five positive units of Social Reflex No. 2 and would be so classified. Social Reflex No. 1 is plotted as the ordinate in most of the graphs that accompany this paper, and may be plotted in terms of time, space, or any combination of time and space. Mass enters these graphs both alone and in combination with time and space as a measure of Social Reflex No. 1.

3. *Analyses in Terms of Physical Concepts.*

a. *Mass alone.* In Figure 6 are shown the plottings at successive intervals of four weeks each of mass as a function of social Reflex No. 2. With the exception of the sixteenth week, there is little linear relationship between mass and Social Reflex No. 2. The relationship at the sixteenth week may be significant. This is the period when "dominance" is being established. After being established, it might persist within limits in spite of variations in mass. This is an interesting possibility that needs investigation.

b. *Time and space in combination with each other and with*

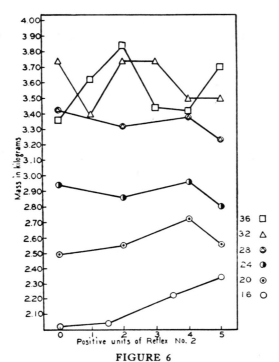

FIGURE 6

MASS IN KILOGRAMS AS A FUNCTION OF SOCIAL REFLEX NO. 2 DURING SUC-
CESSIVE PERIODS FROM THE 16TH TO THE 36TH WEEKS. MALE *Gallus
domesticus*, GROUP D

mass. The meters per second of Social Reflex No. 1 at successive
intervals of four weeks each are plotted in Figure 8. It is seen that
the correlation of the two series would be very high, but that the
linear relationship is not present except at the 20th week. It should be
noticed, however, that the 5th unit of the abscissa shows a surprising linear progression of increasing speed with the passage of time.
This is shown markedly in a special graph, Figure 9. It is not easy
to understand this phenomenon, as the chicks have arrived at their
maximum physical-speed ability by the age of 16 weeks.

The above phenomena are exhibited still further in Figures 10-13,

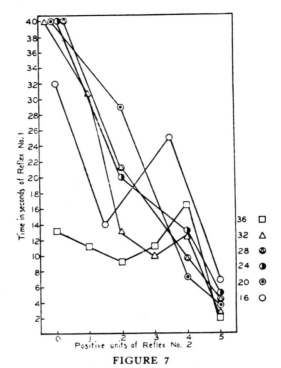

FIGURE 7

TIME IN SECONDS OF SOCIAL REFLEX NO. 1 AS A FUNCTION OF SOCIAL REFLEX
NO. 2 DURING SUCCESSIVE PERIODS FROM THE 16TH TO THE 36TH
WEEKS. MALE *Gallus domesticus*, GROUP D

except that the measurements in the ordinates are momentum and
kinetic energy. These latter two concepts are included not because
of any belief on my part that they may be applied in the field of social
phenomena, but simply for the purpose of demonstrating that such
concepts can actually be tested in the social field under experimental
conditions.

4. *Analyses in Terms of General Rational Concepts.*

a. Time alone. Figure 7 shows the plottings at successive
intervals of four weeks each of the time of Social Reflex No. 1 as a
function of Social Reflex No. 2. The linear relationship does not

240

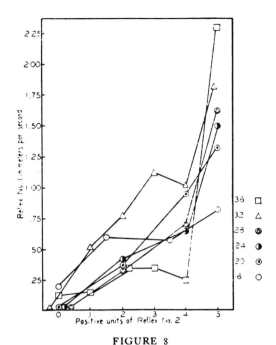

FIGURE 8

METERS PER SECOND OF SOCIAL REFLEX No. 1 AS A FUNCTION OF SOCIAL REFLEX
No. 2 AT SUCCESSIVE PERIODS FROM THE 16TH TO THE 36TH WEEKS.
MALE *Gallus domesticus*, GROUP D.

exist at the beginning and at the end of the experimental period, but
is very marked at the 20th, the 24th, and the 28th weeks. If one
will examine Figures 2-5, one will observe that the period of marked
linear relationship coincides with the long period of social stability
indicated in Figure 3. Time as a linear function of Social Reflex
No. 2 becomes greatly disturbed during a period of rearrangement
for this reflex. This is indicated in the 16th, 32nd, and 36th weeks.

b. *Space alone.* In Figure 14 the average of Social Reflex No.
1 in meters at successive intervals of four weeks is plotted as a
function of Social Reflex No. 2. Considering that these measurements were taken over a period of approximately six months with
rapidly growing and pugnacious young roosters, this function is

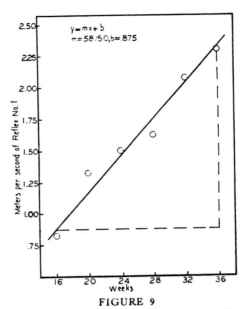

FIGURE 9

VELOCITY PER SECOND OF SOCIAL REFLEX NO. 1 OF THE MAXIMUM POSITIVE
CLASS OF SOCIAL REFLEX NO. 2 AS A FUNCTION OF TIME IN UNITS
OF FOUR WEEKS

amazingly close to true linearity. Most of the deflection from true linearity is caused by the chick in the 0 unit of the abscissa. This chick refused to move out of its tracks for a long period of time because of a foot infection. As soon as this chick recovered from the foot infection and defeated the next two roosters in two successive fights, the straight-line dominance of Figure 5 is reflected here in the almost exact linear function at the 36th week, Figure 15.

A still more interesting function is shown in Figure 16. Here we have a progressively increasing slope to the functional relationship at three widely separated intervals during the experimental period. This figure should be compared with Figures 8-13.

SOME THEORETICAL DIFFICULTIES

1. *The Theoretical Values of "m" and "b."* Let us refer once more to Figure 14. In that graph m has a value of 47/50 while

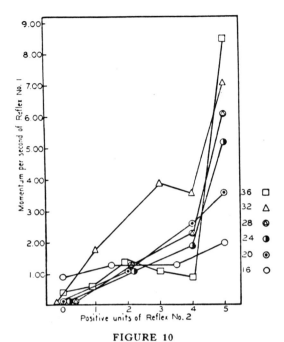

FIGURE 10

MOMENTUM PER SECOND OF SOCIAL REFLEX NO. 1 AS A FUNCTION OF SOCIAL
REFLEX NO. 2 AT SUCCESSIVE PERIODS FROM THE 16TH TO THE 36TH
WEEKS. MALE *Gallus domesticus*, GROUP D.

b has a value of .5. We are dealing with a social hierarchy of six
individuals. Suppose we had a different number of animals in this
hierarchy, some number that might vary anywhere from 2 to some
number greater than 6. What effect would this have on the values of
m and *b?* We answer this question with a family of theoretical
functional curves in Figure 17. Whatever the number of individuals
in the hierarchy, if we are dealing with the equation $y = mx + b$,
we may be quite certain that $m = \dfrac{y - b}{x}$, and that *b* remains a
constant. This hypothesis is limited by the size of hierarchy that
can exist under the life conditions of *Gallus domesticus.*

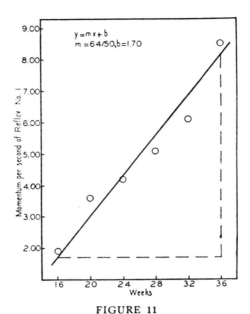

FIGURE 11

MOMENTUM PER SECOND OF SOCIAL REFLEX NO. 1 OF THE MAXIMUM POSITIVE
CLASS OF SOCIAL REFLEX NO. 2 AS A FUNCTION OF TIME IN UNITS
OF FOUR WEEKS

2. *Given that Social Reflex No. 1 is a Function of Social Reflex
No. 2, does $\Delta y = k\Delta x$?* The units of Social Reflex No. 1 are continuous experimental units, while the units of Social Reflex No. 2
are only assumed to be continuous. It is only fair to question whether
we are justified in this assumption. Our reply to this difficulty is
contained in graphical form in Figure 18, and is there explained.
On the theoretical assumption that Social Reflex No. 1 is a true
linear function of the true units of Social Reflex No. 2, the function is
plotted with abscissa units all equal. The experimental points are
then all moved into the straight line with ordinate units unviolated.
Perpendiculars are dropped to the abscissa from these points. The
intersections of these perpendiculars with the abscissa are the theoretically true locations. We believe that this is a valid theoretical
solution of a theoretical difficulty. The form of the theoretical

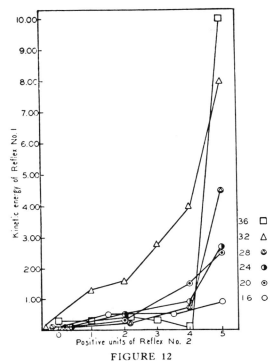

FIGURE 12

KINETIC ENERGY PER SECOND OF SOCIAL REFLEX NO. 1 AS A FUNCTION OF
SOCIAL REFLEX NO. 2 AT SUCCESSIVE PERIODS FROM THE 16TH TO THE
36TH WEEKS. MALE *Gallus domesticus*, GROUP D.

construction fits all the empirical data. The unusual distance from
(4) to (5) is supported by Figures 8-13 and the tables upon which
those figures are based. The unusual distance from (1) to (2)
is explained by the fact that the chick that occupies position (2)
has just defeated in successive fights the two chicks that occupy
positions (0) and (1). Making these theoretical corrections in the
hypothetical values of the abscissa, empirical data may be plotted
almost on the straight line day after day until this special dominance
situation becomes smoothed down or breaks out in a new reversal
and new adjustment. Theoretically there is just one conclusion

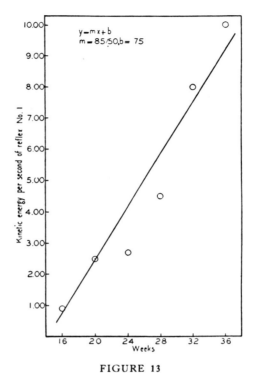

FIGURE 13

KINETIC ENERGY PER SECOND OF SOCIAL REFLEX No. 1 OF THE MAXIMUM
POSITIVE CLASS OF SOCIAL REFLEX No. 2 AS A FUNCTION OF TIME IN
UNITS OF FOUR WEEKS

to this discussion, and that conclusion is that Social Reflex No. 2 may be determined best as a perfect linear function of Social Reflex No. 1. Further, it follows that Social Reflex No. 1 may be accepted as a more accurate measure of dominance even than is fighting. Fighting to a decision is the most simple way to determine rank, but the series of fights do not carry any indication of the relative sizes of the dominance units when plotted on the abscissa of a functional curve. If we have not erred in our argument, we have exhibited a method of determining these relative sizes.

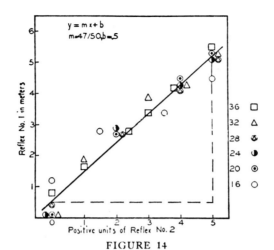

FIGURE 14

SOCIAL REFLEX NO. 1 AS A FUNCTION OF SOCIAL REFLEX NO. 2, THE MEAS-
UREMENTS FOR BOTH SERIES BEING TAKEN DURING THE 16TH, 20TH,
24TH, 28TH, 32ND, AND 36TH WEEKS. MALE *Gallus
domesticus,* GROUP D

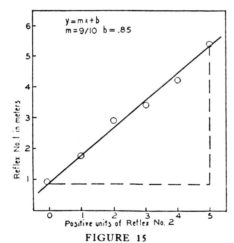

FIGURE 15

SOCIAL REFLEX NO. 1 AS A FUNCTION OF SOCIAL REFLEX NO. 2 DURING THE
36TH WEEK. MALE *Gallus domesticus,* GROUP D

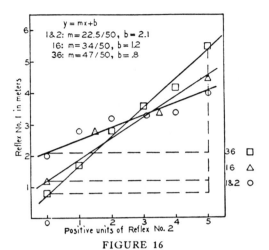

FIGURE 16

SOCIAL REFLEX NO. 1 EXPRESSED AS A FUNCTION OF SOCIAL REFLEX NO. 2
DURING SUCCESSIVE INTERVALS OF TIME (2ND, 16TH, AND 36TH
WEEKS), THE LINEAR FUNCTION SHOWING A PROGRESSIVELY
INCREASING SLOPE AND PROGRESSIVELY DECREASING
VALUE OF THE Y INTERCEPT. MALE *Gallus*
domesticus, GROUP D

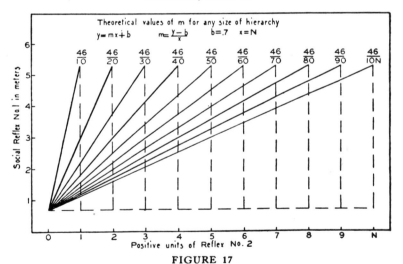

FIGURE 17

SOCIAL REFLEX NO. 1 AS A FUNCTION OF SOCIAL REFLEX NO. 2 IN A THEORET-
ICAL HIERARCHY OF ANY SIZE THAT CAN EXIST UNDER THE LIFE
CONDITIONS OF *Gallus domesticus*

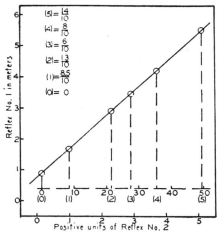

FIGURE 18

ON THE THEORETICAL ASSUMPTION THAT SOCIAL REFLEX NO. 1 IS A TRUE
LINEAR FUNCTION OF THE TRUE UNITS OF SOCIAL REFLEX NO. 2, THE
FUNCTION IS PLOTTED WITH ABSCISSA UNITS ALL EQUAL. THE
EXPERIMENTAL POINTS ARE THEN ALL MOVED INTO THE
STRAIGHT LINE WITH ORDINATE UNITS UNVIOLATED. PER-
PENDICULARS ARE DROPPED TO THE ABSCISSA FROM THESE
POINTS. THE INTERSECTIONS OF THESE PERPEN-
DICULARS WITH THE ABSCISSA ARE THE THEO-
RETICALLY TRUE LOCATIONS

GENERAL SUMMARY

As Spektorsky, Sorokin, and others have pointed out, ambitious
workers in the field of social science have always proclaimed their
destination to be the formulation of social law. Few have arrived
even near the goal, and none beyond all doubt.

The history of social physics has been a history of attempts to
apply the laws and concepts of physics to the phenomena of 'social
behavior. Such methods are classified as physical analogies.

If the social scientist should be satisfied to work with the general
logical concepts of time, space, and their subsidiary categories, he
could not be accused of dealing in analogies, and social law would
seem to be inevitable. These categories, plus observations on physical phenomena, were all the physicist had to start with. Plus ob-

servations on identified quanta of social phenomena, they are all the social scientist needs.

Simple social quanta that are easily identifiable are the movement of social objects towards each other, the fighting of social objects to a decision, the sex reflex, the inference that another social body is greater than or less than oneself, and the inference that another social body is greater than or less than some other social body. The first two of these social quanta have been identified in *Gallus domesticus,* have been measured in terms of time and space categories, have been subjected to the technique of the covariable, and have been exhibited in terms of space alone as satisfying both empirically and rationally the equation $\Delta y = k \Delta x$.

REFERENCES

1. BINGHAM, H. C. Size and form perception in *Gallus domesticus.* *J. Anim. Behav.,* 1913, **3**, 65-13.
2. ————. Visual perception of the chick. *Behav. Monog.,* 1922, **4**, No. 4, 1-104.
3. BIRD, C. The relative importance of maturation and habit in the development of an instinct. *Psychol. Bull.,* 1924, **21**, 96-97.
4. ————. The effect of maturation upon the pecking instinct of chicks. *J. Genet. Psychol.,* 1926, **33**, 212-234.
5. ————. Maturation and practice: their effects upon the feeding reactions of chicks. *J. Comp. Psychol.,* 1933, **16**, 343-366.
6. BREED, F. S. The development of certain instincts and habits in chicks. *Behav. Monog.,* 1911, **1**, No. 1, 1-78.
7. ————. Reactions of chicks to optical stimuli. *J. Anim. Behav.,* 1912, **2**, 280-295.
8. COLE, L. W. The relation of strength of stimulus to rate of learning in the chick. *J. Anim. Behav.,* 1911, **1**, 111-124.
9. DUNLAP, J. Organization of learning and other traits in chickens. *Comp. Psychol. Monog.,* 1933, **9**, No. 4, 1-55.
10. DUNLAP, K., & MOWRER, O. H. Head movements and eye functions of birds. *J. Comp. Psychol.,* 1930, **11**, 99-113.
11. FISCHEL, W. Beiträge zur Soziologie des Haushuhns. *Biol. Zentbl.,* 1927, **47**, 678-695.
12. FLETCHER, M. M., COWAN, E. A., & ARLITT, A. H. Experiments on the behavior of chicks hatched from alcoholized eggs. *J. Anim. Behav.,* 1916, **6**, 103-137.
13. GOODALE, H. D. Note on the behavior of capons when brooding chicks. *J. Anim. Behav.,* 1916, **6**, 319-324.
14. HAR, K. D. Social laws. Chapel Hill: Univ. North Carolina Press, 1930. Pp. xii + 256.
15. HESS, C. Ueber Dunkeladaptation und Sehpurpur bei Hühnern und Tauben. *Arch. f. Augenhk.,* 1907, **57**, 298-316.

16. ———. Untersuchungen über den Lichtsinn und Farbensinn bei Tagvögeln. *Arch. f. Augenhk.*, 1907, **57**, 317-327.

17. ———. Untersuchungen über das Sehen und über die Pupillenreaktion von Tag und von Nachtvögeln. *Arch f. Augenhk.*, 1908, **59**, 143-167.

18. HUNT, H. E. Observations on newly hatched chicks. *Amer. J. Psychol.*, 1897, **9**, 125-137.

19. JOHNSON, H. M. Visual pattern discrimination in the vertebrates. I. Problems and methods. *J. Anim. Behav.*, 1914, **4**, 319-339.

20. ———. Visual pattern discrimination in the vertebrates. II. Comparative visual acuity in the dog, the monkey, and the chick. *J. Anim. Behav.*, 1914, **4**, 340-361.

21. ———. Visual pattern discrimination in the vertebrates. III. Effective differences in the width of visible striae for the monkey and the chick. *J. Anim. Behav.*, 1916, **6**, 169-188.

22. ———. Visual pattern discrimination in the vertebrates. IV. Effective differences in direction of visible striae for the monkey and the chick. *J. Anim. Behav.*, 1916, **6**, 189-204.

23. KATZ, D., & RÉVÉSZ, G. Ein Beitrag zur Kenntnis des Lichtsinns der Hühner. *Nachr. d. Gesellsch. d. Wiss. Göttingen, Math. Phys. Kl.*, 1907, 406-409.

24. ———. Experimentelle-psychologische Untersuchungen mit Hühnern. *Zsch. f. Psychol.*, 1907, **50**, 93-116.

25. KLINE, L. W. Methods in animal psychology. *Amer. J. Psychol.*, 1899, **10**, 265-277.

26. KUO, Z. Y. Ontogeny of embryonic behavior in Aves. I. The chronology and general nature of the behavior of the chick embryo. *J. Exper. Zool.*, 1932, **61**, 395-430.

27. ———. Ontogeny of embryonic behavior in Aves. II. The mechanical factors in the various stages leading to hatching. *J. Exper. Zool.*, 1932, **62**, 453-489.

28. ———. Ontogeny of embryonic behavior in Aves. III. The structure and environmental factors in embryonic behavior. *J. Comp. Psychol.*, 1932, **13**, 245-271.

29. ———. Ontogeny of embryonic behavior in Aves. IV. The influence of prenatal behavior upon postnatal life. *J. Comp. Psychol.*, 1932, **14**, 109-122.

30. ———. Ontogeny of embryonic behavior in Aves. V. The reflex concept in the light of embryonic behavior in birds. *Psychol. Rev.*, 1932, **39**, 499-515.

31. ———. Ontogeny of embryonic behavior in Aves. VI. Relation between heart beat and the behavior of the avian embryo. *J Comp. Psychol.*, 1933, **16**, 379-384.

32. LASHLEY, K. S. The color vision of birds. I. The spectrum of the domestic fowl. *J. Anim. Behav.*, 1916, **6**, 1-26.

33. MILLS, W. The nature and development of animal intelligence. New York: Macmillan, 1898. Pp. xii + 307.

34. MORGAN, C. L. Habit and instinct. London: Arnold, 1896. Pp. 351.

35. ————. Instinct and intelligence in chicks and ducklings. *Nat. Sci.*, 1894, **4**, No. 25.

36. ————. The habit of drinking in young chicks. *Science*, 1896, **3**, 900.

37. ————. Instinct *vs.* experience in newly hatched chicks. *Nature*, 1900, **62**, 590.

38. ————. The animal mind. London: Arnold, 1930. Pp. xii + 275.

39. MOSELEY, D. The accuracy of the pecking response in chicks. *J. Comp. Psychol.*, 1925, **5**, 75-97.

40. MUNN, N. L. The relative efficacy of form and background in the chick's discrimination of visual patterns. *J. Comp. Psychol.*, 1931, **12**, 41-75.

41. PEARL, R. Studies on the physiology of reproduction in the domestic fowl. VII. Data regarding the brooding instinct in relation to egg production. *J. Anim. Behav.*, 1914, **4**, 266-288.

42. PREYER, W. The senses and the will. New York: Appleton, 1890. Pp. xxv + 346.

43. ROMANES, G. J. Mental evolution in animals. New York: Appleton, 1884. Pp. 411.

44. SCHJELDERUP-EBBE, T. Beiträge zur Sozialpsychologie des Haushuhns. *Zsch. f. Psychol.*, 1922, **88**, 225-264.

45. ————. Soziale Verhältnisse bei Vögeln. *Zsch. f. Psychol.*, 1922, **90**, 106-107.

46. ————. Das Leben der Wildente in der Zeit der Paarung. *Psychol. Forsch.*, 1923, **3**, 12-17.

47. ————. Weitere Beiträge zur Sozial- und Individualpsychologie des Haushushns. *Zsch. f. Psychol.*, 1923, **92**, 60-87.

48. ————. Zur Sozialpsychologie der Vögel. *Zsch. f. Psychol.*, 1924, **95**, 36-84.

49. ————. Fortgesetzte biologische Beobachtungen des Gallus domesticus. *Psychol. Forsch.*, 1924, **5**, 343-355.

50. ————. Sociale tilstande hos utvalgte inferiore vesner. (Social behavior in selected animals and primitive peoples). *Ark. f. psyckol., og. ped.*, 1926, **5**, 105-220.

51. SHEPARD, J. F., & BREED, F. S. Maturation and use in the development of an instinct. *J. Anim. Behav.*, 1913, **3**, 274-285.

52. SOROKIN, P. Contemporary sociological theories. New York: Harper, 1928. Pp. xxiii + 785.

53. SPALDING, D. A. Instinct, with original observations on young animals. *Macmillan's*, 1873, **27**, 284.

54. SPEKTORSKY, E. The problems of social physics in the seventeenth century. Vol. I, Warsaw, 1910. Vol. II, Kiev, 1917.

55. STAVSKY, W. H., & PATTIE, F. A. Discrimination of direction of moving stimuli by chickens. *J. Comp. Psychol.*, 1930, **10**, 317, 323.

56. THORNDIKE, E. L. The instinctive reaction of young chicks. *Psychol. Rev.*, 1899, **6**, 282-291.

Clark University,
Worcester, Massachusetts

13

Reprinted from *J. Social Psychol.*, **6**, 172–175, 179–181 (1935)

THE EXPERIMENTAL MEASUREMENT OF A SOCIAL HIERARCHY IN *GALLUS DOMESTICUS:* V. THE POST-MORTEM MEASUREMENT OF ANATOMICAL FEATURES[1]

From the Psychological and Biological Laboratories of Clark University

CARL MURCHISON, C. M. POMERAT, AND M. X. ZARROW

With the appearance of each successive paper in this series, it becomes increasingly important that each paper be considered in sequence with the previous papers. For the convenience of the reader it seems necesary to continue in each paper a brief summary of the findings of the previous papers (4, 5, 6, 7).

> Beginning at sixteen weeks of age, six young roosters are arranged in a hierarchy of dominance, the order being determined by the number of individuals in the group that each rooster is able to defeat in physical combat (Social Reflex No. 2). This order of ranking is revised at intervals of four weeks from the 16th to the 36th weeks. Beginning immediately after being taken from the incubator, these individuals had been tested at frequent intervals in the Social Reflex Runway. This test consisted simply of releasing two individuals simultaneously from opposite ends of the runway, and then observing the time spent and distance traversed by each in running to the other (Social Reflex No. 1). Various operations involving the concepts of physics were applied without great success to these data. Then simple measurements of time and space were applied. When plotted as a function of Social Reflex No. 2, it was found that Social Reflex No. 1, plotted in terms of space alone, was almost truly linear. A theoretical correction of the abscissa units, which agreed with the empirical data, satisfied the requirements of linear function.
>
> Social discrimination in *Gallus domesticus* is identified as it

---

[1]Beginning with the section headed *Procedure* and continuing through the section headed *Special Groups of Characters,* this paper was written by C. M. Pomerat. The other sections of the paper were written by Carl Murchison. C. M. Pomerat planned the measurements, and he and M. X. Zarrow executed the measurements.

Editor's Note: A row of asterisks indicates that material has been omitted from the original article.

is measured in the Social Discrimination Cage. When the discriminations are plotted against Social Reflex No. 2, a relationship appears which approaches a linear form. When the discriminations are plotted against Social Reflex No. 1, true linearity is approximated at the 36th week. The analyses show that male discriminations for pairs of males are away from dominance, and that female discriminations for pairs of males are in the direction of dominance. The constancy of this trait in males makes it possible to measure social discrimination in *Gallus domesticus* in units of space.

Social Reflex No. 3 (the sex reflex) is measured in terms of the total treadings in which each individual engages during a period of time. As so measured this reflex in male *Gallus domesticus* is a linear function of Social Reflex No. 1 and Social Reflex No. 2, while in female *Gallus domesticus* it is a function of Social Reflex No. 1. Observations seem to indicate that such subjective matters as sex favoritism, social insults, and social integration may eventually be exhibited as linear functions of such measurable quanta as Social Reflex No. 1, Social Reflex No. 2, and Social Reflex No. 3.

Under the conditions of this experiment the net loss in body weight over a divided period of 48 hours of starvation containing equal increments of inactivity and social activity approximates true linear functions of the three social reflexes and of social discrimination, but raises no question of metabolic causality.

The object of the present study is to determine whether the three social reflexes can be explained in terms of physical superiority as expressed by body measurements, weight of endocrine glands, size of secondary sexual characters, chemical composition of the blood, or blood count.

Procedure

After having made a series of erythrocyte counts the animals were carefully weighed and then killed by bleeding. Blood was taken for chemical analysis at this time in test tubes containing potassium oxalate. The pituitary was removed under binocular by cutting the stalk, opening the sella (dorsal approach), removing the dura, and lifting out the gland. The anterior lobe was then quickly dissected free and weighed to tenths of a milligram. The thyroids, adrenals, and testes were then dissected out and weighed separately. Uniform procedure was followed to insure speed and accuracy. In spite of this care the anterior pituitary of animal *Green* and the right thyroid of *YY* were damaged and, therefore, are not included in the table which follows.

The measure of intestinal length was made according to the technique employed by Riddle (10). As soon as peristaltic action had ceased the mesenteric folds binding all curvatures were cut and the intestine lying between the gizzard and the anus was laid on a meter stick and measured.

Body measurements, such as interorbital width, length of spur, etc., were made with the use of dividers and calipers. The structures measured and techniques employed were based on Baldwin's *Measurements of Birds* (1). The comb and lappets were dissected free in as nearly uniform a way as possible. The area of the lappet was determined by means of a Willis planimeter.

THE RAW DATA

The measurements which were obtained have been grouped in four tables: general body measurements, weight of endocrine glands, blood chemistry values, and erythrocyte counts. All measurements are given in the metric system. Values, with the exception of those dealing with the blood, have been grouped both as to the position of the birds in the dominance hierarchy and according to the size or the weight of the individual character. In those instances where two or more birds were found to have exactly the same size, they have been grouped together in the "size, weight" portion of the tables. All of these measurements are given in Tables 1-4.

* * * * * * *

THEORETICAL DISCUSSION

None of the anatomical measurements correlate any better with the social reflexes than does the single measurement of gross body weight. This negative result is highly significant, as both popular and expert opinion would have been in the opposite direction.

This result minimizes the importance of anatomical structure in these situations, but does not eliminate the question of physiological function. As far as these experiments are concerned, adequacy of physiological function may vary independently of the size of anatomical structure.

It would seem that the next step is to vary experimentally the

TABLE 1
BODY MEASUREMENTS

No.	Character	Order in dominance hierarchy					Order arranged according to size and weight of character					
		YY	Blue	Green	Red	White	I	II	III	IV	V	
1	Final body weight	3401.10	2989.07	3071.13	3001.15	2792.16	YY	G	R	B	W	1
2	Exposed culmen length	27	27	26	22	26	(YY,B)		(G,W)		R	2
3	Bill length from gape	42	45	42	40	41	B	(YY,G)		W	R	3
4	Height of bill at base	15	16	15	15	14	B	(YY,G,R)			W	4
5	Width of bill at gape	28	31	28	29	28	B	R	(W,G,YY)			5
6	Length of base of comb	64	59	62	59	54	YY	G	(B,R)		W	6
7	Greatest length of comb	122	111	123	129	113	R	G	YY	W	B	7
8	Height of comb	59	56	63	57	68	W	G	YY	R	B	8
9	Weight of comb	40.714	28.301	37.94	35.45	44.784	W	YY	G	R	B	9
10	Comb weight / body weight	0.01197	0.00947	0.01248	0.01181	0.01603	W	G	YY	R	B	10
11	Area of lappets in sq. mm.	4853.86	4922.60	5687.14	3671.64	2650.87	G	B	YY	R	W	11
12	Inter-orbital width of head	31	31	31	32	31	R	(YY,B,G,W)				12
13	Length of head from nostril	66	69	69	62	64	(B,G)		YY	W	R	13
14	Length of keel	138	146	145	137	136	B	G	YY	R	W	14
15	Length of forearm	112	118	119	117	112	G	B	R	(YY,W)		15
16	Length of hand	102	108	113	103	105	G	B	W	R	YY	16
17	Length of leg	435	451	480	450	532	G	B	W	R	YY	17
18	Tibia length	173	182	187	175	177	G	B	W	R	YY	18
19	Diameter of mid-tarsus	18	18	22	20	20	G	(R,W)		(YY,B)		19
20	Length of middle toe	55	57	66	60	51	G	R	B	YY	W	20
21	Length of hind toe	21	21	22	23	24	W	R	G	(YY,B)		21
22	Length of spur from scale (Rt)	27	20	24	28	33	W	R	YY	G	B	22
23	Length of spur from scale (Lt)	24	19	23	27	30	W	R	YY	G	B	23
24	Greatest width of spur (Rt)	13	10	12	12	10	YY	(R,G)		(W,B)		24
25	Greatest width of spur (Lt)	13	10	12	12	10	YY	(R,G)		(W,B)		25
26	Inter-trochanteric span (dorsal)	111	110	117	109	99	G	YY	B	R	W	26
27	Intestinal length	171	137	163	155	146	YY	G	R	W	B	27

secretions of the endocrine glands. That sort of thing, in order to be convincing to one skilled in dealing with social hierarchies, must be done on groups in which straight-line dominance has been established and has become stable.

GENERAL SUMMARY

The five surviving members of a social hierarchy, male *Gallus domesticus,* were subjected to various physical measurements and blood analyses. The physical measurements especially concerned the head, the beak, the comb, the wattles, the legs, the toes, the spurs, the intestines, the endocrine glands, etc. The blood analyses consisted of a blood count and the chemical determination of non-protein nitrogen and of creatinin. None of these measurements indicated an orderly relationship with the social reflexes. It seems very unlikely that size of anatomical structure alone has any important relationship with the social reflexes. This result does not eliminate the question of physiological function in relation to rank in a social hierarchy, and that question becomes open for experimental investigation. It is suggested that such experiments be done only on those groups in which straight line dominance has become established and stable.

REFERENCES

1. BALDWIN, S. P., OBERHOLSTER, H. C., & WORLEY, L. G. Measurements of birds. Cleveland: Mus. Nat. Hist., 1931. Pp. ix+165.
2. FORKNER, C. F. Blood and bone marrow cells of domestic fowl. *J. Exper. Med.,* 1929, **50**, 121-142.
3. FREEMAN, W. The weight of endocrine glands. Biometric studies in psychiatry, No. 8. *Human Biol.,* 1934, **6**, 489-523.
4. MURCHISON, C. The experimental measurement of a social hierarchy in *Gallus domesticus*: I. The direct identification and direct measurement of Social Reflex No. 1 and Social Reflex No. 2. *J. Gen. Psychol.,* 1935, **12**, 3-39.
5. ————. The experimental measurement of a social hierarchy in *Gallus domesticus*: II. The identification and inferential measurement of Social Reflex No. 1 and Social Reflex No. 2 by means of social discrimination. *J. Soc. Psychol.,* 1935, **6**, 3-30.
6. ————. The experimental measurement of a social hierarchy in *Gallus domesticus*: III. The direct and inferential measurement of Social Reflex No. 3. *J. Genet. Psychol.,* 1935, **46**, 76-102.
7. ————. The experimental measurement of a social hierarchy in *Gallus domesticus*: IV. Loss of body weight under conditions of mild starvation as a function of social dominance. *J. Gen. Psychol.,* 1935, **12**, 296-312.
8. PONDER, E. The erythrocyte and the action of simple haemolysins. London: Oliver & Boyd, 1924. Pp. x+192.
9. RIDDLE, O., & FLEMION, F. A sex difference in intestinal length and its relation to pituitary size. *Endocrinol.,* 1928, **12**, 203-208.
10. RIDDLE, O., & NUSSMANN, T. C. A sex difference in pituitary size and intestinal length in doves and pigeons. *Anat. Rec.,* 1933, **57**, 192-204.

Clark University
Worcester, Massachusetts

14

Reprinted from *Physiological Zoöl.*, **12**(4), 412–430, 436–440 (1939)

MODIFICATION OF THE SOCIAL ORDER IN FLOCKS OF HENS BY THE INJECTION OF TESTOSTERONE PROPIONATE[1]

(One figure)

W. C. ALLEE, N. E. COLLIAS, AND CATHARINE Z. LUTHERMAN

Whitman Laboratory of Experimental Zoölogy, University of Chicago

MODERN scientific recognition of the existence of a definite social organization within small flocks of birds dates from the work of Schjelderup-Ebbe (1922) on the so-called "peck-order" in the common domestic fowl (*Gallus domesticus*). He has also observed the phenomenon which he calls "despotism" in a large number of birds, and he summarized his work on the subject in 1935. Schjelderup-Ebbe's original description for chickens[2] has been essentially verified, among others, by Masure and Allee (1934) and by Murchison in his 1935 series of papers.

This social organization is based primarily on the outcome of an initial pair-contact, or a series of such contacts, in which one bird loses a fight or submits passively without fighting. Thereafter, when these two fowls meet, the one which has acquired the peck-right (that is, the right to peck without being pecked in return) exercises it frequently. The subservient bird of an original pair often avoids close contact with her superior and, in event of a revolt, usually does not again fight back vigorously. Revolts occur rarely among hens, and successful revolts that result in a change in social status are still more rare.

Factors which make for dominance between members of the same sex and species of birds have been reported to include the following (see Schjelderup-Ebbe, 1935; Shoemaker, 1939):

1. Other things being equal, the stronger bird usually wins.

2. One bird acts as though intimidated by the appearance of a strange and apparently unfrightened individual and so gives way without fighting.

3. Both birds act as though frightened, but one recovers more rapidly and so wins the contact reaction.

4. An individual which is normally aggressive and victorious may be temporarily out of condition (tired or ill or molting severely) when it meets a newcomer, and so loses when it might, from other relations, be expected to win.

5. Mature hens usually dominate immature chickens.

6. The location of the first contact is important; birds, like many mammals, fight better in their home territory.

7. Even in territory strange to both members of a contact pair a bird wins more readily if surrounded by others with which it has associated.

---

[1] This work has been supported in part by a grant from the Committee on Research in Endocrinology of the National Research Council. We are indebted to Professors C. R. Moore and F. C. Koch and to Drs. L. V. Domm and Walter Hoskins for advice concerning the application of hormones to our particular problem and to certain of them for reading this manuscript.

[2] The *Standard Dictionary* defines "chicken" as "the young of the common fowl; loosely, a fowl of any age. . . . ." The word is used here in this looser sense.

Editor's Note: A row of asterisks indicates that material has been omitted from the original article.

8. Birds with young fight more fiercely than they would at others times. With some birds—canaries, for example—intensity of fighting varies with the different phases of the reproductive cycle.

Two main types of social order have been described for birds. Among chickens, for example, once social dominance is established, it tends to be maintained unless there is a decided revolt. The dominant birds seem to have an absolute peck-right over their subordinates. With several other species—of which pigeons, doves, and shell parakeets may be mentioned—the result of any given pair-contact can seldom be predicted with certainty, even though, in the long run, one of the pair wins more frequently than does the other. Masure and Allee (1934) called this a "peck-dominance," in contrast with the peck-right found among chickens.

It is readily seen that with animals that have an almost absolute peck-right any reverse pecks which may be observed are much more significant than they would be if the society were organized about peck-dominance. Hence, for the present studies the use of the common domestic fowl was indicated. Also, since cocks have a more unstable peck-order than hens, the latter were used exclusively.

A description based on the pecks administered or received does not give a very intimate picture of the actual social relations in a flock. As a result of observing birds, all of which are plainly marked to permit individual recognition, one finds definite personality traits which are not revealed by the more formal statistical summaries of the results of pair-contacts. A given individual may be hesitant, apparently fearful, or aggressive in different degrees; and its behavior may or may not vary in relation to the different social ranks of its associates. Dominant hens are characterized by the lack of any attempt to avoid other members of the flock; they may ignore the others, or they may chase and peck them. Only exceptionally are they pecked in return.

Subordinates low in the social order tend to avoid birds of higher rank. They retreat when pecked or threatened by high-ranking birds, and may give distinctive sounds, as though in pain, when pecked severely. The lowest-ranking birds in three of the first four pens observed in this study showed objective signs easily interpreted as fear of their superiors. Such behavior depends, often, but not always, on the behavior of the superior birds and grades from a mere quickening of pace to a strong avoiding reaction, which may include dodging backward, turning, and speedy retreat with some vocalization, as though in fear.

The severity of the peck, when given, varies widely. Blood may be drawn from the comb; feathers may be plucked; or, on the other hand, the peck may be light. The attack is often aimed at the comb or the top of the head; but, as the subordinate usually dodges the blow, it frequently lands on the neck or even on the back or shoulders or not at all. Under these conditions the superior hen often renews the attempt to deliver an effective blow. The fowls are usually close together when pecking, but threats may be given from a distance of even 2 feet or more. Pecks are rarely attempted when the subordinate is too far away to be hit.

Pullets and hens do not always react to each other when they come close together. They may be feeding busily about the food hopper and may pay no attention to the presence of another individual. Similarly, when sunning or otherwise resting during the day and after they settle on the roost at night, pecking is usually suspended. Frequently contacts can be stimulated by placing some mixed grains or similar food among the litter on the floor of the pen.

VOL. XII, No. 4, OCTOBER, 1939]

Some birds, usually high in the pecking-order without being at the top, show little negative reaction to their superiors and, when pecked, generally give only a slight avoiding reaction. Some individuals react strongly to being pecked. They may retreat to the other end of the pen, the roost, or the top of the nest boxes, or may hide in some corner, stand still with the head held low, or even crouch low with wings half lifted and the tail depressed, i.e., squatting, as hens do before a rooster. It would be easy to read human emotions into these reactions, a temptation which not all writers on the subject have resisted successfully.

Revolts and reverses are rare among hens. When an inferior pecks a superior, the latter, if it reacts at all, appears "surprised" rather than "frightened." It does not retreat, except to gather itself for attack; it draws itself up, leaps forward, and administers vigorous return pecks. The subordinate hen, which initiated the revolt, retreats as a rule but may also stretch up as high as possible, in characteristic fighting pose. Then each hen may leap forward, beak toward beak with neck elongated and arched and neck feathers raised, and so attempt to peck down on the head or seize the comb of its opponent. Noble, Wurm, and Schmidt (1938) in their study of the social organization of the black-crowned night heron report what may be a foreshortened variant of this type of behavior.

The frequency of pair-contacts varies greatly with different individuals and also with the time of day for the same birds. Lorenz (1935, p. 361), speaking presumably of jackdaws, says that there is a certain tension between birds which stand close to each other, particularly between those which stand near the top animal, and "on the other hand birds high in the peck order are good-natured to the low ranking birds." We have some evidence that this is true in our flocks. With some exceptions, we have found that hens high in the social order tend to peck their more immediate inferiors more often than they do those lower in social rank. A good part of the explanation, but not all, is that the lower birds, especially those near the bottom of the peck-order, tend to avoid their dominant pen mates when these come too near. Then, too, birds that rank near together are more often in close competition with each other.

Some of the individual differences mentioned in the preceding paragraphs are illustrated by observations made in preparation for the injections to be reported below. In one pen, BB, the α-bird pecked its subordinates an average of 41 times each over a period of 42 days; while in another pen during the same time, BR, also an α-bird, pecked its inferiors an average of 126 times. BR had fewer subordinates to peck; and one of them, GG, always kept away from the mash hoppers until the other birds had left. Another evidence of the difference between these two top birds was that BR frequently drove all other birds away from the mash hopper in the mornings; BB was not seen to do this.

Other complexities, which are poorly revealed in the simple tabulation of pecks, include cases of apparent antipathy between individuals. In one of the preliminary studies, BR pecked GG 131 times and its only other subordinate, which, furthermore, was its immediate inferior, only 64 times, although it had more opportunity to peck the latter, since GG, the bottom bird of its flock, commonly kept at some distance from the other birds. RW was seen to peck RY 122 times, while during the same observation periods it pecked RR only 57 times; yet RR stood just below RY in social rank. This could be interpreted as an example of the suggestion made by Lorenz which has just been cited; if so, the α-birds are able to draw a fine line in their behavior toward closely ranking subordinates. Discrepancy of treatment is also indicated by the observation that some superiors allowed some inferiors, but not all, to peck food from their bills.

Sickness limits the social activity of a bird and may result in a loss in social position. Unfortunately, we have had a considerable amount of experience with this factor. Of the original 24 top-crossed pullets which were received October 19, 1937, only 3 are now living. The majority of those that died have been autopsied by the veterinary department of Iowa State College,[3] and, of these, almost all died of fowl leucosis, with which the animals were infected when first received.

Evidence has been accumulating of a partial hormonal control of the aggressive reactions on which the social position of birds is based. This led Allee in 1936 to suggest the use of sex hormones in analyzing the underlying physiological basis of the social order in flocks of birds. Briefly, some of the evidence is: With domestic fowls, and with many other dimorphic species in which the male is larger and more showy, the male normally stands above the females in the social order. Domm and Van Dyke (1932) had found that injection of hebin, an extract of the anterior pituitary which contains LH and FSH as well as thyrotropic hormone, into young male leghorns causes great precosity in the development of head furnishings and also induces crowing (9 days), fighting, and treading by the age of 13 days. Capons certainly do not fight as vigorously as do cocks; in fact, Domm's experience (1927) led him to believe that a "gonadless" male neither crows, treads, nor fights. Benoit (1929) cites a decided exception to this general rule, and one of us has seen brief capon fights in Professor Koch's capon colony. Shoemaker, as early as 1936, had evidence that the social position of an individual canary tended to change during the progress of its breeding cycle.

These and other indications suggested that female aggressiveness might be enhanced by the injection of male hormone. The work of several investigators had shown (see Witschi and Miller, 1938, for a summary) that hens produce male hormone; and Gustavson (1932) prepared an extract from hen feces which stimulated comb growth in capons, a reaction now generally agreed to be produced only by androgenic stimulation. The injection of more male hormone would not mean the introduction of a totally strange material into the female system.

EXPERIMENTAL MATERIAL

Four lots of pure-bred white leghorn pullets or hens have been used in these studies. Their constitution and previous history follows:

Flocks A 1, A 2, A 3, and A 4.—These were top cross white leghorns procured from Dr. N. F. Waters, of Iowa State College. Each of these A flocks was composed of full sisters, the offspring of a random-bred hen mated with an inbred sire. The inbreeding coefficients of the sires used were 52.5, 67.0, 52.1, and 76.8 per cent, respectively. This means that the amount of inbreeding ranged from approximately four to something less than eight generations of brother-sister matings. Their breeding history is given by Waters and Lambert (1936). These pullets were approximately 6 months old when received in October, 1937. They were housed, 6 to a pen, in newly rebuilt and carefully fumigated pens in the Whitman Laboratory chicken-house. The pens were 14.5 feet long by 4.5–5.0 feet wide and about 8.0 feet high. Each pen is provided with raised roosts over a solid board platform to catch the droppings. The cement floor of the pens is kept covered with litter, which is changed weekly. Small doorways open onto sun porches on the south side of the buildings. These porches have wire bottoms as well as sides and top, so that the animals are held well above the ground. Light comes to the

---

[3] We are indebted to Professor N. F. Waters and to Dr. E. F. Waller for making and reporting these autopsies.

VOL. XII, No. 4, OCTOBER, 1939]

pens from a window at the south end of each; and, in addition, a 60-watt bulb hangs over the center of each pen. These are operated on a time switch and provide extra illumination during morning and evening hours during the winter. Winter temperature was controlled by a thermostat and ranged between 10° and 15° C. Our four pens were completely separated from other parts of the chicken-house and were on separate thermostatic and light controls. Our compartment was entered from the outside by its own door, which opened into a narrow corridor that gave direct access to each pen.

Each bird carried numbered wing and leg bands. For our purposes these were supplemented by one or more colored leg bands the colors of which gave the designations of the different individuals. GY, for example, wore a green and a yellow leg band. The subscript which is added at times indicates the flock number.

Almost immediately after receipt, flocks of the A series were placed under observation to determine the normal peck-order. These preliminary observations lasted from October 24 to December 24, 1937. Between that date and April 7, 1938, they were used in a set of experiments to determine the effects of injection of epinephrine.

As a result primarily of fowl leucosis, only 16 of the original 24 birds were alive on April 8. At or before that date the remnants were combined into two flocks, which were appropriately designated "A 1, 2" and "A 3, 4." The former contained 9 and the latter 7 individuals; further deaths reduced these numbers to 6 and 5, respectively, by the end of the first experiments with testosterone propionate, hereafter referred to as TP.

Flocks A 1, 2 and A 3, 4 were given some weeks under daily observations until the peck-order became stable. Injections of TP lasted from June 6 through July 29 (54 days in all), after which postinjection observations were continued until August 16. The 9 healthy birds still remaining were combined into flock A 5 and watched for 2 weeks; and selected birds were injected for 51 days, beginning August 30. After 3 weeks of postinjection observation only 5 were left. These were combined with 2 old hens recently received from an Indiana flock that had been line bred for 18 years without the introduction of new stock. This combination will be called flock "A 6."

Flocks B 1 and B 2.—These were originally composed of 24 inbred white leghorns from Iowa State College equally divided between their families 4 and 6 (Waters and Lambert, 1936). The individuals hatched out between January 3 and 31, 1938, and were received by us in two lots on March 19 and April 9, respectively, when they were 2–3 months old. Family 4 had at that time an inbreeding coefficient of between 68.0 and 73.6, and Family 6 had one of 59.3. Each of our B flocks contained 6 pullets from each family. In 1936 the birds in these two families had an *inter se* relationship of 62; the full sibs of Family 4 had a relationship of 88; and of Family 6, one of 74. The B flocks were housed 12 together in pens vacated by combining the A flocks. They were placed under observation immediately on receipt in order to study the establishment of the peck-order. Half of flock B 1 was removed for ovariotomy on June 22 and 23 and was permanently returned to the flock on July 19.

Injections of TP were made into selected individuals of flock B 2 from October 19 to December 21, 1938. On January 13, 1939, only 6 of this flock were still alive. Those autopsied showed no fowl leucosis. There were diagnoses of ulcer of the gizzard or proventriculus or erosion of the gizzard (4 cases) and an infected air sac (1 case). In flock B 1, 4 operated and 4 unoperated controls were alive on the same date. Those autopsied showed: acute hepatitis with fatal hemorrhage from rupture of the liver; abscess on breast muscle and a discolored liver, apparently secondary to abscess, since this bird was never injected.

[PHYSIOLOGICAL ZOÖLOGY

Flocks C 1, C 2, and C 3.—These were also composed of pure but random-bred white leghorns. They were battery reared in our laboratory from newly hatched chicks which we received September 19, 1938, from the Illinois hatchery, of Metropolis, Illinois. Later they were housed 6 to a battery 40 inches long by 24 inches wide by 18 inches high (C 1 and C 2) and six more in one 28 inches long by 22 inches wide by 24 inches high (C 3). The former were kept at 20°–21° C.; the latter at 6° C. They had not been used in previous experiments before the present work began. The chickens received TP injections from December 10, 1938, to January 30, 1939.

Flock D 1.—This flock was originally composed of 12 aged hens of which 2 were taken to make flock A 6. They were received October 22, 1938. Something of the ruggedness of the stocks and the adequacy of our treatment is seen by the fact that, beyond the death of 1 bird in each of the C and D flocks almost immediately on receipt, all the rest were alive and apparently well in late February, 1939.

Care and feeding.—The A, B, and D flocks were fed and cared for by Mr. Collias except for brief periods, when Miss Lutherman acted as substitute. These two also made all the primary observations. While no other person has handled or cared for these flocks, Dr. L. V. Domm and his experienced caretaker, Mr. Emil Stange, have been extremely courteous in keeping watchful eyes on the welfare of our chickens and in giving advice concerning their care. The diet and treatment have been carefully based on the experience of Dr. Domm with his much more extensive colony.

Briefly, they were fed moist mash containing cod-liver oil in the morning and coarser, mixed grain in the afternoon. A hopper of dry mash was always present in each pen. Green food (cabbage or lettuce) was constantly present off the floor at the side of the pen. Water was before the birds at all times, and the containers were cleaned daily. Oyster shell, crushed limestone, and charcoal were always available to the birds.

The C flocks have been constantly under the care of Miss Lutherman. They have been fed Conkey's starting mash, growing mash, and mixed grains and have also had water constantly before them. In place of cod-liver oil they have been irradiated from 1 to 2 hours daily with light containing about the amount of ultra-violet found in noon sunlight.

Methods of observation.—In the A and B flocks behavior records throughout have been made by an observer sitting quietly, notebook in hand, in a neighboring pen, partly screened from the sight of the birds by the solid wooden partition that makes up the lower part of the walls of each pen. Only large-meshed chicken wire was present to interfere with the view of the near-by birds. Although easily disturbed when first placed under observation, the successive flocks were readily conditioned to the presence of the observer, as was shown by the resumption of normal activities even in the pen in which the observer was seated.

All pair-contacts were initiated by the birds themselves without stimulation other than by the fresh distribution of food. During the period of these experiments the birds were under observation 3–5 hours daily except that on Saturdays, when the pens were cleaned, and on Sunday observations were curtailed or omitted. Observations were made during one or two morning periods between 8:00 and noon and an afternoon period between 2:00 and 6:00. The B flocks were observed by the same method; also, they, together with flocks A 5 and A 6, were placed together in a neutral pen, pair by pair, for observation of initial contacts. The details of this procedure will be given later. Flocks C 1 and C 2 were watched from a distance of 4–6 feet; under the conditions that prevailed one could see the bands on the birds in the two middle batteries. So far, up to 6 months of age, it has been possible to observe all birds of both pens at once and to record all pecks. Periods of observation of these C flocks have varied from 15 minutes to an hour, with 15–20 minutes per day being the more usual time spent. The first peck was observed when the birds were about 2.5 months old. During the early weeks, before pecking started, there was little to observe that is of interest from the present point of view. Filling the feed troughs just before the observation period is of little importance with these battery-housed birds, since they do not "fight" over the opportunity to feed. It was difficult to make accurate observations in the cold room because of the small space there.

OBSERVATIONS

Flock A 1, 2.—As constituted on April 8, this flock contained 9 hens. These were observed for 7 weeks before TP injections were begun; during the last 6 weeks of this preliminary period the social order had remained stable and is summarized in Table 1. This

TABLE 1

THE SOCIAL ORDER IN FLOCK A 1, 2 BEFORE TREATMENT

Individual	Number Pecked	Individuals Pecked							
BB..........	8	RG	RR	GG	BR	YY	BY	GY	RW
RW.........	7	RG	RR	GG	BR	YY	BY	GY	....
GY..........	6	RG	RR	GG	BR	YY	BY	....	....
BY..........	5	RG	RR	GG	BR	YY	....	....	....
YY..........	4	RG	RR	GG	BR	....	....	....	....
BR..........	3	RG	RR	GG	....	....	....	....	....
GG.........	2	RG	RR	....	....	....	....	....	....
RR..........	1	RG	....	....	....	....	....	....	....
RG..........	0	....	....	....	....	....	....	....	....

flock had organized itself in a pure-line system or simple hierarchy. The birds indicated by bold-faced type were selected for injection with the male hormone.[4] All injections were given daily and were made intramuscularly into the pectoral muscle. The locations of the punctures were varied systematically. The hormone was administered as received in sesame oil, and all other birds in the same flock were similarly injected daily with 0.1 or 0.2 cc. of sesame oil. In the present experiment **RG** received 1 mg. and **BR** 0.5 mg. of TP daily. As a check on the effectiveness of the hormone, Dr. L. V. Domm kindly consented to make independent injections of our TP into one of his long-time capons. According to expectations, there was prompt and vigorous comb growth.

Four of the hens died during the period of the experiment: BB on June 7, of fowl leucosis; GG on June 19, of impaction of crop and gizzard, RR on July 3, of fowl leucosis; and GY on August 1 (no diagnosis). BY died approximately 3 weeks after injections stopped. The reversals over these birds that died and even over BY may have been made more certain by reason of the ill-health of the losing bird. This cannot always have been

[4] The testosterone propionate (TP) was donated by the Ciba Pharmaceutical Products Company, Lafayette Park, Summit, N.J., and by the Schering Corporation, Bloomfield, N.J. Without their kind co-operation this work could not have been done. We detected no difference between the hormone furnished by the two companies.

an important factor, however, for BB maintained her position at the top of the pre-injection order up to her death. The ranking of the reduced flock on August 1, 2 days after injections stopped, is given in Table 2.

The basic observational data on which earlier tables are based are given in Table 3 in the form of summaries of the observed pecks for stated intervals which are arbitrarily chosen to give the maximum of information without making the table too long and complicated. Another observer was added during the fourth week after injections were started. The unsuccessful revolts that were observed are shown by asterisks, and the approximate date of each reversal can also be estimated. None of the reversals was actually observed. All the return pecks that are recorded for the first period actually were given on the first day that flocks A 1 and A 2 were united. The dramatic thing shown by the tables is the rise of **RG** from the bottom to the top position in the flock. This hen stood fifth among the 5 survivors of flock A 1. There is evidence that it did not submit passively to the lowly position which it occupied in flock A 1, 2. After the first day, however, no

TABLE 2

THE SOCIAL ORDER IN FLOCK A 1, 2 AFTER 54 DAYS OF INJECTIONS

Individual	Number Pecked	Individuals Pecked				Former Rank*	Formerly Pecked*
RG.............	3	BY	....	YY	RW	5	0
RW.............	3	BY	BR	YY	....	1	4
YY...	2	BY	BR	....	....	3	2
BR.............	2	BY	....	....	RG	4	1
BY.............	0	....	....	....	....	2	3

* Among those still living.

overt indication of revolt was seen in the long period before injections started, and it was not until these had been going on for 4 weeks that **RG** began its social climb. **BR** successfully revolted against GY almost 5 weeks before the death of the latter, against BY 6 weeks before she died, maintained her dominance over **RG**, and staged one observed unsuccessful revolt against RW.

Flock A 3, 4.—On March 19 the remnants of former flocks A 3 and A 4 were combined into a flock of 7 individuals. No reversals in position occurred in the 11 weeks preceding injection of male hormone, which, as with the preceding flock, began on June 6. The dominance-subordination pattern of the flock at that time is shown in Table 4. The social order in this flock illustrates another basic type from that shown in Table 1. Here the top bird, GY, dominated all the others except RG, which, although it stood third from the bottom, had managed to get and hold superiority over GY. As shown again by bold-faced type, the lowest-ranking hens were injected daily with TP, **GG** with 0.5 mg. and **BG** with 0.25 mg. As always, the others were similarly injected with sesame oil. Two birds died during the injection period;[5] the ranking of the remaining members of the flock on August 1 is shown in Table 5.

Again no successful revolts were actually seen, but the daily observations permitted these to be dated with confidence. The unsuccessful revolts that were observed and the

[5] RG on June 16, of fowl leucosis; GY on July 19, no diagnosis but with "no enlargement of organs as expected in leucosis."

TABLE 3

SUMMARY OF OBSERVED PECKS IN FLOCK A 1, 2 BEGINNING APRIL 8

Hens injected with hormones and pecks in the reversed direction are shown in bold-faced type. Double vertical lines inclose the period of injection. Asterisks mark unsuccessful revolts.

DOMINANT	SUBORDINATE	DAYS SUMMARIZED								PECKS	REVERSED PECKS
		10	21	28	14	14	14	12	18		
RW	GY	1	4	8	25	15	11	3	Dead	67	0
	BY	3 6	*10	17	4	3	5	5	13	63	3
	YY	2	3	3	3	1	5	4	9	30	0
	BR	1	12	14	*13	13	11	1	19	84	2
	GG	2	8	8	6	Dead	......	......		24	0
	RR	1	11	5	9	5	Dead	......		31	0
	RG	1 1	12	4	0	4	1	2	62	22	65
GY	BY	1	3	4	15	8	0	6	......	37	0
	YY	6	2	6	14	3	0	4	GY	35	0
	BR	5	2	8	4	1 46	26	8	Dead	20	80
	GG	5	1	1	3	Dead	......	......		10	0
	RR	1	4	1	24	1	Dead	......		31	0
	RG	6 2	0	2	*7	6	1	3	......	19	10
BY	YY	14	23	15	38	15	5 9	15	8	110	32
	BR	9	14	8	8	8	12	1	9	47	22
	GG	7	11	2	0	Dead	......	......		20	0
	RR	5	24	9	24	7	Dead	......		69	0
	RG	1 8	7	10	6	5	1 26	11	20	37	8
YY	BR	5	36	5	19	10	20	12	10	117	0
	GG	3	16	6	9	Dead	......	......		34	0
	RR	6	8	8	13	2	Dead	......		37	0
	RG	3 2	8	6	9	5	3 19	25	58	33	105
BR	GG	3	11	12	2	Dead	......	......		28	0
	RR	4	34	26	37	8	Dead	......		109	0
	RG	2	13	18	6	27	3	1	6	76	0
GG	RR	3	31	4	5	GG	Dead	......		43	0
	RG	2 2	13	12	0	......	......	......		27	2
RR	RG	2	1	3	8	1	RR	Dead	......	16	0
RG		0	0	0	0	......	0	0	0	0	0
Totals	......	127	323	227	311	194	158	101	214	1,276	379

TABLE 4

THE PECK-ORDER IN FLOCK A 3, 4 BEFORE TREATMENT

Individual	Number Pecked	Individuals Pecked						
GY	5	BG	GG	....	BY	BR	RR	....
RR	5	BG	GG	RG	BY	BR	....	....
BR	4	BG	GG	RG	BY	....	....	....
BY	3	BG	GG	RG	....	....	....	....
RG	2	....	GG	....	....	....	....	GY
GG	1	BG	....	....	....	....		
BG	1	....	....	RG	....	....	....	....

other basic data are summarized in Table 6; as in Table 3. In this flock again both injected birds moved up in the social scale: **GG** went from next to the bottom to the top,

TABLE 5

THE PECK-ORDER IN FLOCK A 3, 4 AFTER 54 DAYS OF INJECTIONS

Individual	Number Pecked	Individuals Pecked				Former Rank*	Formerly Pecked*
GG..............	4	BY	BR	**BG**	RR	4	1
RR..............	3	BY	BR	**BG**	....	1	4
BG..............	2	BY	BR	....	....	5	0
BR..............	1	BY	....	....	....	2	3
BY..............	0	....	....	....	....	3	2

* Among those still living.

TABLE 6

SUMMARY OF OBSERVED PECKS IN FLOCK A 3, 4 BEGINNING MARCH 19

Hens injected with hormone and pecks in the reversed direction are shown in bold-faced type. Asterisks mark unsuccessful revolts. Double vertical lines inclose the period of injection. BG was added on April 30.

DOMINANT	SUBORDINATE	DAYS SUMMARIZED									PECKS	REVERSED PECKS
		8	28	14	28	14	14	14	12	18		
GY........	RR	19	17	15	47	29	46	16	GY	.....	189	0
	BR	18	17	20	19	11	*20	7	Dead	.....	112	0
	BY	12	20	4	17	25	16	11	.....	.....	105	0
	GG	15	13	7	21	10	*16	*10	.....	.....	92	0
	BG	.....	.....	16	41	19	18	35	.....	.....	129	0
RR........	BR	***30	33	18	27	16	13	2	11	22	172	0
	BY	40	24	19	33	41	38	18	19	2	234	0
	RG	31	32	9	38	19	Dead	.....	.....	.....	129	0
	GG	18	17	15	24	21	15	*9	7	113	126	**113**
	BG	.....	.....	27	65	22	55	76	82	55	382	0
BR........	BY	6	10	5	6	7	*1	12	15	8	70	0
	BG	24	51	25	21	12	Dead	.....	.....	.....	134	0
	GG	5	17	8	17	18	*1 **22**	**21**	**37**	**57**	67	**137**
	BG	.....	.....	36	57	19	3 **11**	**18**	1 **33**	1 **52**	117	**114**
BY........	RG	20	15	3	6	14	Dead	.....	.....	.....	58	0
	GG	0	0	3	1	*8	**12**	**2**	**8**	**23**	12	**43**
	BG	.....	.....	2	5	4	**22**	**44**	**54**	**5**	11	**125**
RG........	GY	24	21	5	21	22	RG	.....	.....	.....	93	0
	GG	14	10	7 1	*19	18	Dead	.....	.....	.....	68	1
GG........	BG	.....	.....	39 2	46	28	23	14	25	40	215	**2**
BG........	RG	.....	.....	2	10	20	Dead	.....	.....	.....	32	0
Totals..	......	276	297	288	542	383	332	295	293	376	2,547	**535**

and **BG** staged successful revolts against BY and BR. It is worth noting that GY maintained its dominant position until its death and was able to put down attempted revolts which were made on three different occasions by **GG**; also, that BR was broody during the injection period, and its comb, and presumably its supply of male hormone, became much reduced.

Flock A 5.—This flock was composed of the 9 remaining birds from the original A flocks. They were placed together on August 17 and were watched then for the first 5 hours, which is the period most crucial for the establishment of the social order. As it happens, in this flock there were no reversals after this initial period for the first 4 weeks that

TABLE 7

THE SOCIAL ORDER IN FLOCK A 5 BEFORE TREATMENT

Individual	Number Pecked	Individuals Pecked								
RG	7	BY	RY	BG	RR	GG	......	YY	RW	......
RW	7	BY	RY	BG	RR	GG	BR	YY	......	......
YY	6	BY	RY	BG	RR	GG	BR	......	......	......
BR	6	BY	RY	BG	RR	GG	......	......	......	RG
GG	4	BY	RY	BG	RR	......	......	......	......	......
RR	3	BY	RY	BG	......	......	......	......	......	......
BG	2	BY	RY	......	......	......	......	......	......	......
RY	1	BY	......	......	......	......	......	......	......	......
BY	0	......	......	......	......	......	......	......	......	......

TABLE 8

THE PECK-ORDER IN FLOCK A 5 AFTER 51 DAYS OF TREATMENT OF
THE HENS SHOWN IN BOLD-FACED TYPE

Individual	Number Pecked	Individuals Pecked									Former Rank	Formerly Pecked
BY	7	**BG**	RR	GG	.....	BR	YY	RW	RG	.....	9	0
RG	6	**BG**	RR	GG	**RY**	.....	YY	RW	.....	.....	1	7
RW	6	**BG**	RR	GG	**RY**	BR	YY	.....	.....	.....	1	7
YY	4	**BG**	RR	GG	.....	BR	.....	.....	.....	.....	3	6
BR	4	.....	RR	GG	**RY**	.....	.....	RG	.....	.....	3	6
RY	4	**BG**	RR	.....	.....	YY	.....	BY	.....	.....	8	1
GG	3	**BG**	RR	.....	**RY**	.....	.....	.....	.....	.....	5	4
RR	1	**BG**	.....	.....	.....	.....	.....	.....	.....	.....	6	3
BG	1	.....	.....	.....	.....	BR	.....	.....	.....	.....	7	2

the hens were together—and then only by injected individuals. All the birds that received testosterone gave 1 or more reversals. There were no deaths during the progress of the experiment. The peck-order on August 30, when injections began, is shown in Table 7.

BG was given 0.75 mg., **RY** 1.00 mg., and **BY** 1.25 mg. of TP daily. **BG** was removed on October 4 (ninth period) because of illness, and her injections were discontinued. Otherwise the hormone was administered for 51 days. The flock organization at that time is shown in Table 8: Again the dramatic feature of the experiment is the climbing of **BY** from the bottom to the top of the flock, where she dominated all except **RY**, which

[PHYSIOLOGICAL ZOÖLOGY

had also received substantial injections of male hormone. **RY** went up from eighth place to a tie for fourth, with reversals over RR and YY and **BG**. Despite illness, **BG** lost only to her pen mates that were also receiving the hormone, and gained a reversal over BR.

Two successful revolts were observed. On September 13 **RY** attacked her superior, YY; they fought briefly, and YY was beaten and retreated. Thereafter **RY** frequently asserted her superiority over YY. Fifteen days later **RY** engaged in a more elaborate encounter with her superior, RR. She fought with RR just after the latter was returned to the pen following injection with sesame oil. RR turned away and retreated, an action which marks a reversal if the fight is decisive. A little later **RY** leaped on RR, and they fought vigorously for a moment, then RR ran. **RY** kept picking up and dropping a small feather and calling. Later **RY** again jumped on RR in fighting pose and pecked her severely; RR retreated and ran. Still later **RY** gave RR another vigorous peck, and this time RR flew to a roost. After a time RR came down to the floor only to receive another peck from **RY**. All this happened between 4:00 and 4:30 P.M. Although these two birds had not been seen to fight that morning, RR had threatened **RY** during the earlier observation; hence, this probably was the complete story of the reversal, while the action between **RY** and YY may have been merely the final scene in a longer struggle, the early phases of which were not observed.

An unexplained pair-contact reaction was seen on October 10 at about 5:00 P.M. Earlier in the period of watching, GG had pecked **RY**, as was usual; then near the end of the observation period, **RY** and GG fought actively for 5 or 10 seconds, during which time **RY** seized GG by the comb and threw her down. GG got away and retreated beneath the roosts, where she remained for some time. **RY** did not follow but flew to the top of the nest boxes and stayed until the observer left. Apparently this had been an entirely successful revolt, but the next day GG was again pecking **RY** as though nothing had happened. The fact that the revolt occurred late in the day and that **RY** did not immediately follow up her advantage may have had something to do with the matter. Of course, there may have been another unwitnessed and crucial encounter which **RY** lost, especially since later observations of another reversal, not in connection with this experiment, showed that a series of revolts occurred before the dominant bird relinquished the peck-right to a revolting subordinate. During the series of combats still another bird, dominant to both, kept joining in and fighting against the initial rebel, almost as if aiding the defending dominant bird, which, nevertheless, in this case eventually lost its position. This illustrates one of the unavoidable difficulties of our work; in order to know all that happens, the hens should be under constant surveillance and yet, as has been indicated earlier, there would be hours and hours of such watching that would ordinarily be wasted.

The basic observations with flock A 5 are summarized in Table 9, which is arranged as were Tables 3 and 6.

Since the other experiments to be reported involved other stocks of birds, at least in part, and used other techniques, a summary of the effects produced in these tests which were made on the A flocks is given in Table 10.

During the same period of time, of the 12 individuals similarly injected with sesame oil alone, only 1 reversal was given. This was by YY over BY in flock 1, 2 and took place after BY had lost to its 2 hormone-injected flock mates; BY died just after the close of the experiment. Of the 22 reversals given by the hens receiving hormones, only 5 are less important, because they were made over a bird which may have been ill. On the other hand, BG in flock A 5, the only one of these injected individuals to become ill while receiving hormone, gained and kept 1 reversal shortly before she was removed because of illness.

TABLE 9

SUMMARY OF OBSERVED PECKS IN FLOCK A 5 BEGINNING AUGUST 17

Hens injected with hormone and pecks in the reversed direction are shown in bold-faced type. Double vertical lines inclose the period of injection. Asterisks mark unsuccessful revolts.

DOMINANT	SUBORDINATE	DAYS SUMMARIZED								PECKS	REVERSED PECKS
		1	12	13	7	7	7	17	18		
RG	RW	22	20	5	2	1	0	6	1	57	0
	YY	17	25	26	4	11	11	5	8	107	0
	GG	1	8	12	15	9	2	16	3	66	0
	RR	.....	17	8	9	1	6	4	3	48	0
	BG	9	19	16	10	13	13	7	0	87	0
	RY	1	11	10	8	4	1	*11	6	52	0
	BY	.....	13	12	30	15	*6	1 **98**	62	77	160
RW	YY	7	2	13	2	10	7	20	8	69	0
	BR	9	6	8	14	27	6	36	5	111	0
	GG	7	6	5	12	1	1	20	5	57	0
	RR	12	*9	11	3	2	4	2	1	44	0
	BG	7	1	11	5	4	8	2	0	38	0
	RY	12	5	10	5	2	1	15	5 **3**	55	**3**
	BY	10	3	14	27	16	5	8 **42**	34	83	76
YY	BR	7	15	13	2	18	31	65	13	164	0
	GG	3	1	0	0	5	1	36	5	51	0
	RR	5	13	6	0	3	4	3	2	36	0
	BG	8	*8	5	5	13	10	9	6	58	0
	RY	4	11	3	3 **4**	**21**	**13**	**27**	**18**	21	**83**
	BY	4	2	3	2	**13**	**3**	**60**	**27**	11	**103**
BR	RG	2	3	12	5	4	4	22	1	53	0
	GG	4	15	13	2	9	18	82	13	156	0
	RR	44	9	*5	2	6	16	12	4	98	0
	BG	9	6	7	1 **1**	2	**8**	,1 **1**	0	26	10
	RY	7	14	7	3	23	24	47	17	142	0
	BY	4	25	12	25	35	9 **1**	**102**	**52**	110	155
GG	RR	5	3	12	8	8	17	3	2	64	0
	BG	.....	1	15	6	21	48	14	0	105	0
	RY	.....	2	7	16	17	21	*46	15	124	0
	BY	.....	4	6	29	14	16	133	19	198	23
RR	**BG**	.....	2	3	3	4	19	5	0	36	0
	RY	.....	5	2	*7	7	12 **8**	**3**	**6**	33	17
	BY	.....	3	5	*6	9	**10**	**5**	**2**	23	17
BG	RY	2	5	3	2	**6**	**23**	**1**	0	12	30
	BY	.....	4	3	8	1	**4**	0	0	15	5
RY	BY	.....	3	7	7	10	14	71	62	174	0
BY	....	0	0	0	0	0	0	0	0	......	......
Totals	....	222	299	310	293	365	405	571	402	2,661	682

In general, as Table 10 shows, there were more reversals as the dosage became larger. **BG** again is an exception on this point; it was also the only individual receiving male hormone which allowed other hormone-injected birds to rise above it in social status. Both

BR in flock A 1, 2 and **RY** in flock A 5 retained the peck-right over their injected associates **RG** and **BY,** respectively, which otherwise went to the top of their social groups.

As the more detailed tables show (3, 6, and 9), there was a latent period before the injected birds began to stage revolts, and the beginning of reversals came still later. Reversals were usually made first over the fowls low in the peck-order, and only near the end of the prolonged injection periods, if at all, over the ranking birds in the respective flocks.

As was to be expected, TP produced other physiological effects besides those reflected directly by changes in social position. Certain of these will be discussed later, when all such effects will be considered together.

Flock B 2.—These birds were placed under continuous daily observation about the time that they were first beginning to peck each other. During the early months of observation, social position in this particular flock of highly inbred birds was more unstable

TABLE 10

THE EFFECT OF MALE HORMONE ON SOCIAL
POSITION IN THE A FLOCKS

Flock	Bird	Daily Dosage (Milligrams)	Number of Reversals
A 3, 4...............	BG	0.25	2
	GG	0.50	3
A 1, 2	BR	0.50	2
	RG	1.00	4
A 5................	BG*	0.75	1
	RY	1.00	3
	BY	1.25	7
Totals..........	7	....	22

* Became ill during experiment.

than that of any other flock of domestic fowls of which we have record. The last spontaneous reversal occurred during the twenty-seventh week of observation; the next to the last happened 4 weeks earlier. Thereafter the social order remained stable. Eleven of the 12 original birds were alive at the beginning of injections. Three control members of the flock and two that were receiving injections (**BB** and **GG**) died during the injection treatment. The peck-order of those that survived throughout and all those injected with TP is shown in Table 11 at the beginning of injections. As before, the individuals listed at the left in bold-faced type were given the male hormone daily; the others received 0.2 cc. of sesame oil at the same time. **GG** was given 1.25 mg. of TP from October 19 through December 4; **BB** received 1.0 mg. daily for the same period. **RW** was injected with 0.75 mg. each day until December 4, when its dosage was doubled for the following 17 days. **GG** and **BB** died about a week after their removal from the experiment on December 4.[6]

No reversals were given by any of the birds in flock B 2 during the period of injection. Four unsuccessful revolts were seen: 3 by birds that were receiving hormone, and 1 by a

[6] Autopsies at Iowa State College showed erosion of the gizzard and extreme emaciation of both; BB also had an egg free in her body cavity.

VOL. XII, No. 4, OCTOBER, 1939]

control. Nineteen reverse pecks were seen; 7 were delivered by 2 of the 3 pullets injected with TP, and 12 by 5 controls. In other effects, as will be seen later, injections in this flock produced physiological results similar to those in the A flocks, and, as will be told in the next section, the injected birds showed definite behavior effects as a result of receiving the hormone. Under these conditions we have no complete and certain explanation of the failure to induce social reversals such as had been given by each of the previously injected birds. Since crowing was induced early in the bird receiving the highest dosage and was discontinued in a short time, it is possible that these young birds became somewhat refractory as a result of continued treatment with heavy dosage. The primary difference between these birds and the others appears to be in age; and it is to be noted that previous investigators (Hamilton, 1938), working with young chicks, did not get a crowing response in the female by injection of male hormone. Since there were no reversals, there is no need to exhibit the basic data. Approximately 4 months later we were able to produce a typical set of reversals in this flock by another series of injections (see p. 434).

TABLE 11

THE SOCIAL ORDER IN FLOCK B 2 BEFORE TREATMENT

Individual	Number Pecked	Individuals Pecked						
BG	7	GG	RW	BB	YY	RR	BR	GY
GY	5	GG	RW	BB	YY	....	BR	....
RR	5	GG	RW	BB	YY	....	....	GY
BR	4	GG	....	BB	YY	RR	....	....
YY	2	....	RW	BB	....	....	....	....
BB	2	GG	RW	....	....	....	....	....
RW	2	GG	....	....	....	....	BR	....
GG	1	....	....	....	YY	....	....	....

INITIAL CONTACTS

The observations reported so far were made by watching the reactions of the birds when engaged in their normal activities in the flocks of which they were constantly members and in which each had a well-established social niche. A revolt or reversal was therefore made in the face of all the social inertia inherent in such a flock organization. As has been shown in the introductory pages, social position among these birds is established by the outcome of initial contacts and, once established, is rarely reversed.

In connection with other studies we devised a system of experimentally induced initial contacts. The typical procedure for Series I was to scatter some scratch food on the floor of the observation pen; then one person caught one of the birds, while another caught the other. Each individual bird came from a separate flock; in the cases we are now considering, only the injected birds had ever been in contact with members of the other flock, and with them 5 weeks or more had elapsed before the contacts were repeated. Each fowl was weighed, the state of its plumage and general health was noted, position in the social order in the home flock was recorded, and standard comb measurements were taken. Length of comb was measured along the greatest length of the blade; height was measured for the point which stood highest and most perpendicularly above the head. Successive measurements on the same individual were made on the same point. Finally a prediction was entered concerning the outcome of the pair contact which was about to be staged. One observer took a bird under each arm, while the other prepared to make and record observations. The chickens were introduced into a strange pen, which was purposely reduced to half the standard size to insure closer association.

In Series I, five sets of such initial contacts were staged. Set I was between fowls in flock B 2 which had been selected for injection and all those in flock B 1. This latter flock was a wholly similar flock of inbred birds of the same age; the only difference was that 5 of the 10 birds then in flock B 1 were poulards. Set II was between all members of these two flocks and was begun 32 days after injections were started. By that time the comb approached its maximum height. Set III consisted of pair-contacts between the survivors of A 5 and the 2 old Indiana hens which were to be combined with them to make flock A 6. They were given these controlled initial contacts on neutral territory; and, when permanently placed together 5 days after the first pair-contact of the series, they took the position in the social order which was indicated by the outcome of the staged initial encounter. Set IV was a series of staged encounters between members of flock A 6 and the B flocks. Set V includes all staged contacts in the C flocks.

Series II was run by one observer alone with as few modifications as feasible from the routine in Series I. The principal difference was that in the earlier series the same bird was never introduced with another on successive days, while in Series II most birds were so introduced every day. Their opponents, however, always varied from day to day. The contacts in Series II were staged from January 11 to February 27 in connection with studies on the effect of estradiol on the social order and were begun 21 days after the last injection of TP.

Although available in tabular form, the full details will not be given. Many of the data may be used later in another connection, and only those are presented here which bear directly on the role of male hormone in determining social position. Originally this series of observations was begun to test whether there is a relation between the size of the comb among closely inbred normal pullets and the result of the socially important first encounter between two strange birds. The observations were extended to include all other normal individuals together with operated and hormone-injected birds, to allow as complete a test as possible. The results are summarized in Table 12. Some of the general findings are: Poulards with small combs, though often weighing more and standing higher in the social order of their home flock, lost all but 2 encounters. These 2 instances were won by the poulard BB, which had the smaller comb, was heavier, had the higher social rank, and won by passive submission of the other bird.

Pullets injected with TP, though low in social position and light in weight, won all but 4 of their pair-contacts. Such birds had, on the average, larger combs and presumably, therefore, had more male hormone in their system than did the birds to which they lost. The number of actual fights in these staged contacts was small; of the 124 encounters which were staged first, only 24 were settled by fighting. In the remainder, 1 bird retreated without a struggle, i.e., there was passive submission.

In 16 of the 74 cases in which poulards were involved, the bird with the larger comb lost. Two of these involved the poulard BB mentioned above, when for some unknown reason the normal birds submitted passively. The 11 such contacts listed in the table were between the poulard GG and the birds of flock B 2. At the time of these experiments GG resembled a masculinized poulard or a pullet on full injection dosage with TP, since its comb was much larger than that of its pen mates. On closer inspection, small cocklike hackles could be detected on the neck and flanks, and small spurs were present. Later observations showed continued masculinization.

The contacts between this GG₁ and YY₂ and between GG₁ and BR₂ were unusual in that the first reactions were inconclusive. The third time GG₁ and YY₂ were placed together, the latter won. With BR₂ it was not until the fourth staged encounter that we decided that GG₁ was the loser. The protocols of these encounters are given in Table 13. Both these normal birds acted more warily in the presence of GG than when placed with other birds which they dominated. To human eyes at least, GG was not as aggressive as was to be expected from its appearance. These observations raise another interesting

TABLE 12

THE OUTCOME OF INITIAL CONTACTS ANALYZED ON THE BASIS OF SIZE OF COMB AS AN INDICATOR OF THE AMOUNT OF MALE HORMONE PRESENT

SERIES	CONTACT PAIRS*	BIRD WITH LARGER COMB		MEAN SIZE OF COMB*†	
		Won	Lost	Won	Lost
1................	Normal:Normal	32	13	128	115
2................	Normal:Normal	33	12	150	133
1................	TP:Normal	12	4	143	131
2................	TP:Normal	22	2	148	133
1................	TP:Poulard	12	2	156	89
2................	TP:Poulard	3	0	156	107
1................	Normal:Poulard	37	11	108	87
2................	Normal:Poulard	6	3	184	103
2................	TP:TP	0	2	142	146
2................	Other combinations	81	35	145	131
Totals or true means.......	....	238	84	140±0.9	120±0.9

* One pair of birds, not included here, had combs equal in size.   † Length plus height.

possibility. The relatively sudden growth of turgid, hypertrophied furnishings on the head of a relatively slight hen present some handicaps for successful fighting. The delivery of effective pecks is probably rendered more difficult, and the possibility of being pecked on the comb is certainly increased. Many of the problems suggested by the changed appearance of injected hens, or of such operates as GG, both psychological and physiological, could be effectively attacked by the use of dubbed birds.

Of the four contacts lost by hormone-injected birds and won by normal controls, 2 fights each were won by 2 normals from the same 2 injected pullets. These normal individuals (BW and BY) had smaller combs than their injected opponents; they were, however, able and enduring fighters. One of the injected birds (BB) was losing weight rapidly at the time, and the other (GG₂) seemed to be blind in one eye; both were ill and died after about 3 weeks. Neither had the endurance of BW or BY, although each won all its remaining staged contacts with normal birds.

[PHYSIOLOGICAL ZOÖLOGY

These experimental contacts provide another indication of the increase of aggressiveness in these fowls following TP treatment. **BB, RW,** and **GG** of flock B 2 lost 9 of their 12 contacts with BW, RW, RR, and BY of flock B 1 before injection started. These were the only ones tried at that time. **BB** and **GG** lost all 4 contacts, and **RW** lost 1 of 4. After 32 or more days of injection, and despite the fact that **BB** and **GG** were not very well at the time, **RW** won all 4 and the others each won 2 of their 4 contacts with the same birds. An adverse score of 9 losses and 3 wins before treatment was turned into a favorable one of 8 wins and 4 losses. Further, the operate BB$_1$ had won from **BB**$_2$ and GG$_2$ before injections were begun but lost to these as well as to **RW** while they were being given male hormone.

TABLE 13

A SYNOPSIS OF EXPERIMENTAL PAIR-CONTACTS BETWEEN THE POULARD
GG$_1$ AND THE NORMAL HEN, BR$_2$

	I	II	III
Date............	10-7-38	10-23-38	10-28-38
Weight, pounds...	3.6:3.6	3.3:3.3	3.2:3.3
Comb, millimeters.	55+101:42+76	56+97:32+67	56+103:30+67
	Large, red:Normal	As before	As before
Social position....	8 :5	As before	As before
Molting..........	No : Yes	As before	As before
Start, P.M........	4:13	3:53	4:23
Behavior.........	GG avoids BR	GG feeds	BR threatens GG
	BR half-threatens GG	GG apparently avoids BR	GG avoids BR
	4:33, out, no contest	BR half-threatens GG	BR threatens GG
	5:04, in again	BR feeds	4:27, out
	GG avoids BR	BR threatens GG	BR judged to have won
	BR feeds	GG feeds	
	GG avoids BR	4:13, out, no contest	
	BR half-threatens GG	NOTE: BR is pecked by RW,	
	5:22, out, no contest	which has large comb, and	
		acts as though afraid of	
		GG's appearance	

In general, we can conclude from these observations on the B flocks that, while the injections did not cause the lowest 3 birds to climb in the social order of their home flock, it did decidedly increase their aggressiveness and success in encounters with their near relatives in a parallel flock. Also, these staged initial contacts between normal pullets showed that, while the amount of male hormone present, as measured by comb size, is not an infallible guide to the winner of such contacts, it is a good indicator—in fact, one of the best single indicators we have found—of the probable winner.

The C flocks.—The whole treatment of the C flocks differs radically from that which we have been describing. These were battery reared in our laboratory. One of our interests with them was to study the onset of pecking in two flocks of 6 birds in separate batteries at 20°–21° C., as compared to those in another battery which, since their thirteenth week of age, had been kept at 6° C. The temperatures were chosen in connection with another experiment. The birds were hatched on September 18. RG, of flock C$_2$ in the warmer room, was seen to peck a battery mate on December 3. A week later, during which time no pecking had been seen, injections of TP were begun into **RR** in the warm room and **GG** in the cool room. **RR** was seen to peck another bird a week later (December 17) and continued to do so steadily. Injections were continued until January 20, 1939, when **RR** became the α-bird of the now firmly established peck-order.

BB of the same flock was first seen to peck another bird on December 17; it was not found pecking again until 4 weeks later. Between December 23 and 29, 5 more of the 11 controls in the warm room were observed pecking other birds. The smaller space in the cold room made observation difficult. No pecks were recorded for these pullets until after injections were begun; and then **GG**, the injected bird, did the first pecking that was seen. **GG** started a few days after **RR** began in the warm room, and has in general behaved much the same as **RR,** at least as nearly as can be detected. We recognize that these observations on the C flocks are inconclusive. For one thing, the injections did not begin until the most precocious bird had started to peck its fellows. However, the behavior is at least what would be expected from the observations on the other flocks.

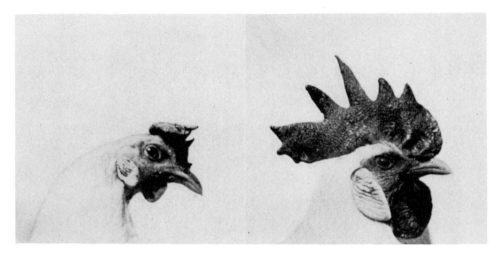

FIG. 1.—**BY** of flock A 5 before (*left*) and after 51 daily injections with testosterone propionate

*   *   *   *   *   *   *

DISCUSSION

In order to begin to understand the foundations of the social order in flocks of the common domestic fowl, one must have data concerning two radically different aspects, namely, those factors which initially determine the position of a given individual in the peck-order and those which make for the maintenance of the position, once it is established. These two sets of factors are not necessarily mutually exclusive, although they do present contrasting emphases. The initial position in the social order is largely determined by the relative aggressiveness of the individuals which compose the flock. Maintenance of position is associated with the factors which make for social inertia; such factors as memory and habit reinforce, and may entirely replace, the aggressive behavior patterns which are so important in the origin of a social order in hens.

In the present experimental analysis we have been investigating primarily the physiological basis of aggressiveness which determines which of two hens or pullets will dominate in a first contact with a stranger from another flock. Also, after the social order is established, we have been particularly interested in analyzing the factors which produce successful revolts that lead to a change in the established order. Although we had indications at the beginning and have since then accumulated more evidence that psychological factors are involved in aggressiveness, we have been directly concerned with a more obviously physiological approach.

The presence of male hormone is a potential source of aggressiveness. Cocks are more aggressive than hens or capons, and cocks dominate hens in a heterosexual flock. Hens apparently secrete male as well as female hormone; that is, they appear to be functional endocrine hermaphrodites, since the combs of ovariotomized fowl drop to a capon level (Domm, 1927). In normal hens both the medulla, which secretes androgenic material that affects the comb, and the cortex, which secretes estrogenic substance that affects the oviduct, respond to gonadotropic agents (Domm, 1937). Hence, the introduction of additional male hormone means increasing the amount of some material originally present rather than the introduction of a wholly foreign substance. Our choice of testosterone propionate (TP) for injection was determined by its chemical purity, its known physiological action on the nervous system as well as on other organ systems (Bize and Moricard, 1937), its relative efficiency (Breneman, 1939), and its availability. The results obtained have justified the selection.

In three flocks of adult hens the TP-injected individual moved from at or near the bottom to the top of the social system, and each of the hormone-treated hens showed one or more social reversals over normal, healthy flock mates. Such a change was shown by none of the controls which were similarly injected with sesame oil. In one flock of relatively young, highly inbred pullets the treated birds did not rise in their social order within the flock but they did win 75 per cent of their first contacts with strange, untreated normal pullets and lost only to 2 aggressive, experienced fighters—and probably then because of illness. Six months later, one of these inbred birds gave an almost diagrammatic response to treatment with TP. Finally, with young pullets in which the peck-order was just being established, the 2 birds receiving TP became dominant in their respective flocks.

Although it is now generally considered that comb growth in capons, when it occurs, is a result of androgenic stimulation, the presence of the appropriate androgen in suitable quantity is not the only factor that determines the size of a fowl's comb. Different breeds vary in this characteristic as in others; even our different flocks of white leghorns had more or less definite variations in comb size. Within limits the amount of sunlight to which an individual is exposed affects the size of head furnishings (Domm, 1930). With these facts in mind, it is not surprising that the observed correlation between size of comb and aggressiveness in first combats was no higher than we found it.

The increased comb growth which we obtained was to be expected from the work of others (Hamilton, 1938; Breneman, 1938; Emmens, 1938), as was also the evidence of a decrease in the rate of egg-laying (Schoeller and Gehrke, 1933) and the lack of maintenance of normal rate of growth in young birds (Breneman, 1939). The greater weight of the older TP-treated hens is in keeping with the experience of Bulliard and Ravina (1938) with ducks about 2 months of age and is opposed to the usual result with growing chicks, our own included.

Our findings concerning the stimulation of crowing differ from those of Hamilton (1938) with very young female chicks and afford a demonstration of the conclusion reached by Myers (1917), and confirmed by Appel (1929), that there is no apparent anatomical reason why the female leghorn fowl should not crow, provided it has the instinct to do so properly developed. Incidentally, these observations throw some light on the question of the ontogenetic origin of instinctive behavior.

Benoit (1929, p. 380) found, by using testicular grafts, that the threshold of hormone stimulation in the cock ran as follows, with the last mentioned having the highest threshold: "(1) Epididyme et canal déférent; (2) Organes érectile; (3) Inhibition du plumage; (4) Instinct sexuel; (5) Chant; (6) Ardeur combative." Benoit suggested that changes which involve the intervention of the nervous system have a higher threshold than do changes in plumage or in comb and wattle growth.

In our experiments with females, changes in comb size were most readily and certainly effected; a change in the rate of egg-laying typically came next; the initiation of crowing, next; and finally changes in social position. It should be noted that in the last, and for our purpose the most important, aspect of these studies we have not investigated the dosage that will cause a given hen to become more aggressive in staged initial contacts; our data regarding dosage apply only to changes in the well-established peck-order of hens that had been living together for some time.

These observations that changes in behavior have a higher threshold than do changes in structure run counter to the early conclusions of ecologists (Shelford, 1911) that with animals it is easier to effect modifications in behavior than in anatomy. We have at present nothing to say about the relative ease of producing changes in behavior and structure in the so-called "lower" animals; obviously, the generalization can no longer be applied to the vertebrates.

There are many indications of the complexity of the factors which initiate and maintain the social order among flocks of hens. One of these indications is found in behavior following the cessation of injections. The sequence of events was as follows: the comb began decreasing in size, courtship vanished, crowing ceased,[7] and after a long interval egg-laying was resumed. In these respects the hens apparently returned to the *status quo ante*. However, the changes won in social position persisted as long as we followed the respective flocks—in the case of flock A 5, for at least 5 months.

Since social position in flocks of hens is not easily affected and changes come only after long treatment, this does not provide the physiologist with a satisfactory criterion for assay of hormone effectiveness. When the hormone begins to produce an increase in aggressiveness, the first revolts against social superiors are likely to fail even in cases when a few more injections will bring successful reversals. The first reversals shown by the low-ranking birds are likely to be the result of successes gained over those just above them in social rank and likewise relatively low in the social order. This is in part a result of the fact that birds near each other in social rank have more mutual contacts and in part because of the lesser aggressiveness one frequently finds in low-ranking hens. Only after a long period of injections, if at all, did our treated hens rise to the top of their respective flocks, and usually the last reversals were over the birds with the highest social status. It is entirely consistent with this picture that the social positions gained by the treated birds were maintained in the postinjection periods. The psychological factors which retarded her rise now helped maintain the hen in her newly acquired status.

It is fairly obvious, from what has been said, that the hen's relation to TP cannot be regarded as similar to that of chemicals in a test tube to which sufficient catalyst is added to cause the reaction to proceed to its logical conclusion. Among the other complicating factors are indications that recent experience in winning or losing encounters affects the outcome of the next pair-contact. These and other so-called "psychological factors" will be discussed in more detail in a later publication.

It is apparent that we have by no means solved the problem of the development and maintenance of social position even under the relatively simple conditions that exist in flocks of hens. We have shown that one physiological factor, the increase in the amount of male sex hormone present—at least to the extent that natural male hormone acts as does TP—if continued for some weeks, increases the likelihood that individuals so treated will better their position in the social order of the flock.

Our experiments were directed toward this one end, and we have not undertaken the arduous task of investigating the physiological mechanism(s) which finally produce changes. For that matter, we do not know with certainty the effect produced on an experienced hen by the physiognomy of a strange bird which possesses the high comb and general appearance of other birds which have recently held the peck-right over her. As an example, the behavior of BR_2 in contact with the large-combed but nonaggressive GG_1, as outlined in Table 13, suggests that this possible factor is not negligible.

---

[7] Domm (1935) reports that the precocious crowing of male chicks stopped within 48 hours after the cessation of daily injections of hebin.

[PHYSIOLOGICAL ZOÖLOGY

Finally, we have been considering some aspects of the social order in flocks of pullets and hens primarily because of our interest in the behavior and social life of birds. We have the distinct impression, however, that in working with this problem we are making an attack on some of the fundamental aspects of group living and graded social position as found in many vertebrates.

SUMMARY

1. Testosterone propionate (TP), injected into low-ranking individuals in these flocks of white leghorn hens, produced a rise in social status in each adult that was treated; and an injected individual eventually occupied the top position in each flock.

2. With one flock of younger pullets, in which the peck-order had just become fixed after an extended period of fluctuations, similar injections produced no changes in the social position within the flock. However, the hormone-treated pullets were more aggressive and successful in initial contacts with strange birds than they were before the injections.

3. In two flocks of battery-reared young pullets studied while the social order was beginning to be established, the injected birds came to dominate each flock.

4. In initial pair-contacts with strange birds, the hen or pullet with the larger comb had decidedly the better chance of dominating the situation; comb size is generally considered to be an index of the amount of male hormone present.

5. Other observed effects of TP include increased size of comb, retardation or suppression of egg-laying, initiation of crowing, and, with three hens, of courtship. These changes vanished soon after the cessation of treatment. Higher social position, once won, was retained.

LITERATURE CITED

ALLEE, W. C. 1931. Animal aggregations. Chicago: University of Chicago Press. Pp. 431.
———. 1936. Analytical studies of group behavior in birds. Wilson Bull., 48:145–51.
———. 1938. The social life of animals. New York: Norton. Pp. 293.
ALLEE, W. C., and COLLIAS, NICHOLAS. 1938. Influence of injected male hormone on the social hierarchy in small flocks of hens. Anat. Rec., 72: Suppl., p. 60.
APPEL, F. W. 1929. Sex dimorphism in the syrinx of the fowl. Jour. Morph. and Physiol., 47:497–513.
BENOIT, J. 1929. Le Déterminisme des charactères sexuels secondaires du coq domestique. Etude physiologique et histologique. Arch. zool. expér. et générale, 69:217–499.
BIZE, P. R., and MORICARD, R. 1937. Modifications psychiques provoquées par l'injection de testostérone chez les jeunes garçons. Bull. Soc. de pediat. d. Paris, 35:38–49.
BRENEMAN, W. R. 1938. Relative effectiveness of testosterone-propionate and dihydroandrosterone-benzoate in the chick as indicated by comb growth. Endocrinology, 23:44–52.
———. 1939. Variations in the reactions of chicks to different methods of administering androgens. Ibid., 24:55–62.
BULLIARD, H., and RAVINA, A. 1938. Effets de la testostérone chez Cairina. Compt. Rend. Soc. biol., 127:525–28.
DOMM, L. V. 1927. New experiments on ovariotomy and the problem of sex inversion in the fowl. Jour. Exper. Zoöl., 48:31–150.
———. 1930. A factor modifying growth of head furnishings in Leghorn fowl. Anat. Rec., 47:314.
———. 1935. The effects of daily injections of hebin on the development of sexual characters in Leghorn chicks. Trans. on Dynam. of Devel., 10:67–79.
———. 1937. Observations concerning anterior pituitary-gonadal interrelations in the fowl. Cold Spring Harbor Symp. on Quant. Biol., 5:241–47.
DOMM, L. V.; GUSTAFSON, R. G.; and JUHN, MARY. 1932. Plumage tests in birds. Chapter xiii in Allen's Sex and internal secretions, pp. 584–646.

VOL. XII, No. 4, OCTOBER, 1939]

DOMM, L. V., and VAN DYKE, H. B. 1932. Precocious development of sexual characters in the fowl by daily injections of hebin. I. The male. Proc. Soc. Exper. Biol. and Med., 30:349–51.

EMMENS, C. W. 1938. Responses of fowl to massive doses of oestrone and testosterone. Jour. Physiol., 92:27–28P.

HAMILTON, J. B. 1938. Precocious masculine behavior following administration of synthetic male hormone substance. Endocrinology, 23:53–57.

HAMILTON, J. B., and DORFMAN, R. J. 1939. Influence of the vehicle upon the length and the strength of the action of male hormone. Endocrinology, 24:711–19.

JUHN, MARY; GUSTAFSON, R. G.; and GALLAGHER, T. F. 1932. The factor of age with reference to reactivity to sex hormones in fowl. Jour. Exper. Zoöl., 64:133–76.

LORENZ, K. Z. 1935. Der Kumpan in der Umwelt des Vogels. Jour. f. Ornith., 83:137–213, 289–413.

MASURE, R. H., and ALLEE, W. C. 1934. The social order in flocks of the common chicken and the pigeon. Auk, 51:306–25.

MURCHISON, C. 1935. The experimental measurement of a social hierarchy in Gallus domesticus. I. The direct identification and direct measurement of social reflex No. 1 and social reflex No. 2. Jour. Gen. Psychol., 12:3–39.

MURCHISON, C.; POMERAT, C. M.; and X ARROW, M. X. The experimental measurement of a social hierarchy in Gallus domesticus. V. The post-mortem measurements of anatomical features. Jour. Social Psychol., 6:172–81.

MYERS, J. A. 1917. Studies on the syrinx of Gallus domesticus. Jour. Morph., 29:165–216.

NOBLE, G. K. 1939. The rôle of dominance in the social life of birds. Auk, 56:263–73.

NOBLE, G. K., and WURM, M. 1938. Effect of testosterone propionate on the black-crowned night heron. Abstract (60) in Anat. Rec., 72 (Suppl.):60.

NOBLE, G. K.; WURM, M.; and SCHMIDT, A. 1938. Social behavior of the black-crowned night heron. Auk, 55:7–40.

PAPANICOLAOU, G. N., and FALK, E. N. 1938. General muscular hypertrophy induced by androgenic hormone. Science, 87:238.

SCHJELDERUP-EBBE, T. 1922. Beiträge zur Socialpsychologie des Haushuhns. Zeitschr. f. Psychol., 88:225–52.

———. 1935. Social behavior of birds. Murchison's Handbook of social psychology, chap. xxi, pp. 947–72.

SCHOELLER, W., and GEHRKE, M. 1933. Der Einfluss der Keimdrüsenhormone auf die Legetätigkeit von Hennen. Arch. f. Gynäkol., 155:234–40.

SHELFORD, V. E. 1911. Physiological animal geography. Jour. Morph., 22:552–617.

SHOEMAKER, H. H. 1939. The social hierarchy in flocks of the canary (Serinus canarius). Auk, 56, No. 4.

WATERS, N. F., and LAMBERT, W. V. 1936. Inbreeding in the white Leghorn fowl. Agric. Exper. Sta., Iowa State College, Research Bull. No. 202.

WITSCHI, E., and MILLER, R. A. 1938. Ambisexuality in the female starling. Jour. Exper. Zoöl. 79:475–87.

15

Reprinted from *Physiological Zoöl.*, **15**(4), 485–506 (1942)

SOME EFFECTS OF CONDITIONING ON SOCIAL DOMINANCE
AND SUBORDINATION IN INBRED STRAINS OF MICE

(One figure)

BENSON GINSBURG AND W. C. ALLEE
University of Chicago

DOMINANCE-SUBORDINATION patterns of behavior, such as the so-called "peck-orders" in birds, have been demonstrated for many different groups of vertebrates ranging from fish to man. Interest in the subject now centers mainly on inquiries concerning the extent to which such social organizations exist in nature and upon an analysis of the factors which underlie and control their establishment and maintenance.

Although this type of group organization exists under natural (Odum, 1941) and seminatural conditions (Lorenz, 1935; Jenkins, in press), it has been most studied with caged or penned animals. In these kept groups, individual social status is achieved either by fighting or by the passive acceptance of subordination. The relative social position of any two individuals with regard to each other may be determined largely by the result of their first pair-contacts, as with hens (Schjelderup-Ebbe, 1922); or there may be a long series of such contacts during which dominance within the pair shifts back and forth before stability appears. Under certain conditions and with some animals, dominance never becomes absolute, and the social flock is organized on a give-and-take basis in which the relative social status of any given individual is judged by the ratio of its wins and losses with each member of the flock.

One of us has recently reviewed, in a general way, present knowledge concerning the factors that make for social dominance (Allee, 1942a, 1942b), and we shall return to the subject in the final section of the present paper. The factors include, among others, heredity, on the one hand, and previous experience or conditioning, on the other. In this paper we are concerned mainly with an analysis of some phases of the effect of conditioning on social status. More specifically, the question we asked ourselves and these animals was: Does a recent victory or series of victories tend to condition further successes, and, on the other hand, does defeat tend to produce further defeats? There were indications in the literature (cf. Whitman, 1919, with doves; Uhrich, 1940, with mice; Diebschlag, 1941, with pigeons) and in the experience of one of us with hens (Allee and Collias, 1940) which indicated that an affirmative answer might be expected, providing we were able to secure proper experimental material and to set up appropriate experimental conditions.[1]

---

[1] Hall and Klein (1942), in a paper which appeared while the present article was in press, concluded from their experiments that (1) "there are marked differences in aggressiveness between rats"; (2) "aggressiveness appears to be a fairly stable and consistent trait"; and (3) "rats selectively bred for fearlessness in a strange situation were found to be more aggressive than rats selectively bred for timidity." They did not test for the possibility that aggressiveness might be affected by conditioning. As with our mice, body weight and the home cage were not important correlates with aggressiveness in these rats.

MATERIAL AND METHODS

Mice were selected for initial experimentation, in part because male mice are notably combative. Among others, Crew and Mirskaia (1931), Retzlaff (1938), and Greenwood et al. (1936) have found that fighting is an important cause of death in groups of laboratory mice. In this connection the last-named authors say: "Many, perhaps most, of the deaths in a herd of normal mice are due to fighting, and it would seem that an immigrant requires a few weeks to obtain recognition as a member of the herd to whom average toleration must be extended."

Uhrich (1938) worked for more than 2 years in this laboratory upon the social order among inbred albino mice. As a result of the study of group relations among some 300 male and 150 female mice, he reported critical evidence which showed that small groups of mice, when caged together, exhibited a social order. For the most part, the order was based on the exclusive dominance of a single mouse which was superior to a number of subordinates that were approximately equal in position. Various other orders also existed for longer or shorter times—among them, a linear order such as is characteristic of the peck-order in hens.

Uhrich also concluded (1938, p. 402) that, despite evidence that mice may at times react to another mouse as an individual, "it sometimes appears as though the subordinates do not recognize the dominant [as an individual] but merely flee from any mouse that happens to attack them or assumes a threatening attitude." A social order based on an impersonal recognition of belligerency, real or threatened, seemed more suitable for our experimentation than did one based on recognition of and reaction to individuals as such.

Later Uhrich (1940) combed through his records and found suggestive evidence that the previous fighting history gave some indication of the outcome of future fights. Briefly summarized, this evidence was that, with mice of known fighting history, some individuals with 12 defeats to their joint credit lost 9 and won only 3 of their next 12 fights. Conversely, another set with 32 previous victories to their joint credit won 20 times in their next 32 starts. Uhrich himself recognized the imperfect nature of this evidence. Its greatest weakness lay in the lack of discrimination between native belligerency of the individual mice and a possible conditioning effect of their preceding experience.

With mice such as Uhrich had been using, a successful attack upon this phase of the problem would be tedious. We were helped at this point by the discovery of Scott (1940), who found in his study of prefighting behavior that different inbred strains of mice at the Jackson Memorial Laboratory show clear-cut genetic differences in aggressiveness. These results suggested that we could use such genetic strains and have predictable outcomes of at least some of the interstrain fights. This would enable us to carry on conditioning and later to test the effects of such experience by rematching with other members of the same strain.

Miss Catherine Fales made a preliminary test of the possibilities as a class project in animal behavior and found that it was possible to cause previously superior or equal mice to become subordinate after a series of 15 successive defeats and that inferior mice tended to fight sooner when their former superiors had been defeated repeatedly. At this point Miss Fales decided to leave the problem, and we took up active experimentation with the mice which she had been using.

At the beginning we had available C-57 blacks (aa), C-3H agoutis (wild type), and C (Bagg) albinos (bbcc), which had been given us by the Jackson Memorial Laboratory, to which we hereby express our thanks. In each case these animals were from two litters and were between 41 and 47 days of age upon arrival at this laboratory about 2 months before the start of the present series of experiments. In addition, we had several litters of the so-called P(aabbddpps$_e$s$_e$) stock and several C albino animals of a different substrain which were obtained from Dr. E. L. Green. We also caught some random-bred wild mice in live traps in the laboratory. We worked seriously with the first three stocks only.

All the animals in the first three strains had been bred through at least fifteen generations of brother-sister matings. For fuller descriptions of the stocks and for a summary of the effects of inbreeding see the stock lists of the Jackson Laboratory and the account of Russell (1941).

In all, we observed 110 male mice from three generations. Males only were used in the actual fighting. Unless otherwise specified in comparable tests, the animals were litter mates or were the offspring of litter mates born a few days apart. All animals were kept in a well-ventilated darkroom. The temperature was held between 20° and 23° C., except during experiments designed to test the effects of more extreme temperatures upon fighting.

Unless otherwise stated, each mouse was housed separately in a coarse-meshed wire cage 9.5×9.5×7 inches. Each cage was equipped with a water bottle and a sufficient amount of cotton and excelsior to allow nests to be built. The animals were fed Purina Fox Chow Checkers in excess of the amount they would consume daily. Cages were cleaned once each week.

Combats were carried on in neutral wire cages ($\frac{5}{16}$-inch mesh) with metal bottoms. They measured 9.5×9.5×12 inches and were divided by a galvanized iron partition which could be readily raised. A mouse was transferred to the fighting cage and back to his home cage in the box in which he was routinely weighed. This procedure did not seem to excite the mice as much as manual handling, and it took considerably less time than when the home cage and the fighting pen were brought together for the transfer. Experiments with box transfer and direct transfer from the pen gave the same results. Fights were started by raising the partition between the two mice.

During a combat, ceiling lights were turned off. The fighting cage was removed some 10–12 feet from the home cages. A 60-watt lamp was placed about 7 inches from the top of the cage. This hindered the mice from seeing outside and, particularly, from reacting to the observer. The near-by light raised the temperature within the cage about $\frac{1}{2}$° C. Unless the fights became too severe, they were allowed to run for 20 minutes, longer on occasion.

The mice were weighed from time to time, and, with generations reared in our laboratory, records of growth in the prefighting age were also taken.

OBSERVATIONS

The pattern of activity which constituted a typical fight was as follows:

1. Mutual sniffing
2. Independent activity, exploration of the cage, accompanied by fluffing of the hair and rapid breathing
3. Lashing of the tail
4. *a*) An attempt to bite back when sniffed at (more common in first encounters)
   b) Direct attack by one member of the pair (common after conditioning)

5. A series of hard and rapid attacks and counterattacks
6. Intermittent chasing and biting
7. The chased mouse rears on its hind legs, draws one foreleg close to the body, extends the other stiffly, remains motionless, and squeals when touched by the other mouse

Point 7 will hereafter be referred to as "submission" and is one important criterion used to judge active or passive defeat. Significant variations from the above pattern will be mentioned separately. The position of the forelegs during "submission" was decidedly different from the poised stand which was frequently assumed just before an attack.

INTERSTRAIN DIFFERENCES

In order to provide a basis for testing the effect of successive wins and losses on the outcome of future battles, it was necessary to assay the differences between the strains. If these were consistent, then conditioning, if effective, should produce reversals in the order of interstrain dominance. The results of calibrating tests are recorded in Table 1.

TABLE 1

A SUMMARY OF THE OUTCOME OF PRELIMINARY TESTS OF THE RELA-
TIVE FIGHTING ABILITY OF THREE STRAINS OF MICE

	C-57 black C-3H agouti	C-3H agouti C albino	C-57 black C albino
Winner.	C-57 black	C-3H agouti	C-57 black
Loser.	C-3H agouti	C albino	C albino
Number of animals.	10	10	10
Number of battles.	133	137	144
Percentage of victories:			
Active.	66.2	40.9	61.1
Passive.	9.8	24.8	4.2
Percentage of "no aggression".	18.8	24.1	31.9
Percentage of defeats:			
Active.	0.0	4.4	0.0
Passive.	5.2	2.2	2.8
Percentage of draws.	0.0	3.6	0.0

Active defeats and victories are determined by battle, while passive ones result from submission of one animal without fighting. It is evident that, whenever a fight took place, the blacks were clearly superior to the other strains and the agoutis were superior to the albinos. In some cases the agoutis displayed more aggressive behavior prior to the actual fight than did the blacks.

All these mice had previously been fought under the supervision of Miss Fales (1940), who had obtained similar results. A period of isolation of 3 weeks' duration had ensued. The first 5 fights between the blacks and agoutis were initial contests for both strains during the present experimental period. The contests were subsequently extended until every mouse had met every other mouse, round-robin fashion, in order to detect individual upsets if any should occur; none did. The behavior of one mouse (Bl₃) ran counter to the expected trend in that it showed no aggressiveness during any of its encounters. This animal had been very badly beaten during the experiments made by Miss Fales and, as will be shown later, had probably been conditioned so that it showed aberrant behavior as compared with other members of its strain.

In the agouti-albino series the albinos may have been favored by the fact that the agoutis had recently been defeated by the blacks. The agoutis, however, proved to be

[PHYSIOLOGICAL ZOÖLOGY]

consistently superior; and, characteristically, the albinos put up a series of poor defenses against the more agressive animals.

The P stock was erratic and highly variable in fighting behavior and was not used in this group of experiments. Perhaps the large number of mutations carried by this stock influenced its vigor and behavior, or, more likely, the fact that it was less highly inbred may account for these variable results.

The data given in Table 1 indicate that there are distinct behavior differences between the three strains tested and that, under our conditions and treatment, the blacks were most successful, the agoutis were less so, while the albinos were most pacific. We thought it wise to retest these relationships with mice which we had reared and whose early experience had been carefully controlled and observed. Accordingly, a second series of 88 interstrain contests were staged between each of the three strains; the animals used had been isolated after weaning. The results will not be summarized in detail, since they support the data already given. The only major deviation is that these younger, less experienced mice showed more instances in which no fight resulted from placing two males together.

Another similar series was arranged which consisted of 54 fights between second-generation blacks and agoutis. This time the young mice were reared with their litter mates until transferred to the fighting cage for the first battle. Again the results are in entire agreement with those already described. No effect of communal rearing was noted other than a possible tendency for the mice to be less excited in the presence of another individual.

As another variant, litters borne by black and by albino females on the same day were split shortly after birth, so that each dam reared both black and albino offspring together. Some of these (a) were isolated after weaning and were kept so until they were mature enough to fight; others (b) were kept together with the same number of males in each lot; the number of females differed with each group. They were removed as soon as it was possible to determine sex accurately. We were also able to make exploratory tests of interstrain fighting between individuals (c) that had the same foster-mother and to compare the results with the outcome of other fights (d) in which the foster-mothers had been different. The numbers available for these tests were as follows: (a) 4; (b) 4; (c) 2; (d) 2.

In all these cases the result was still the same as has already been reported; and, while the numbers are too small to allow a final statement, there was no indication of differences in the outcome of interstrain fighting which could be related to early experience. It will be remembered that we attempted to allow these fights to continue until there had been an actual defeat. This means that our criterion of success combines the factors of aggressiveness in starting a fight with that of success in fighting. Table 2 summarizes the results from all these preliminary interstrain contests between mice that had been reared in our laboratory.

On the basis of the statistical data presented in Tables 1 and 2 we conclude that, in our hands and under the described conditions, the C-57 black mice used in this study were definitely superior to the C-3H agoutis and the C albinos and that the latter were inferior to the agoutis in ability to win fights. The mice were assorted according to strains in their first few fights. Other observations, which are not readily summarized statistically, support these conclusions. Consistent changes in this order which are correlated with

VOL. XV, NO. 4, OCTOBER, 1942]

conditioning may, if environmental conditions are kept constant, be ascribed to conditioning.

Differences other than superiority and inferiority in fighting were also observed between the strains. These will not be reported in detail. So far as they go, however, they tend to support the hypothesis advanced by Scott (1942) and others that there is a genetic basis for such traits. It is pertinent to our main interest to report that the black males were more sensitive to handling and were more easily made highly nervous than were the males of the other strains studied. Unless the blacks were handled gently, they often failed to fight when placed with a stranger, and they squealed every time the cage was moved. Similar treatment did not cause as severe a response on the part of any of the males of the other strains.

Agouti males started fights most frequently in initial encounters, but chiefly by feinting or nipping when sniffed at. Neither the blacks nor the albinos were so likely to make false feints and lunges as were the agoutis.

TABLE 2

THE RESULTS OF FURTHER TESTS OF THE RELATIVE FIGHTING
ABILITY OF THREE STRAINS OF MICE

	C-57 black	C-3H	C-57 black
Winner	C-57 black	C-3H	C-57 black
Loser	C-3H agouti	C albino	C albino
Number of animals	14	8	20
Number of battles	142	88	208
Percentage of victories:			
Active	61.3	40.9	33.9
Passive	10.5	21.6	25.4
Percentage of "no aggression"	28.2	37.5	40.7

It is evident from the data that not every initial meeting of two male mice results in a battle. There may be much random behavior, mutual sniffing and grooming, and no display of aggressiveness. Females are regularly less aggressive than the males; and, while they will fight between themselves, such fights are more rare than when two males are concerned. What, then, are the immediate causes of a fight between males?

It is impossible to give a satisfactory answer on the basis of observed battles. There is, however, one conjecture which is suggested after watching even a few fights. This suggestion is that initial fights develop on occasion from a modification of normal sex behavior. Strange males that have been kept in isolation and are meeting another male for the first time since weaning begin to sniff and nudge him as though he were a female. This is not long sustained, since such behavior is usually sufficient cause for the other mouse to nip at the more aggressive individual, and a fight commonly ensues.

Where mice have had previous fighting experience, such exploratory behavior is usually omitted. An experienced, aggressive male, when he meets another male, shows increased activity, fluffs his hair, thrashes his tail, and exhibits other phases of "threatening" behavior. If an experienced male is pacific, he tends to keep to himself or tries to groom the other mouse. In certain other circles we would say that he was either an isolationist or an appeaser.

If an unaggressive male is allowed to copulate and is then placed with another male, which may or may not be aggressive, the second pursues the first rather insistently

[PHYSIOLOGICAL ZOÖLOGY

(probably attracted by odors incident to copulation). This method has been used in these experiments in attempts to provoke fights between nonagressive animals and was somewhat successful.

INTRASTRAIN DIFFERENCES

In addition to knowing the interstrain differences in fighting ability, it was essential to find the intrastrain status of each individual mouse. Accordingly, an extended series of intrastrain fights were staged in order to "calibrate" every mature male mouse in the colony against the other mature males of his own strain and general age class. With such information at hand we could then detect relatively subtle changes in the order of dominance which might arise during the process of conditioning.

For these intrastrain contests the mice were treated exactly as they were for the interstrain battles except that now the members of each strain were fought against each other every fourth day until a fairly stable order of dominance was reached. Care was taken to fight the mice in a different order in each successive round.

THE C ALBINO STRAIN

The first series of staged pair-contacts between a lot of five males of this strain will serve to illustrate in some detail the characteristic behavior of this particular stock. Each set of symbols—W_1, for example—represents a single animal.

W_1 vs. W_2: The mice were transferred to the fighting cage and left in separate compartments until they showed no signs of undue excitement. The partition was then rattled and lifted. Each mouse ran into the half of the cage which was newly opened to it. For the next 3 minutes each explored the cage independently. Then they touched vibrissae in passing. Mutual sniffing followed, and they subsequently groomed each other. There was no evidence of hostility or fear. At the end of 20 minutes the partition was lowered, and the animals were transferred to their home cages. The encounter was recorded as "no battle."

The behavior of the following contact-pairs in their initial meeting in this series did not differ significantly from that of the pair just described: W_3 vs. W_5; W_6 vs. W_1; W_3 vs. W_1; W_5 vs. W_1.

W_2 vs. W_3: W_3 rattled its tail once during the 20-minute period. There was no other sign of hostile activity.

W_5 vs. W_6: After 5 minutes of typical random behavior, W_6 nipped at W_5, which immediately squealed. W_6 then attacked, and W_5 gave a typical submission reaction. The entire conflict lasted 17 seconds.

W_2 vs. W_5: After 1 minute of random behavior W_2 attacked, and W_5 submitted at once.

W_3 vs. W_6: As soon as the partition was lifted, both animals showed signs of excitation. Their breathing was rapid and labored; the dorsal hair of each was erected; movements were quick and jerky; and both mice frequently rattled their tails in passing each other. W_6 attacked, and a series of spirited skirmishes followed. Each attack was fast and vicious. Single battles lasted up to 46 seconds and were followed by periods of from $\frac{1}{2}$ to 2 minutes of independent activity. After 7 minutes of such behavior, each mouse retired to a separate corner and groomed itself. Then they began all over again. After 11 minutes each retired to his previously occupied corner, but this time they both lay prone on the bottom of the cage for 3 minutes in seeming exhaustion. Another series of battles followed. At the end of 20 minutes neither mouse had shown signs of submission, and the encounter was recorded as a draw.

W_2 vs. W_6: W_6 showed typical signs of aggressiveness (hair-fluffing, tail-beating, etc.) and attacked almost immediately. W_2 made no attempt to fight back but squealed at each attack. After 9 minutes of such behavior, W_2 gave the submission reaction and the fight was stopped. W_6 was, of course, recorded as the winner.

Each mouse had its own peculiar fighting habits, and changes in relative aggressiveness between the members of a contact-pair were easily observed. As the series of battles progressed, some mice (notably W_5) submitted passively before being attacked. Such passive submission was usually to an aggressive, and almost never to a pacific, mouse. Aggressive mice from another substrain of albinos, which did not differ noticeably in behavior from those most used, were introduced from time to time in an effort to determine whether the submissive animal "remembered" a specific mouse by which it had formerly been beaten or whether it reacted to the general deportment of an aggressive individual. The latter seemed to be the case. Further evidence on this point is furnished by the behavior of these mice when matched against the aggressive black strain. Under these conditions the albinos did not usually submit to Bl_3, which never attacked, but did submit passively to all the other blacks. They did not, therefore, react to the strain color, or to a possible strain odor; and we have no evidence that they remembered individuality. The whole behavior complex can be adequately interpreted on the supposition that these mice were responding to the deportment and activity of their opponents in the fighting cage.

For the purpose of ranking the animals a victory following a fight was considered more important than a victory by passive submission, since the most aggressive animals attacked regardless of what the other mouse did. The lack of any fighting within the 20-minute period was taken as an indication of at least temporary pacifism, and continued lack of such fighting was thought to reveal a generally pacific or passive "character."

The most aggressive mice (as already noted) typically attacked opponents almost immediately. The least aggressive typically submitted passively almost as soon as the partition had been lifted and the presence of an opponent was sensed. Mice in the middle of the order seemed to watch for behavior clues; they attacked, if at all, only after some time had elapsed, and they seldom submitted until an aggressive opponent actually began his attack.

Figure 1 summarizes the observed changes in the order of dominance during a series of 200 intrastrain fights among the albinos. The order of dominance is indicated by the ordinates, and the number of fights which had occurred in this series is shown by the abscissae. At times two or more individuals occupied the same rank.

The available data indicate that the order of dominance is not established by the initial encounters and that a relatively stable order emerges only after a considerable number of battles. Fluctuations at the top and the bottom of the social order are more rare than are changes in the mid-ranks.

The system of obtaining social rank by considering the number of active and passive victories and defeats was tested after 130 such pair-contacts had been staged. All five of the albinos were placed in the fighting cage simultaneously. Under these conditions W_1 was dominant and W_5 was most subordinate. At the end of 200 fights the others were intermediate in about the same way in which they are shown in Figure 1. The

[PHYSIOLOGICAL ZOÖLOGY

ranking system, as used, was accordingly judged to be a fairly accurate index of the true order in group dominance.

Mouse W_1 showed no agressiveness in his initial battles. He fought back when attacked and, at first, did not pursue the attacker after he had been driven off. Evidently prefighting aggressiveness in unconditioned mice is not necessarily correlated with success in fighting. As the series progressed, W_1 became more aggressive and at length became and remained the alpha-mouse of this lot. W_2 showed similar tendencies but was not so successful as W_1. After a series of 7 defeats by the latter, W_2 dropped to a position of parity with W_3 and W_6.

The changes in aggressiveness of W_3 and W_6 after 50 or 60 contact reactions with the other mice of this lot did not seem to be a result of actual defeats but rather of the long series of hard-fought battles. This is an exception to our general experience. Both of these mice were relatively aggressive as compared with other members of this strain; they fought longer and harder than any of the other albinos. They were never defeated with the same decisiveness with which they defeated their opponents; yet, after a few relatively mild defeats, both mice began to lose status, which they did not regain. This drop was sharply paralleled by increased aggressiveness on the part of W_1 and W_2 and by reduced passiveness

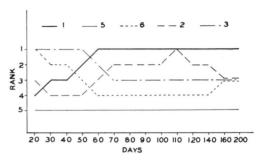

Fig. 1.—The developing social status during 200 intrastrain staged fights among the C albinos.

on the part of W_5 when opposed either by W_6 or by W_3. Even relatively subtle changes in attitude are apparently detected by other mice when the animals in question have frequent contact.

W_5 showed no signs of aggressiveness and was the only one of these mice which never contested the position of the mice with which he was matched. He was 4.5 gm. lighter than any of the other albinos at this period and gained no weight during the entire course of the experiments. He finally showed definite signs of weakness and eventually died. Autopsy revealed a greatly enlarged heart and marked peritoneal edema. The behavior of W_5 can, therefore, scarcely be considered as normal.

THE C-3H AGOUTI STRAIN

In our hands this stock seemed most lethargic in its general behavior. Any one of them might sit in the fighting cage with eyes almost closed, while it shook its head from side to side and paid no attention to its opponent. At times this behavior was so extreme as to suggest impaired vision. In the first encounters these mice were cautiously aggressive and indulged in feints rather than proceeding to actual skirmishes. The early fights were not so full of action or so long or frequent as were the fights between the albinos. It was not until after about 20 encounters that real battles became usual during the 20-minute combat periods. When they did fight, the agoutis either bit harder or were less able to defend themselves, for, in general, their fights resulted in more frequent injuries to one or both antagonists.

Vol. XV, No. 4, October, 1942]

Throughout the series of 200 preliminary contact reactions Br_5 was either definitely first or was tied for that position. At the other extreme, Br_6 either occupied or was tied for the last place except once, early in the series, when it ranked next to last. In the final standing, after 200 contacts, Br_5 was slightly more successful than Br_4; and both of these were decidedly above Br_1, Br_2, and Br_6, which were judged to be approximately even in fighting prowess.

THE C-57 BLACK STRAIN

These mice were in better condition and were seemingly more vigorous than either of the preceding stocks. In general, they were more sensitive to handling and more restless. They were also more rapid in their movements. They sometimes showed less aggressiveness than either of the other two stocks; usually, however, they fought more readily, longer, and more rapidly than did either of the others. Prior to conditioning, no one of our black mice, with the possible exception of Bl_3, which had been severely defeated during Miss Fales's experiments, was defeated in actual combat by a mouse from either of the other strains. The fights among themselves were frequent and fierce from the very beginning of the testing series. No feinting was observed, and each of the 200 contacts staged in the present preliminary series resulted in one or more active attacks. The readiness to fight at the first opportunity was in marked contrast to the other stocks. With this general picture it is somewhat surprising that before conditioning in our hands Bl_3 submitted passively most of the time.

A linear order was established after 80 fights, which remained stable through the following 120 battles. It was as follows: Bl_2, Bl_5, Bl_1, Bl_4, Bl_3. There was an indication that the last three mice were less aggressive than the other two, which was apparent by the end of the first 10 fights. The only change that occurred after the first 50 combats was that Bl_2 replaced Bl_5 as the alpha-mouse. It is worthy of note that all the battles between these two aggressive black mice were fiercer and more intense than were any others in these testing series.

A close study of the behavior of Bl_2 and Bl_5 suggests that the latter may actually have been physically superior as a fighter. He was, however, perhaps more easily intimidated. The first 50 intrastrain combats between the blacks included 5 encounters between these two. The fights were always severe. Bl_2 tired more rapidly than its opponent. One or two added vigorous onslaughts by Bl_5 after Bl_2 was tired would usually suffice for a submission by the latter. In the meantime Bl_2 had been defeating all its other opponents and was becoming more and more aggressive. During the sixth round robin of intrastrain battles (there were 10 fights to a round robin) Bl_2 began to attack as soon as the partition was lifted. In previous pair-contacts there had been some time spent in mutual threats and observations. From that time on, Bl_2 consistently defeated Bl_5, which never attained this degree of aggressiveness. Some reversals of form were observed after short periods of isolation, which extended up to a week in length. In each of these instances Bl_2 did not immediately attack his opponent; on such occasions Bl_5 was the winner until Bl_2 again displayed his usual sharp aggressiveness.

With all three strains the order, once set up, remained relatively stable at both extremes but fluctuated somewhat in the mid-ranks. Unlike the albinos and the agoutis, the blacks soon established a linear order which was maintained. In fact, the differences between the subordinate black mice tended to become intensified as the fights proceeded. With the other two strains, and particularly with the agoutis, the differences in the lower

[PHYSIOLOGICAL ZOÖLOGY]

ranks tended to become less extreme as time went on. It appeared as though the agoutis particularly became more completely submissive to their dominants as a result of successive defeats; and, as they did so, they showed decreased hostility among themselves.

EFFECTS OF CONDITIONING

THE INFLUENCE OF SUCCESSIVE LOSSES ON SUBSEQUENT FIGHTING BEHAVIOR

The C albinos—W_1 had been the alpha-mouse among the albinos for 140 battles when the conditioning experiments were begun; he, himself, had fought 28 times. There was no indication of a threatened shift in dominance; rather, his social status was becoming more firmly established. If this status could be changed following a series of defeats, we should have good evidence that the loss of dominance was a direct result of such experience.

There was a complicating factor which we attempted to overcome: isolation is important in determining fighting behavior. Mice which are regularly submissive often become aggressive if they have been kept isolated from their fellows for varying periods of time. Analogously, dominant animals are not quite so prone to attack a former subordinate after similar isolation. For this reason it is not entirely valid to assume that relative changes in status result from conditioning a particular mouse if the individual in question has had no contact with those against which he is finally tested during the conditioning period, provided that the period is relatively long (several weeks or more).

In order to circumvent the effects of isolation, W_1 was fought against Bl_2 twice a day for 8 days. Bl_2 almost invariably attacked its opponent as soon as the partition was lifted, and he showed signs of aggressiveness (tail-beating, etc.) whenever the partition was rattled. W_1 showed active resistance during the first 3 battles but was never able to hold out against Bl_2 for more than 15 seconds at most, despite the fact that both animals were of approximately equal weight and were each at the top of his own social order. After the twelfth defeat W_1 submitted passively as soon as Bl_2 attacked actively before the end of the exposure period, and the conditioning was thus regular and relatively severe. On the ninth day the albinos were fought among themselves in the manner described for the earlier intrastrain battles.

During the first round robin of intrastrain contacts W_1 submitted passively to every opponent, including even W_5, which had first submitted to W_1. None of these relatively passive albinos showed aggressiveness; hence the uniform submissiveness of W_1 was the more striking. Submissions were usually given only in the presence of an aggressive individual. Bl_3 and W_5 were exceptions to this general rule; they often submitted passively to the rattling of the partition or the proximity of the weighing box or even to a female shortly after her introduction into the cage. The conditioned submission of W_1 was similarly nondiscriminatory. These seem to be unoriented fear reactions rather than typical submission.

After two more series of pair-contacts—a total of 30 such encounters during the same day—W_1 was no longer completely submissive; he still showed no aggressiveness. On the following day he was threatened by W_2 with appropriate tail-beating and hair-raising and replied in kind. After 30 more passive but unsubmissive contacts W_2 again threatened, and he was attacked and promptly beaten by W_1. The battle was severe and lasted 50 seconds. After this fight the old order of dominance was quickly established.

VOL. XV, No. 4, OCTOBER, 1942]

At this point, W_1 was exposed to Bl_3, which, as usual, showed random, nonaggressive behavior. W_1 remained in its corner of the cage in a rigid attitude while Bl_3 explored the cage, and W_1 reared on its hind quarters as Bl_3 approached. Bl_3 also reared and squealed and went through all the signs of submission. During the remainder of the 20-minute period both mice explored the cage freely but reared on their haunches each time they approached each other. This suggested that W_1 might be reacting to Bl_3 as a member of the black strain. It was, therefore, placed with the nonaggressive agouti, Br_6, with which it had had no recent contacts. The same cautious pattern of behavior was shown by both animals. In the presence of the albinos, however, W_1 reassumed its old dominant attitude.

The foregoing behavior suggests that severe conditioning over a short period causes random submissive behavior, which can be interpreted as an unoriented fear reaction. After such conditioning, the mouse was able to readjust itself in a relatively short length of time and to assume its former dominance. If it had met with more active resistance from its own group during the first few postconditioning encounters, it would probably not have reassumed its aggressive attitude so quickly. The effect of conditioning W_2 in the same manner described for W_1 points to this conclusion. W_2 was attacked by W_1 during the early postconditioning intrastrain battles and retained its submissive attitude until after a series of passive encounters with the ultra-meek W_5, which covered a period of 10 days and began 2 weeks after the 8-day conditioning period.

If such an intermediate mouse does not meet any of its superiors after a series of defeats, it, too, is able to readjust fairly rapidly. Thus W_3 was given the usual series of 16 defeats by Bl_2 in 8 days and was then exposed only to the unaggressive W_5 until it resumed its normal behavior to the latter. W_3 was then able immediately to take its former position in the group despite threats from the alpha-mouse, W_1.

W_7 was the dominant animal in a group of four albino males of the following generation. It was treated to the customary series of 16 defeats in 8 days. The results paralleled those described for W_1 in most details. It took slightly longer for W_7 to reachieve its former position after the downward conditioning.

At this point an attempt was made to bring about a lasting change in the order of dominance among the oldest group of albinos. In order to avoid complications resulting from isolation from other members of the group, these mice were fought daily in round-robin fashion. In addition, the alpha-mouse, W_1, was made to meet several different black or agouti mice each day. The number of such pair-contacts varied from day to day. They were always continued until one or two severe defeats were suffered daily by W_1. Intrastrain contacts were arranged each morning; interstrain combats were staged in the afternoon.

Under this sytem of conditioning, the behavior of W_1 varied considerably. Most of the time he showed pronounced pacific tendencies in the presence of black or agouti animals; though he seldom attacked, he usually fought vigorously against an attacker. He seemed to show no discrimination between individual mice. After 2 weeks the other albinos, with the exception of W_5, which was ill, ceased to submit passively to W_1. At this time W_5 was replaced by $W_{5'}$, an albino from a different substrain. $W_{5'}$ had earlier been defeated 3 times in encounters with W_1, but now the entire social order was upset. $W_{5'}$ replaced W_1 as the dominant mouse, and the latter was now on a par with the others.

$W_{5'}$ was removed in a few days in order to simplify the situation, and the encounters continued otherwise according to schedule. W_1 was still being defeated regularly by black or agouti mice and now submitted passively without showing resistance. W_3 became the dominant albino in this lot, and W_1 submitted passively to all his former subordinates. This conditioning series lasted for 35 days, during which time, W_1 had been defeated at least 47 times. In the 2 months that followed, W_1 did not regain his former status and showed no signs of aggressiveness. At first he submitted passively when in the presence of another mouse. He stopped doing so after a time but submitted immediately in the face of any slight show of aggressiveness by another mouse.

The summer quarter intervened. During this period of 4 months W_1 remained isolated. When he was again fought against the members of his group in the autumn, his former aggressiveness had been regained.

The C-3H agoutis.—Br_5, the most successful of the agouti strain, was given the standard series of defeats (16 severe losses in 8 days) and then fought again against its fellows. Br_4 became dominant; it had already been a close second, and Br_5 did not recover its former status until Br_4 was experimentally lowered in the scale.

Br_4, the new dominant, was consistently beaten in contacts with the blacks without interrupting its daily exposure to the other members of its group of agoutis. At the end of 3 weeks of such treatment he was submitting passively to all except the nonaggressive Br_6, which usually submitted faster than he could. Br_5 became dominant once more, and the order remained stable during 10 more days of experimentation.

As with the albinos, an attempt was made to repeat the results with agoutis of the succeeding generation. The outcome was obscured because Br_8, which had previously been judged to be unaggressive, suddenly changed its behavior and dominated the entire group, including Br_7, the former dominant, which was in process of being conditioned downward. The increased meekness of Br_7 after partial conditioning may have stimulated Br_8 to display his aggressiveness. Uhrich's (1938) observations indicate that reversals do occur at times in such groups of mice without known causation; hence, in the absence of more critical evidence the suggested explanation remains a guess.

The C-57 blacks.—The ordinary procedure was modified somewhat because none of our available mice were capable of administering defeats to the dominant blacks. For this reason, Bl_5, which was in beta-position in the social order of the blacks, was fought against the dominant Bl_2, once daily for the first 4 days. It was quite an effort for Bl_2 to defeat Bl_5 during their initial battles. After the defeats became more routine, the two were matched twice daily. After the fifth day, intrastrain fights were also staged daily between all these blacks with the exception of the omega-individual, Bl_3; and the formerly aggressive Bl_5 went to the bottom of the group with which it was being fought in 15 days of such treatment.

Less direct evidence of the effect produced by defeats is furnished by our experience with Bl_2, the most aggressive mouse in our entire colony to date. Despite some relatively serious injuries and a rigorous fighting regime, Bl_2 could always be counted on to attack an opponent fiercely the moment the partition was raised in the fighting cage. If the tip of an opponent's tail chanced to protrude beneath the partition, Bl_2 would attack that. In nearly 200 battles, Bl_2 lost only some 10 times; he was always less prone to attack after having been defeated. Bl_5 showed similar tendencies.

Interstrain tests.—Before conditioning, Bl_5 invariably fought all the males of the C-3H and C albino strains and was victorious in all encounters except one, which was accidentally disturbed. After conditioning he was submissive in the presence of W_3, W_6, Br_5, and

Br_2 and pacific in the presence of W_1, W_2, and Br_6. Br_2 was the only mouse which made an active attack in these postconditioning, interstrain contacts, and Bl_5 submitted passively both before and after the attack.

The effect of successive defeats cannot be ascribed to fatigue or injury. When at all able to do so, Bl_2 was normally fighting from 4 to 6 times daily during this period in addition to intrastrain battles and had, moreover, been injured on several occasions. Three of its injuries necessitated several days of complete inactivity. None of the mice which were showing the effects of defeat were forced through so rigorous a schedule, nor did any of them sustain appreciable, inactivating injuries.

A victorious, favorably conditioned animal will fight successfully against heavy odds, while an oft defeated one of similar constitution and stamina is psychically handicapped to such an extent that it seldom does well even under favorable conditions. Psychic, rather than physical, factors seem most important here.

THE INFLUENCE OF SUCCESSIVE VICTORIES ON SUBSEQUENT FIGHTING BEHAVIOR

The albinos.—The most effective way of testing the effect of successive victories would be to see whether a change in status could be brought about in one of the completely subordinate animals and, if so, how great a change was possible. The difficulty in such experimentation was that, for the most part, such animals would not fight or even threaten an opponent. In an attempt to surmount this difficulty, W_5, the passive omega-mouse of its lot, was placed in a neutral cage with W_6 for several hours daily. W_6 was nonaggressive, and W_5 soon stopped submitting to it. At this point W_5 died.

W_{15} was the omega-mouse in a group of five second-generation albinos bred from our original stock. As usual with these second-generation mice, all those grouped together were litter mates. W_{12} was second lowest, and neither animal was aggressive. They were caged together for several hours daily, as has just been described, with no apparent change in behavior toward each other. At the end of a week W_{15} was mated. The next encounter between the two males came just after an observed copulation. W_{12} pursued W_{15} all around the fighting cage and sniffed insistently at his hind quarters. W_{12} finally nipped at W_{15}, which promptly submitted. At this point, W_{13}, which was approximately on a social par with W_{12}, was introduced. It also pursued W_{15}, which submitted. W_{12} tried to crowd the newcomer away, and a fight ensued. The entire group was matched in round-robin fashion with the exception of W_{12}. W_{13} was beaten in a routine fight and was rematched immediately with W_{12}, which attacked after W_{13} had first submitted passively.

On subsequent days this practice was continued. W_{12} was omitted from the intrastrain battles but was matched with a mouse of its group just after the latter had been beaten. After a week and a half of such a "build-up" W_{12} became quite aggressive and defeated W_{11} (on a par with W_{13}), W_{15}, and W_{13} regularly, regardless of whether they had been beaten previously. At the end of 18 days of such treatment W_{12} defeated the dominant mouse of its group (W_{14}), from which it had been isolated for the entire period. It must be noted that W_{14}, for all its dominating position, was not generally aggressive. It dominated by threats after a very few battles. The fight with W_{12} was the first severe combat in its history. Even so, the results show that the aggressiveness of W_{12} was strongly influenced by its run of successful fighting.

The C-3H agoutis.—Our even more striking experience with conditioning Br_6, the least aggressive agouti, so that he became a vigorously aggressive individual, has been

[PHYSIOLOGICAL ZOÖLOGY

told elsewhere (Allee, 1942a, 1942b). He not only became the dominant mouse in his own group as a result of some 6 weeks of careful social manipulation, but he also became aggressive enough to give Bl_2 a long hard fight—so hard, in fact, that Br_6 died as a result.

The C-57 blacks.—All of the techniques which were used with the albinos and agoutis were tried with Bl_3, which was the least aggressive of these blacks. This animal was sensitive to handling and did not fight under any of the conditions mentioned. Finally, he was isolated for 5 weeks. His first encounter after isolation was with Bl_4, which had formerly ranked just above him. Both moved about the cage for several minutes: Bl_3 then sniffed at his relatively mild opponent, which rose on his haunches and squealed. Both mice explored opposite corners of the cage for awhile. Then they touched vibrissae; and Bl_4 squealed, ran, and reared up in a corner. Bl_3 approached and groomed it for several seconds. Again Bl_4 ran, squealed, and reared up as before. Bl_3 followed cautiously, waited until Bl_4's back was turned, and attacked. Submission was prompt; Bl_3 attacked once more before the partition was interposed between them.

Bl_3 then attacked and defeated Bl_1 when the two were matched after the latter had recently been beaten by the dominant Bl_2. In the meantime, Bl_2 suffered one of its rare defeats by Bl_5, and Bl_3 was matched with him. There were 3 severe fights in which both mice attacked practically simultaneously. These fights were stopped before a decision was reached, in order to shield Bl_3 from possible early defeat. Later in the day Bl_3 defeated Bl_5 in a single severe battle.

A week later the battles were repeated, with the same results. Then the animals were fought daily, and Bl_3 again went to his earlier place at the bottom of the social order after 4 more days of fighting. He was then isolated for 8 weeks and was again gradually "built up" by being matched with low-ranking agoutis, a method which had proved effective with Br_6.

After only 3 days of such opposition he again became obviously aggressive. He was matched against each of the three bottom blacks of his group and beat them all. At the end of the week, Bl_3 was again put in with Bl_5, which had just been beaten by his superior, Bl_2, and won. The interval between the defeat of Bl_5 and the latter's being matched with Bl_3 was gradually increased to 3 hours, and Bl_3 still won. Bl_3 was then matched with the potent Bl_2 approximately 15 minutes after the latter had had a winning fight with Bl_5. Bl_3 was fresh and won. Two days later Bl_3, which had fought and won from all the blacks in the meantime except Bl_2, was again matched with this hard-fighting mouse. Both were fresh, and there was a real fight. Bl_3 had become extremely aggressive and defeated Bl_2 in a series of pitched battles.

The next series was fought 2 days later. It involved 30 minutes of intermittent fighting between Bl_3 and Bl_2. Both animals appeared to be completely exhausted several times during the encounters. Bl_2 finally won a complete victory. Bl_3 was passively submissive to any mouse of any strain with which he was matched from that time to the end of the experimental period 16 days later. At the time of his winning the dominant position, Bl_3 was the heaviest mouse in its group by a slight margin.

Isolation, or some other device to overcome the reluctance of low-ranking animals to fight, appears to be a necessary prerequisite for their upward conditioning. The normal reluctance may be partly a matter of memory and of fear, both produced by previous defeats. It is far easier to reduce the aggressiveness of a superior animal by a series of defeats than it is to get an animal at the bottom of his social order to show signs of aggressiveness. The latter process is slow and painstaking, and the experimenter must be continually alert and must modify his procedure to meet suggestions given by the slight behavior clues which indicate the degree of progress his subject is making.

DISCUSSION

It is difficult to summarize adequately such complex behavior patterns as those which we have been considering. The different categories shade off from their modes. There are exceptions to be evaluated and individual variations of predominant patterns to be considered, even when they are not described in the final account. Probably no one except the observer himself could get a clear picture or even an approximately accurate impression of the whole matter at hand, even though the account were to be spun out far beyond publishable length. Such troubles increase when attempts are made to tabulate observations in definite classes. We hope that readers will appreciate our difficulties. We have tried to be conservative in our statements, and we have seen many fights, the results of which are barely suggested in the foregoing pages. Despite this extensive experience, we feel that we have merely made a modest sampling of the fighting behavior of these mice, not that we have reached definitive conclusions.

This study is one of an informal series, with which the second author is concerned, in which an analysis is being made of factors which make for social dominance as contrasted with social subordination. Without going into details, which would differ with the animals studied, we know that the following factors are involved, at least in many cases:

1. *Physiological state.*—This factor includes such matters as fatigue, which has been seen to be effective in some of the fights among these mice—so much so that when we were trying to condition a mouse to be more aggressive it was standard practice to pit it against another that had just been in a hard fight. Illness also tended to reduce aggressiveness, as did severe injuries with most mice.

Under the conditions of these experiments starvation for 12-, 24-, and 36-hour periods made no difference in aggressiveness. Only once was there a fight over food, and that did not occur during or soon after a starvation test. Witholding water for 6 or 8 hours from isolated or communally penned mice and then introducing a water bottle with a small nozzle, so that only one mouse could drink at a time, did not stimulate fighting.

Temperature is another easily controlled factor which affects the physiological state of mice as well as other animals. We found that fights were possible over a temperature range from 18° to 28° C. The combats were more sluggish and shorter at the higher temperatures; and inexperienced, unaggressive mice did not fight at 27°–28° C. Wider variations in range were not tried. There was no differential response to temperature on the part of the various strains that was detected by our methods.

The action of various hormones may also affect the physiological state and so have a controlling action on aggressive behavior. This had been demonstrated for the androgen, testosterone propionate, with a number of different kinds of animals (cf. Allee, Collias, and Lutherman, 1939). Such hormones supposedly affect aggressiveness by acting through the nervous system. These strains of mice which differ in innate aggressiveness appear to offer promising material for hormonal experimentation.

2. *Weight.*—Uhrich (1940) reported suggestive evidence that differences in weight tended to make an effective difference in the winning of fights. The heavier mouse won, in his series, 6 times and lost 1 time when weight differences were as much as 5 gm. (approximately 15–20 per cent).

[PHYSIOLOGICAL ZOÖLOGY

In our studies we have comparable data regarding the relation between weight and success in preconditioned, intrastrain fights for 53 mice. These included 22 blacks, 12 agoutis, and 19 albino mice. The departure of weight from the mean for the strain was as much as -5 gm. with one mouse and -4 gm. with another; the others all ranged between -3 and $+3$ gm. from the average for the given strain. The coefficient of correlation between success and weight in these cases is positive and slight ($r = 0.2852 \pm 0.1356$ S.E.); however tested statistically, this degree of correlation is just on the border line of validity. A larger series of animals is needed for a conclusive test.

3. *Heredity.*—There seems little room for doubt that some of the factors which make for dominance can be inherited. The greater power to win fights by the black strain, as contrasted particularly with the more passive albinos, makes this point. The distinction between either of these and the intermediate agouti strain is also clear-cut. The strain differences which determine success in fighting have not been separated either genetically or, for that matter, by controlled, analytical experimentation. Several factors can be recognized which contribute to success. In addition to possible hereditary differences in physiological state and weight, there are also aggressiveness, ability in the attack, ability in the defense, and general stamina, which are easily seen to be important.

Aggressiveness toward the keeper differs from aggressiveness toward other mice. All three people who have cared for this colony—Miss Fales, Mr. Ginsburg, and now Miss Elizabeth Beeman—have been bitten by the albinos. The blacks are most pacific, according to this criterion; and the agoutis are intermediate. Aggressiveness in mice is a complex quality which needs further careful analysis.

The albinos tended to stop and explore the fighting cage even in the presence of another mouse. To some extent the agoutis showed the same behavior, while, in contrast to both, the blacks usually attacked. Before being conditioned, they had a strong tendency to attack even if low in the social scale and if easily beaten.

In view of the varying fortunes in the series of fights between Bl_2 and Bl_5, which have been given in a previous section and which finally resulted in the emergence of Bl_2 as the dominant animal of our whole colony, it cannot be unequivocally stated that the physically superior mouse becomes the dominant one. If the animals are normal and of comparable weights, even when unconditioned, natural dominance depends on a number of traits. One of these is aggressiveness.

We are not prepared to evaluate fully the role of aggressive behavior in the winning of fights. Its importance as a contributing factor is readily apparent. Aggressiveness, apparently, is an inherited trait as far as initial contacts go. It is, however, subject to great modification over varying periods of time, after conditioning. In general, our evidence shows that unconditioned fighting ability is a complex strain character not necessarily determined by any single gene (since the strain differences are largely multiple-factor differences) and subject to great environmental modification.

It has previously been mentioned that the agoutis sometimes showed more aggressiveness in initial encounters than did the blacks, but the latter invariably won such pair-contacts. This does not refute the importance of the role played by general aggressive behavior, for two reasons: (1) Decisive initial encounters are usually settled through actual battles if the animals are exposed to each other for a sufficient length of time. Submissive responses to aggressiveness do not usually occur until after several defeats, and this is an indication that learning is an important factor in such submissions. (2) The aggressiveness displayed by the agoutis before initial fights was confined to hair-fluffing and tail-beating. The aggressiveness displayed by superior mice after fighting experience consists of a direct and vigorous attack, the results of which, through repeated experiences, may well come to be associated with pain and defeat by the inferior mice. This is the combination which presumably determines the social order in nature, if, in fact, it exists there among populations of mice.

Elton (1942, p. 205) comments on fatal attacks in small caged colonies of voles (*Microtus*) and states: "This social antagonism among voles probably exists in nature, and may be very important at different levels of population, and may prove to explain some of the mortality at higher numbers."

There are other phases of behavior—nest-building proclivities, for example—which differ in the different strains we have studied. This is not the place to discuss further the relations between heredity and behavior in these mice; nor, indeed, is this our problem. We have primarily used the hereditary differences as originally discovered by others as a tool with which to find evidence of the effect of conditioning on dominance-subordination patterns of behavior in these mice.

4. *Conditioning*.—It is somewhat surprising how rapidly the effect of previous failure in fighting can be detected in the fighting behavior of a given mouse. This is more noticeably true with mice which are intermediate in the social order of their group. Such mice are already partly conditioned to win and partly conditioned to lose by their routine contacts; hence it is easier to cause an intermediate mouse to lose status than it is to do the same with a dominant individual, and, on the other hand, it is easier to condition an intermediate mouse so that he advances in his social scale than one which stands at or near the bottom of his own group.

As yet, we have only a small number of cases in which mice have been conditioned to become more aggressive in behavior and more successful in fighting. We are impressed by the results obtained, because the mice which were lowest in the social scale often showed extreme subordination. Their responses were unorganized in the fighting cage and suggested strongly behavior disintegration similar to that described for rats (Anderson, 1941; Hall, 1941). Hence we find the conditioning upward of such mice as Bl_3, for example, particularly impressive and convincing.

Another effect of conditioning is worthy of note. The correlation between aggressiveness, as shown by starting a fight and success in fighting, becomes much higher as a result of conditioning.

As with many other biological processes, including other phases of conditioning, the retention of the effects of training depends in part on its duration. This was best shown by our experience with the effects of continued defeats. If such conditioning was freed from the counteracting effects of isolation by frequent contacts within the group, then long-term treatment had relatively lasting effects. Severe conditioning for a short period altered the behavior of a dominant animal for a relatively short time only. Such short-time conditioning treatment had a more lasting effect on an intermediate animal, since it was really extended, in such an instance, through defeats administered by superior animals in its own group after the experimental conditioning was over. Where this factor was experimentally controlled, an intermediate mouse recovered from the effects of a short series of defeats as rapidly as did a dominant individual.

Conditioning can also be accomplished by rearing young mice together in the same cage. For controls, one contrasts their fighting behavior with that of mice which have been reared apart. Our experience on this point is not extensive, but we do have indications with the albino and the black strains that the fighting tendencies in such communally reared mice are less than if they were reared, from weaning on, in isolation. These indications hold even in the more provocative situations in which fertile females are present.

Two other factors which have been reported to affect social dominance are: (1) what may be called the "home-cage effect," which is, at least in a general way, an aspect of territoriality, and (2) age. We did not find that limited age differences were important when they were divorced from differences in weight of over 5 gm. and also from differences in experience in fighting. Our tests did not include differences of more than 3 or 4 months. In the strains which were regularly tested either in intrastrain or in interstrain pair-contacts, the mice were all nearly of the same age; hence, for the purposes of these experiments this factor was effectively eliminated. The possibility of a home-cage effect was also eliminated, since the contests were staged on neutral ground except in those cases in which, during a conditioning series, we were attempting to favor one of the mice. In our hands and with these mice, we did not find any strong evidence that the home cage exerted great influence on the outcome of a fight unless the home cage contained a female. The presence of such a female, then, becomes the last of the factors that make for dominance with which we have worked in these experiments.

The behavior of these animals before and after conditioning suggests a certain analogy to the conditioned reflex in Pavlov's sense (1927, p. 26), though a more complex category of nervous activity is being dealt with. In Pavlov's words:

The fundamental requisite [to establish a conditioned reflex] is that any external stimulus which is to become the signal in a conditioned reflex must overlap in point of time with the action of an unconditioned stimulus. . . . . It is not enough that there should be over-lapping between the two stimuli, it is also and equally necessary that the conditioned stimulus should begin to operate before the unconditioned stimulus comes into action.

The most generalized unconditioned stimulus for fighting in an aggressive male mouse is the proximity of another male mouse. Aggressive deportment before a battle, if it comes to be associated with defeat, soon elicits the submissive response, even if no battle has taken place. Likewise, the transfer to the fighting cage before the battle also provokes a certain type of fighting behavior, depending on the previous experience of the animal in the fighting cage. If the mice are allowed to remain in the fighting cage under the experimental conditions centering about the battle (i.e., raised partition, top lighting, and the presence of another male mouse) and if they remain pacific during several such encounters, it soon becomes difficult to elicit the fighting response from many of them. The fighting cage has lost its efficacy as a conditioned stimulus for fighting, and the animal responds with a random behavior pattern, or even by the social action of grooming its "opponent." For these reasons, care must be taken to insure active battles for all mice that are to respond belligerently under the experimental conditions, even if short periods of inactivity must be interpolated to accomplish this. In the absence of such precautions the number of active battles is greatly diminished.

Perhaps the analogy to the reflex will help explain several obscure points having to do with fighting behavior under the conditions described. Ordinarily, two animals are promptly separated by the partition after one of them has submitted, and no further fighting can then take place. Infrequently, an intermediate animal will turn on a "superior" adversary directly after having been defeated, and the "victor" may than submit without a battle. During the course of these experiments this was never observed when the victor was a very successful animal at the top of its own order, but it was observed in several instances involving battles between intermediate mice when the partition was not interposed promptly between the contenders. If it is true, as these and other experiments indicate, that mice respond to deportment rather than to individuality as such, it may well be that the foregoing can be accounted for as follows:

The less successful animal submits, either after an actual battle or after exposure to the aggressive behavior which it has come to associate with defeat and which may be loosely analogized to a conditioned stimulus. After submission the victor retires to an unoccupied portion of the cage and ceases to show signs of aggressiveness, having been partly conditioned to this type of behavior by the fact that the animals are normally separated directly after submission. This may provide a weakly conditioned stimulus for the mouse that has just submitted to attack an unaggressive victor, if the former was not too decisively beaten in the previous encounter. This hypothesis has not been tested experimentally, and it is difficult to push the analysis further until more data make it possible to subdivide fighting behavior into various aspects, especially with respect to the following: (1) the effects of handling, (2) prefighting aggressiveness, (3) initiating a fight, (4) the fight proper, (5) submission, and (6) behavior after submission.

In the absence of a more adequately diagnostic term, we propose to refer to the type of behavior and behavior changes that we were able to induce as a "conditioned automatism."

Finally, we need to emphasize that within each strain the mice show decided individuality as regards both aggressiveness and fighting techniques. Some of the individual differences we can trace back to previous conditioning; others we cannot. While evidence on this point is far from complete, there are strong indications that the young mice reared in our laboratory showed individual variations which resulted from differences which the animals possessed at birth. Such differences may, or may not, be truly genetic. We remember that the Dionne quintuplets (Blatz, Millichamp, and Charles, 1937) showed clear evidence of differing among themselves in social status. Presumably, these were an identical set of sisters, each with the same heredity as the others. Their differences may well have arisen from their relations in and to their intrauterine environment. Such differences would also be expected among mice. The extent and importance of such fetal interrelationships and particularly their bearing on adult behavior, especially on social behavior, is an intriguing problem.

With mice which are not litter mates, even with uniform handling, controlled external conditions, and close inbreeding, there are a number of subtle factors which may act to produce individual differences in behavior.

In the guinea pig, Wright (1926) has shown that, though the white spotting pattern is determined largely by one pair of alleles, the age of the mother is an important factor in establishing the extent of the spotted areas. The data of Castle (1936a and b) have demonstrated a decided effect of the size of the mother on birth weight and litter size. Bittner (1937, 1939a) has shown that cancerous strains of mice will transmit a tendency to form mammary cancers to noncancerous strains which they foster. This tendency is probably resident in the mother's milk. Wright (1934) has demonstrated the best-known case for a strict maternal-age effect acting independently of the known gene constitution of an animal. His data demonstrate conclusively that age of the mother, per se, is responsible for the relative changes in frequency of the numbers of digits produced in genetically uniform material.

Thurstone and Jenkins (1931) have made a study of such factors as are expressed in human material. Their data include ten thousand cases, and they seem to have avoided most of the common pitfalls that usually attend such investigations. They found a negative correlation between growth and birth order; a significantly high proportion of cases of pyloric stenosis, tuberculosis, criminality, epilepsy, schizophrenia, feeble-mindedness, albinism, congenital cataract, and genius among the first-born. Mongolism was most frequent among the last-born. There was a general progressive increase of I.Q. from the first- to the eighth-born. The age of the parents did not prove to be a factor in these I.Q. studies.

In general, it should be emphasized that maternal and other extra-genetic influences on genetic and seeming genetic characters are manifold and well established; and, though it seems logical to refer unconditioned behavior traits to genetic differences, it is impossible to postulate the nature of these differences without resorting to a rigorous and exacting demonstration of any alleged premises.

SUMMARY

1. Definite genetic behavior differences exist between highly inbred strains of mice, some of which are consistently demonstrable under specified conditions. In the strains studied, the C-57 blacks were superior, the C-3H agoutis were intermediate, and the C albinos were least effective, as judged by the outcome of actual battles. Other consistent differences were also noted.

2. Superiority in fighting appears to result from a multiplicity of psychic and physical factors. The mice in a small group show a social organization which is based on dominance-subordination relations, determined in the last analysis by superiority in winning fights.

3. The order of dominance in both isolated and communally penned mice varies from group to group and has temporary stability. It is most stable at the extremes. Relatively stable social structures may appear after many battles.

4. Individual mice may be conditioned to be less aggressive as a result of repeated defeats. They can also be made more aggressive as a result of continued victories over submissive mice. Both types of conditioning strongly affect the ability of a given mouse to secure or to maintain high social status.

5. It is far easier to condition a socially superior mouse downward in the social scale by a series of defeats than it is to condition a socially inferior mouse upward. Individuals which are intermediate in their social position are most readily conditioned in either direction.

6. Factors known to be of importance in determining social dominance include: physiological state; differences in weight, if over 5 gm. (15–20 per cent); hereditary differences in aggressiveness and fighting ability; and conditioning.

7. The presence of a female not only stimulates fighting among otherwise passive male mice, but she also tends to stimulate her mate to make a successful defense of his home cage or of any other cage into which the pair was introduced prior to the entry of an intruder.

8. Because of the small number of animals worked with (some forty-five were studied intensively) and because of the standardization of procedures, no finality is claimed for the results which have been presented. Comparison with other work suggests a certain amount of universality.

LITERATURE CITED

ALLEE, W. C. 1942a. Group organization among vertebrates. Science, 95:289–93.
———. 1942b. Social dominance and subordination among vertebrates. A chapter in Levels of integration in biological and social systems. Lancaster, Pa.: Jacques Cattell Press.

ALLEE, W. C., and COLLIAS, N. 1940. The influence of estradiol on the social organization of flocks of hens. Endocrinology, 27:87–94.

ALLEE, W. C.; COLLIAS, N. E.; and LUTHERMAN, C. Z. 1939. Modification of the social order in flocks of hens by the injection of testosterone propionate. Physiol. Zoöl., 12:412–40.

ALLEE, W. C., and GINSBURG, B. 1941. Some effects of recent experience on aggressiveness in male mice: a study in social dominance. Anat. Rec., Suppl., 81:50.

ANDERSON, E. E. 1941. Sex differences in timidity in normal and gonadectomized rats. Jour. Genet. Psychol., 59:139–53.

BITTNER, J. J. 1937. Mammary tumors in mice in relation to nursing. Amer. Jour. Cancer, 30:530–48.

———. 1939. Relation of nursing to the extra-chromosomal theory of breast cancer in mice. Ibid., 35:90–97.

BLATZ, W. C.; MILLICHAMP, D.; and CHARLES, M. 1937. The early social development of the Dionne quintuplets. "University of Toronto Studies of Child Development," Ser. No. 13. 40 pages. Or in Collected studies on the Dionne quintuplets. Toronto: University of Toronto Press.

BRAINERD, BARBARA HALE. Social organization of geese and ducks under semi-natural conditions. Manuscript in hands of W. C. Allee.

CASTLE, W. E. 1936a. Size inheriteance in mice. Amer. Nat., 70:209–17.

———. 1936b. Superior influence of the mother on body size in reciprocal hybrids. Science, 83:627.

CASTLE, W. E.; GATES, W. H.; REED, S. C.; and LAW, L. W. 1936. Studies of a size cross in mice. II. Genetics, 21:310–23.

CREW, F. A. E., and MIRSKAIA, L. 1931. The effects of density on an adult mouse population. Biol. Gen., 7:239–50.

DIEBSCHLAG, E. 1941. Psychologische Beobachtungen über die Rangordnung bei der Haustaube. Zeitschr. f. Tierpsychol., 4:173–87.

ELTON, CHARLES. 1942. Voles, mice and lemmings: problems in population dynamics. New York: Oxford University Press. Pp. 496.

FALES, C. H. 1940. Fighting behavior in three inbred strains of mice. Term paper deposited with W. C. Allee, University of Chicago.

GREENWOOD, M.; HILL, A. B.; TOPLEY, W. W. C.; and WILSON, J. 1936. Experimental epidemiology. Med. Res. Counc. Spec. Rept. Ser., No. 209. Pp. 204.

HALL, C. S. 1941. Temperament: a survey of animal studies. Psychol. Bull., 318:909–43.

HALL, C. S., and KLEIN, S. J. 1942. Individual differences in aggressiveness in rats. Jour. Comp. Psych., 33:371–83.

JENKINS, DALE. Territory as a result of despotism and social organization as shown by geese. (In press.)

KEELER, C., and KING, H. D. 1941. Multiple effects of coat color genes in the Norway rat, with special reference to the "marks of domestication." Anat. Rec., Suppl., 81:48.

LORENZ, K. Z. 1935. Der Kumpan in der Umwelt des Vogels. Jour. f. Ornith., 83:137–213, 289–413.

MOURER, O. H.; KORNREICH, J. S.; and YOFFE, I. 1941. Competition and dominance in rats. Biological Abs., 16, Item 829. Original study a film deposited with the Psychological Cinema Register, Lehigh University at Bethlehem, Pennsylvania.

ODUM, E. P. 1941. Annual cycle of the black-capped chickadee. Auk, 58:314–34.

PAVLOV, I. P. 1927. Conditioned reflexes. New York: Oxford University Press. Pp. 430.

RETZLAFF, E. 1938. Studies in population physiology with the albino mouse. Biol. Gen., 14:238–65.

RUSSELL, W. L. 1941. Chapter in Biology of the laboratory mouse, ed. G. D. Snell, pp. 325–48. Philadelphia: Blakiston Co.

SCHJELDERUP-EBBE, T. 1922. Beiträge zur Sozialpsychologie des Haushuhns. Zeitschr. f. Psychol., 88:225–52.

———. 1940. Hereditary differences in social behavior (fighting males) between two inbred strains of mice. Anat. Rec., Suppl., 78:103.

SCOTT, J. P. 1942. Genetic differences in the social behavior of inbred strains of mice. Jour. Heredity, 33:11–15.

THURSTONE, T. L., and JENKINS, R. L. 1931. Order of birth, parent-age and intelligence. Chicago: University of Chicago Press. Pp. 135.

UHRICH, J. 1938. The social hierarchy in albino mice. Jour. Comp. Psychol., 25:373–413.

———. 1940. The effect of fighting experience on albino mice. Ecology, 21:100–101.

WHITMAN, C. O. 1919. Posthumous works of Charles Otis Whitman. Ed. H. A. Carr. Vol. III: Behavior in pigeons. Carnegie Inst. Pub. No. 257. Pp. 161.

WRIGHT, S. 1926. Effects of age of parents on characteristics of the guinea pig. Amer. Nat., 60:552–59.

———. 1934. An analysis of variability in number of digits in an inbred strain of guinea pigs. Genetics, 19:506–36.

Reprinted from *Amer. Naturalist,* **77**(773), 519–538 (1943)

STATISTICAL ANALYSIS OF FACTORS WHICH MAKE FOR SUCCESS IN INITIAL ENCOUNTERS BETWEEN HENS[1]

DR. NICHOLAS E. COLLIAS

UNIVERSITY OF CHICAGO

THE basic importance of initial encounters, among certain animals at least, in many different major groups of phenomena (Collias, 1943) makes it seem important to analyze the factors deciding the outcome of such encounters. To make such an analysis I have taken advantage of the remarkable fact that the first thing two strange hens will do upon meeting is to settle their future dominance relations either by fighting or by passive submission of one bird. Once established the social order is very stable and is easily ascertained at any time by noting which hens peck which without being pecked in return. The social order has been observed to persist a year or more with no change in the pecking order.

METHODS AND RESULTS OF ANALYSIS

Two hundred pair contacts were staged in a neutral pen using normal moderately inbred white leghorn hens from different flocks. These contests were conducted over a period of three years, and their original purpose was merely to serve as controls in experiments designed to test the influence of various injected hormones on the out-

[1] This article represents part of a thesis submitted to the faculty of the University of Chicago in partial fulfilment of the requirements for the degree of doctor of philosophy. I am indebted to my special advisor, Professor W. C. Allee, for suggesting that I work on this problem for a thesis, for facilities needed and valuable comments on the work. No claim is made by me to professional qualification as a statistician, and without the aid generously furnished by Professor Sewall Wright, who suggested and supervised the statistical procedure, this study would not have been possible. I wish also to knowledge assistance by Miss Catharine Z. Lutherman in gathering some of the data, and to express my appreciation to Dr. A. M. Guhl for permission to cite from his unpublished work.

come of initial pair contacts. The care and housing of the birds and the method of conducting the encounters have been described in a previous publication (Allee, Collias and Lutherman, 1939). Briefly, two hens from different flocks were caught, weighed, state of moult was noted, position in the social order in the home flock was recorded, and standard comb measurements (length plus height) were taken. The birds were then placed simultaneously in a strange pen purposely reduced to half size to insure closer association. The time of the latent period before contact and the length of the fight, if any, were recorded along with notes on general behavior of the birds toward each other. The statistical methods used in analyzing the results are described below.

By artificially increasing certain variables, it is often possible to discover which are real factors in deciding the outcome of initial encounters. This makes it possible to control those factors which can be readily controlled, and by the application of statistical analysis to evaluate the others and so to reach some idea of the relative importance of the different factors in a fairly normal situation. It is the extent to which prediction can be attained that is important, and the finer the discrimination possible, the more nearly will rules apply to the vast majority of individuals which cluster about the modes of species variability rather than merely applying to the less numerous variants at the extremes. The present report describes the progress attained toward this objective as related to success in initial encounters.

A. *Nature of the Encounters*

In a typical fight the birds first become oriented with respect to one another; as they examine each other the face becomes red and flushed, the neck hackles rise, the tail becomes more erect and the wings droop. The birds may then jockey for a favorable position or at once leap up and at each other and slash towards the head with the beak in an attempt to seize and bite their opponent's comb or wattles. During the heat of battle the birds pay absolutely no attention to a human observer. The fight often lasts but a few seconds. The loser retreats and seeks to escape, its face pales, its feathers are depressed, it appears to be confused and panic stricken and looks for a place to hide, especially if closely and viciously pursued. The winner maintains much the same attitude as during the fight and only gradually does its excitement subside. Inhibition of the attack is based on experience, as well as subordination, since aggressive cocks or hens as a rule attack any strange individual, whereas attacks on flock mates are generally confined to relatively mild pecks. Fighting among hens, while often very strenuous, is much less severe than among cocks, and in the present series of encounters dominance relations were more often decided by passive submission of one bird than by an active battle.

B. Controlled Factors

1. *Sex.* As a rule cocks dominate hens when full grown (Schjelderup-Ebbe, 1935). Only hens were used in these experiments.

2. *Territorial defense.* Of 1,428 first meetings of domestic fowl observed over a period of 10 years Schjelderup-Ebbe (1922) noted that the home bird "wished" to fight in 93 per cent. while the strange bird "wished" to fight in only 32 per cent., and the home birds won 62 per cent. of the 476 battles observed. The courage of the new birds seemed lessened by their surroundings and they fought with less vigor. All the encounters reported in the present study were conducted in a neutral pen which was quite similar to a part of the home pens of the birds concerned.

3. *Social facilitation.* There is evidence that the presence of a powerful familiar despot betters the chances of its subordinates in contacts with strange individuals (Dr. A. M. Guhl, personal communication). Since in the present experiment never more than two birds were intro-

duced into the fighting pen at the same time, this factor does not complicate the results.

C. *Statistical Analysis of Uncontrolled Factors*

These factors included the following readily measured features: (1) size of comb as an indicator of male hormone output; (2) weight as a partial indicator of strength, relative size, including "impressiveness" and general physiological state; (3) social rank in the home flock as a presumed indicator of the "psychology of success"; and (4) state of moult.

Previous experience had demonstrated that very unaggressive hens could readily be caused to win all their encounters with normal hens and to rise in the social order by injecting them with testosterone propionate (Allee, Collias and Lutherman, 1939). Possibly size of comb has some "impressive" or bluff value (*ibid.*). However, Dr. A. M. Guhl (unpubl.) in this laboratory recently found that treatment of dubbed hens with testosterone caused such hens to rise in the social hierarchy; local application of this androgen induced considerable increase in comb size of normal hens, but had no influence on social rank. It had also been found that doses of thyroxin large enough to cause moulting and a marked reduction in body weight as well as to inhibit the gonads would considerably reduce the chances of a hen of winning its initial encounters (Allee, Collias and Beeman, 1940).

Size of comb was recorded in millimeters, weight in ounces, moult by arbitrary grades and social rank in terms of numbers of subordinates. For purposes of statistical analysis moult of the winner was simply treated as greater or less than the moult of the loser, while rank was adjusted for difference in size of flock by a transformed scale making use of inverse probability functions. For the latter purpose use was made of the following formula given me by Professor Wright.[2]

[2] Prf $\frac{x}{\sigma}$ is the area of the unit ($\sigma = 1$) normal curve between the mean (0) and x/σ and prf^{-1} is the symbol for the inverse function and gives the devia-

$$\text{Adjusted rank} = \text{prf}^{-1}\left(\frac{\text{Number pecked} + \frac{1}{2}}{N} - \frac{1}{2}\right)$$

We are assuming that the bird is located at a point which dichotomizes a normal distribution in such a way that the subordinates are on one side and the dominants are on the other. The bird itself is assumed to contribute equally to both sides. The proportion of subordinates is thus taken as $\dfrac{\text{number pecked} + .50}{\text{size of flock}}$. The proportion of subordinates between the mean of the hypothetical scale and the rank of the given bird is then $\dfrac{\text{number pecked} + .50}{\text{size of flock}} - \dfrac{1}{2} = \text{prf R}$. The inverse function $(\text{prf}^{-1}\,R)$ gives us the value of the adjusted rank on the hypothetical scale. The following diagrams may help to make this clear.

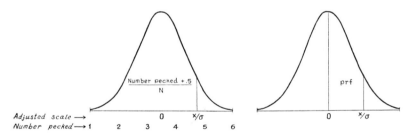

FIG. 1. Diagrams to illustrate statistical adjustment of rank in peck order to size of the home flock.

In a flock of 7 birds (maximum of 6 subordinates for highest bird) the adjusted rank for a bird which pecked 5 subordinates would be $\text{prf}^{-1}\left(\dfrac{5 + .50}{7} - .50\right) = .79$. This formula tends to bunch together the birds near the center of the social scale on the assumption that the underlying factors which make for success would exhibit a normal distribution if a large enough population were considered. All the statistical methods used in this report were based on the same assumption. All modifications of the usual

tion (in terms of σ) that underlies the designated area. Tables of these functions are provided by Davenport and Ekas (1936).

statistical methods which are used were kindly provided to me by Professor Wright.

The first step was to correlate each factor with success. The two hundred fights, rather than the individual birds, have been taken as the units throughout in obtaining correlation coefficients. A preliminary tabulation indicated that even in contests conducted between the same pair of individuals at different times the results were frequently reversed. It must be remembered that these hens came from different flocks and did not associate with each other between these staged fights. If successive pair contacts are well spaced in time, each becomes essentially an initial contact. Hence such variability is not incompatible with the relative permanence of social order in stable flocks of hens. None of the data were grouped.

The variables that have been considered are the differences between the birds involved in the fight, not the absolute measurements. Each fight may be considered from the view-point of either bird as the one whose victory or defeat is to be used in describing the results and from whose grades in the various characters, those of the other bird are to be subtracted to obtain the differences. Thus a fight in which bird A, weighing 3 lbs. 8 oz., defeated bird B, weighing 3 lbs., may be entered in the correlation table relating success and difference in weight either as a "victory" associated with a difference of + 8 oz. or a "defeat" associated with a difference of – 8 oz. As this is an arbitrary matter, it is legitimate to make entries from both viewpoints, thereby making the marginal frequencies of the correlation table symmetrical (as in the case of a fraternal correlation table). In the correlations involving success, the complete table would have 200 entries in the category "victory" and 200 entries in the category "defeat." The total distributions of the differences in weight, comb size, rank and state of moult are necessarily symmetrical about zero.

In determining the signs of the entries, it is assumed that the winner of a fight stood higher than the loser on a hypothetical scale describing the array of factors making for success. Thus "victory" is treated as positive, "defeat" as negative. A difference in rank is treated as positive if the first bird (*i.e.*, the one whose grade is to be treated as minuend) stood higher in the peck order of its flock than the second, according to the scale discussed above. With regard to the state of the plumage, it has seemed best to treat freedom from moult as highest in the scale and extreme moult as lowest. In this way, the differences which one might expect to find associated with success (*viz.*, those that go with greater weight, greater comb size, higher rank and relative freedom from moult) should exhibit positive correlations with success if, as turns out to be the case, this expectation is realized.

The correlations between success and the differences in the graded characters, weight, comb size and rank, were obtained by Pearson's coefficient of biserial correlation (in the form given by Kelley, 1923).

Let S be the difference between a first and a second bird on a hypothetical scale based on the factors making for success in fighting. A positive value implies that the first bird won, a negative value that it lost. It is assumed in the theory that the differences are distributed normally on this scale.

Let W be the weight of the first bird minus that of the second.
Let W_1 be a value of W in a case in which the first bird won.
Let W_2 be a value of W in a case in which the first bird lost.
$\overline{W}_1 = \Sigma W_1/200$, $\overline{W}_2 = \Sigma W_2/200$ are the means.
If each fight is entered in both ways $\overline{W}_1 = -\overline{W}_2$, $\overline{W} = 0$.
Thus $\sigma_w = \sqrt{\Sigma W^2/400} = \sqrt{(\Sigma W^2_1 + \Sigma W^2_2)/400} = \sqrt{\Sigma W^2_1/200}$ is the standard deviation of all differences in weight about zero.

The formula for biserial correlation is as follows:

$$r_{sw} = \left(\frac{\overline{W}_1 - \overline{W}_2}{\sigma_w}\right)\left(\frac{pq}{z}\right)$$

where p and q are the proportions in the two categories with respect to success and z is the ordinate of the unit normal curve at the point of dichotomy. In the present case $p = q = 1/2$, $z = 1/\sqrt{2\pi} = .399$. $r_{sw} = 1.253\,\overline{W}_1/\sigma_w$.

Differences in comb size (C) and rank (R) were treated similarly.

The standard errors of these correlations were obtained from Soper's approximate formula as given by Kelley (1923).

$$SE_r = \frac{1}{\sqrt{N}}\left(\frac{\sqrt{pq}}{z} - r^2\right) = (1.253 - r^2)/\sqrt{200}.$$

The case of moult (differences represented by M) requires further consideration. Three categories were recognized: more, equal and less moult. It is assumed that these represent a trichotomy of a normal distribution of a scale of graded differences. Let a, b and c be the proportions with more, equal and less moult respectively in any distribution. Assume that the threshold between more and equal moult is at $-.50$ on this scale and that that between equal and less moult is at $+.50$. The difference between these thresholds can be expressed as follows in terms of the standard deviation (*cf.* Wright, 1934a).

$$[\mathrm{prf}^{-1}(a+b-.50) - \mathrm{prf}^{-1}(a-.50)]\sigma = 1$$

If a frequency between mean and threshold comes out negative as calculated from $(a+b-.50)$ or $(a-.50)$ the sign of the inverse probability function is to be taken as negative.

Among the winners there were 16% with more moult than the loser, 32% with equal moult and 52% with less moult. This yields for the standard deviation of this category (σ_{M_1}).

$$\sigma_{M_1} = 1/[-\mathrm{prf}^{-1}(.02) + \mathrm{prf}^{-1}(.34)] = 1.059$$

The location of the mean of the winners on our hypothetical scale can now be obtained.

$$\overline{M}_1 = .50 - \sigma_{M_1}\mathrm{prf}^{-1}[a+b-.50] = .50 + 1.059\ \mathrm{prf}^{-1}(.02) = .553$$

Because of symmetry, the mean of the losers $(\overline{M}_2)$ is $-.553$ on this scale.

The variance of the total (σ_M^2) is compounded of the average within the rows (both $1.059^2 = 1.122$) and that between them (which is $.553^2 = .306$). Thus $\sigma_M^2 = 1.122 + .306 = 1.428$, $\sigma_M = 1.195$.

The standard deviation of the total could also be calculated directly from the symmetrical trichotomy of the total ($a = .34$, $b = .32$, $c = .34$).

$$\sigma_M = 1/[\mathrm{prf}^{-1}\ .16 + \mathrm{prf}^{-1}\ .16] = 1.212$$

The discrepancy between these estimates is due to the fact that if the distributions of the separate rows are normal (as assumed in calculating their means) the distribution of the total is more or less platykurtic, instead of normal (as assumed in the second estimate of σ_M). Properly it is this total (based on birds taken in both orders with respect to success and therefore equivalent to ones taken in random order) that should be considered as normal, but for consistency with the calculation of the row means, the estimate $\sigma_M = 1.195$ is used here. The error in assuming normality of the distributions within the separate rows does not appear to be important as indicated in some measure by the comparison of the two estimates of the total standard deviation.

The biserial correlation may now be estimated.

$$r_{SM} = 1.253 \ M_1/\sigma_M = .580 \pm .065$$

This has been assigned a standard error by the same formula as in the preceding cases, although this is undoubtedly somewhat of an underestimate because of the nature of the scale. There is, however, no doubt of the significance of this correlation.[3]

In order to estimate the relative importance of the different factors it was necessary to take into account their correlations with each other. These correlations were obtained by use of the ordinary product moment formula in the case of differences in weight, comb, and social rank.

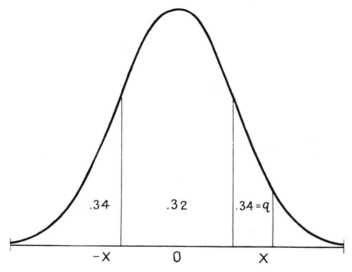

.34   .32   .34 = q

−X   0   X

FIG. 2. Transformation of values for moult to a numerical scale on the normal probability curve.

Pearson's broad category method was used in the cases in which moult is involved. The marginal frequencies (34% more, 32% equal and 34% less moult) are here assumed to trichotomize a normal probability curve with unit standard deviation, a different scale from that used above. The mean of the middle category is obviously zero. The mean deviation (d) for the category "less moult" is given by the following formula in which q is

[3] Professor Wright, personal communication.

the tail frequency (.34) and z is the ordinate at the threshold.

$$d = z/q = .3664/.34 = 1.078$$

The deviation of the category "more moult" is obviously -1.078.

$$\sigma_d^2 = (.34d^2 + 0 + .34d^2) = .790$$

is the variance as estimated from the mean deviations. The true variance is 1.

According to Pearson's formula,

$$r_{MW} = \frac{\Sigma dW}{N\sigma_W \; \sigma_d^2}.$$

The formulae for r_{MC} and r_{MR} are similar.

All these correlations between factors have been assigned standard errors by the usual formula, $\sigma_r = (1 - r^2)/\sqrt{N}$ although this is undoubtedly slightly too low in the cases involving moult, because of the coarseness of the scale.

We assume that the array of factors upon which the result of a fight depends includes cases in which differences in weight, comb size, state of moult and social rank are close indicators. In Figure 3 the correlations are known and are indicated by curved lines. The object is to find the value of coefficients measuring the influence along the paths indicated by arrows. An estimate can be obtained by the method of path coefficients (Wright, 1921, 1934a, 1934b) which in a symmetrical system such as used here is essentially the same as that of multiple regression, except for qualifications from the hypothetical character of the scale in the case of the differences in state of moult, social rank and success. The correlation between success and one of the factors such as weight may be analyzed into a direct contribution measured by the coefficient for the path from weight to success and three indirect contributions measured by the products of the coefficients along the paths from W through C, M, and R respectively to S. The correlations between success and the four factors provide four simultaneous equations which are identical with the normal equations of multiple regression.

$$r_{SW} = p_{SW} + p_{SC}r_{WC} + p_{SM}r_{WM} + p_{SR}r_{WR}$$
$$r_{SC} = p_{SW}r_{WC} + p_{SC} + p_{SM}r_{CM} + p_{SR}r_{CR}$$
$$r_{SM} = p_{SW}r_{WM} + p_{SC}r_{CM} + p_{SM} + p_{SR}r_{MR}$$
$$r_{SR} = p_{SW}r_{WR} + p_{SC}r_{CR} + p_{SM}r_{MR} + p_{SR}$$

Table 1 summarizes the correlation coefficients and the path coefficients resulting from solutions of the normal equations.

TABLE 1

CORRELATIONS AND PATH COEFFICIENTS RELATING DIFFERENCES IN CHARACTERS TO SUCCESS IN FIGHTS FOLLOWING INITIAL ENCOUNTERS

Difference	Correlation with success	Path coefficient relating to success
slightness of moult	.580 ± .065	.417
comb size	.593 ± .064	.354
weight	.474 ± .073	.111
rank in own flock	.262 ± .084	.209

Multiple correlation = $\sqrt{\Sigma rp}$ = .748

TABLE 2

CROSS CORRELATIONS AMONG THE CHARACTER DIFFERENCES

Character differences	
Slightness of moult and comb	.379 ± .061
Slightness of moult and weight	.387 ± .060
Slightness of moult and rank	− .064 ± .070
Comb and weight	.440 ± .057
Comb and rank	.156 ± .069
Weight and rank	.220 ± .067

TABLE 3

DEGREE OF DETERMINATION OF SUCCESS BY INDICATED FACTORS

Moult	.174
Comb	.125
Weight	.012
Social rank	.044
Moult and comb, joint residual	.112
Moult and weight, joint residual	.036
Moult and rank, joint residual	− .011
Comb and weight, joint residual	.035
Comb and rank, joint residual	.023
Weight and rank, joint residual	.010
Total determination (= .748² =)	.560
Residual determination =	.440
	1.000

These results need some interpretation. The correlation coefficients measure the total influence direct and indirect, of the various factors. The multiple correlation coefficient (.748) indicates the correlation between the best linear function of all of the factors and success.

Because of inter-correlations, the correlation coefficients do not however give a reliable evaluation of direct effects of the factors. The relatively high correlations between moult, weight, and comb size may be an expres-

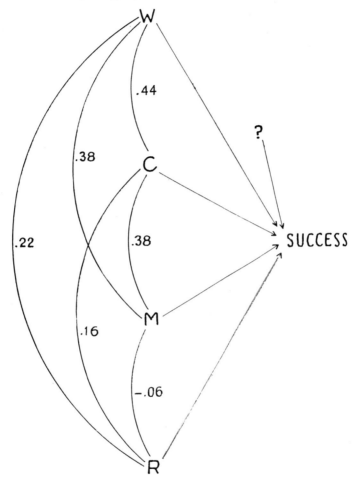

Fig. 3. Interrelations of factors influencing result of initial encounters.

sion of pituitary action but they can be more directly explained as being due in large part to a common factor which is presumably the amount of thyroxin which is present since this hormone in fairly large doses will induce moulting, reduce weight and suppress the gonads, all of which are more or less associated with loss of combat. The correlation between social rank and weight (.220) between social rank and comb size (.156) are low. The correlation that does exist may be a result of the fact that differences in weight and androgen which helped decide the original rank have persisted in part. Social rank once decided is much less variable than is weight, size of comb, or state of moult, hence the correlation is low. Then, too, birds high in the peck order have precedence to food and probably thrive better. The very low correlation between moult and social rank ($-.064$) is probably of no significance.

Inspection of the tables indicate that according to the path coefficients, moult (absence) was the most important factor deciding success, followed by male hormone indicated by comb, social rank and weight in that order. The small influence of weight is noteworthy. Moult exerts the greatest relative influence, but exactly how it does this is not known since it has been statistically freed from its actions through weight or the gonads. Perhaps it tends to make the bird more retiring since moult is really a kind of physiological illness, perhaps it makes the bird more sensitive to blows received, and perhaps the loss of feathers renders a bird less able to maintain equilibrium and to fight effectively. Size of comb is one of the best convenient indicators of male hormone output which we know at present. Hens secrete male as well as female hormone and injection of androgen has been shown to have a very marked and stimulating effect on the outcome of initial encounters unlike injection of estrogen (Allee, Collias and Lutherman, 1939; Allee and Collias, 1940). Social rank is perhaps a somewhat objectionable criterion since it may conceal factors of the very nature we are trying to uncover. However, it was the most convenient indicator of the "psychology of success" available, and furthermore this objection is obviated to a substantial degree by the low correlations of moult, comb and weight with social rank as compared to their relatively high correlations with success in initial encounters. The rather small importance of social rank as an independent factor as indicated by its path coefficient is also noteworthy in this respect.

The degrees of determination of the variance by the various factors reduce the results to a percentage basis and give a good idea of how the different factors influence variability in degree of success, acting singly and in combination. Acting singly, they necessarily exhibit the same order as their path coefficients since they are merely the squares of the former. Joint determination represents the contribution to variability (plus or minus) due to correlated occurrence of the two factors under consideration. The order of joint determinations obviously may differ from the order of the intercorrelations since the latter do not measure degree of influence on success. For example, moult and weight are more closely correlated than moult and comb, yet the joint determination by the latter pair is almost four times as great as that by the former pair which involves weight, a relatively ineffectual factor. One of the important properties of the degree of determination is that the sum of the components must be equal to 1. The general equation is $\Sigma d_{x \cdot A} + d_{x \cdot \overline{AB}} = 1$, which is read, the sum of the determination of x due to A and the sum of the determination of x due to the correlation between A and B must be equal to 1 (Wright, 1921). This gives a means of estimating the percentages of unknown factors (residual) by subtraction. In the present case 56.0 per cent. of the factors are accounted for by the single or correlated action of moult, comb size, weight and social rank, and presumably the underlying factors of which they serve as more or less imperfect indicators are responsible for a somewhat similar percentage. It follows that 44 per cent. of the factors were not measured in this analysis or were not known.

D. Unmeasured and Unknown Factors

One possible way to account for the unknown factors lies in the probability that more accurate indicators of the underlying factors would have given higher values for the paths, cutting down the amount assigned to residual factors. This is not to assume that still other factors do not exist which are unrelated to these indicators.

There are certain known but unmeasured factors which influence success and may have complicated the results to an unknown extent in the present case. Among these are age, the behavior of the other bird, and previous fighting experience.

Young pullets up to at least 9 months of age are somewhat refractory to injected androgen in terms of the stimulation of aggressive behavior (Allee, Collias, and Lutherman, 1939). Some of the birds used were barely older than this and included flock-mates of the treated birds. When encounters involving birds less than one year of age with older birds were omitted, the multiple correlation coefficient fell slightly from 0.75 to 0.70. The path coefficients were almost identical except that that for the comb size fell to 0.31 from 0.35. Possibly difference in general size weakened the accuracy of the comb as an indicator since the older hens tended to be somewhat heavier and larger. On the other hand, senility may play a role. Degenerative changes occur in the pituitary of old fowl (Payne, 1941). There is some recent evidence which suggests the possibility that senile animals may become less sensitive to injected androgens (Hoskins, Levine and Bevin, 1939). One flock of our hens was approaching senility, but were still laying, during the last set of encounters in which they were used. However, no effect of age differences on success is evident from the gross data; of the 200 encounters, 36.5 per cent. were won by older birds, 39.5 per cent. by younger birds, and in 25 per cent, of the cases the opponents were of about the same age.

The behavior of the other bird is undoubtedly important. A hesitant individual will often take heart and attack when it observes that its opponent seems frightened. In only 33.5 per cent. of the encounters did an actual battle ensue, and 78 per cent. of the encounters were won by the bird that started the encounter. The longer the latent period before contact began the shorter a fight was likely to be and the greater the probability that one bird would submit passively.

Experience in winning or losing can be an important factor at times, and by purposeful application of this factor the results of a later encounter between a given pair of birds can often be reversed in a later encounter. It is a little difficult to extract suitable test cases in encounters of hens. Seven cases were tested in which the original loser had won from two to five encounters while the original winner had lost from three to six encounters just before the second meeting, and the results were reversed in every case. Ginsburg and Allee (1942) have recently brought out the marked importance of this factor of conditioning in mice by a series of carefully controlled experiments. Whether conditioning played much of a role in the present series of encounters is somewhat debatable, but a conscious effort was made to control this factor by not fighting any bird more than once a day. The more immediate history of a bird might perhaps be expected to exert the greatest influence; however, it is interesting to note in this connection that the correlation between winning a given encounter and success in the preceding encounter was exactly the same as the correlation with the second preceding encounter, that is, 0.66. This tetrachoric correlation was conveniently secured by use of computing diagrams prepared by Thurstone and Chesire.

Endurance is probably a factor when an actual fight takes place. By fighting the same tough hen over and over again in succession Schjelderup-Ebbe (1922) was able to make her yield and be subordinate to weaker birds. During a battle it is obvious from the increasing slowness of movements of the birds that they are becoming more and more tired. Since none of the hens were fought more than once a day in the present series of encounters some measure of control over this factor was afforded.

The nature and importance of more completely unknown factors can merely be speculated on at present—differences in fighting skill, chance blows, differences in sensitivity to hormones, wildness, mild indisposition akin to illness, resemblance of the opponent to a despot or subordinate in the home flock, the details of past history, minor external disturbances, slight differences in handling, and errors in measurement perhaps have an aggregate effect of significant dimensions.

APPLICATIONS OF THE METHOD AND RESULTS

The significance and general implications of the problem will be developed elsewhere (Collias, 1943). The general importance of the statistical method employed is that it provides something of a model whereby the relative importance of the various factors which decide social dominance relations in various situations of ecological significance may be quantitatively evaluated. The present statistical method of evaluating the factors which make for high social level would have to be modified for each particular case, but this would probably not entail insurmountable difficulties.

One of the biggest complications and not adequately dealt with in this report since its influence was experimentally eliminated, is the interrelationship of territory. Schjeldrup-Ebbe's data indicate the possibility that in chickens, which have not generally been regarded as territorial birds, this factor may often outweigh all the others. However, even in nature, birds may on rare occasions be displaced from their territories by stronger rivals. Perhaps territorial influence could be measured by the distance from the territorial center.

More specific parallels of the general results to natural situations, at least cases where one factor apparently dominates the situation, can be cited even in the present inadequate state of investigation of social hierarchies in the vertebrates. At first thought, the size and weight factor is perhaps most obvious, *e.g.*, a general order of precedence to food in rough order of size differences is very common, for example, in birds of different species at a feeding shelf (Nice, 1929), or in different species of ducks and geese on a park lagoon (Jenkins, in press). The factor of male hormone is probably the important one in many situations. Males usually dominate females, and Lorenz (1931) states that young jackdaws rise in the peck order as they become reproductively activated, and much the same thing has been found by Shoemaker for canaries (1939). Coveys of quail and winter flocks of many other birds break up with the onset of the breeding season and the parallel rise in gonad activity and individual aggressiveness. Fighting is very common in groups of young male mammals (Alverdes, 1935). Examples of the action of thyroxin are more obscure and lie on more unstable assumptions, because of the complexity of its effects. In general, birds when moulting tend to retire to dense cover and hence to avoid conflict with other members of their species. The same retiring habit is seen during incubation when gonad activity is likewise reduced. Schjelderup-Ebbe (1922, 1935) has described a decided increase in aggressiveness of broody hens and such accords with the popular idea, but some evidence has been gathered in this laboratory which, while inadequate in itself, suggests the need for caution in relating such things as the defense of her chicks by a broody hen to position in the social hierarchy of the flock. The effective stimuli for types of aggressive behavior with different physiological bases may be very different.

SUMMARY

Initial encounters lie at the basis of the social order in flocks of chickens, as is known to be the case with a number of other vertebrates. To gain some insight into the factors which decide the outcome of initial encounters 200 pair contacts were staged in a neutral pen using normal, moderately inbred white leghorn hens from different flocks. Controlled factors included sex, territorial familiarity, and social facilitation. Evaluation of the less easily controlled factors was made mainly by the use of a modified biserial correlation formula. The degrees to which success in encounters was controlled by the more important factors were determined by the method of path coefficients which in the present instance was the same as the method of multiple regression.

Factors of major importance were male hormone output as indicated by comb size and thyroxin secretion as indicated by the complex of changes which accompany moulting. Social rank in the home flock had much less influence, and weight was of only small importance.

The multiple correlation coefficient of success with the four factors analyzed was 0.75. Forty-four per cent. of all factors were unknown or unmeasured in this analysis.

LITERATURE CITED

Allee, W. C. and N. E. Collias.
    1940. *Endocrinology*, 27: 78–94.
Allee, W. C., N. E. Collias and E. Beeman.
    1940. *Endocrinology*, 27: 827–835.
Allee, W. C., N. E. Collias and C. Z. Lutherman.
    1939. *Physiol. Zool.*, 12: 412–440.
Alverdes, F.
    1935. ''The Behavior of Mammalian Herds and Packs.'' (A chapter in Murchison's ''Handbook of Social Psychology''). Worcester, Mass.: Clark Univ. Press.
Collias, N. E.
    1944. *Physiol. Zool.* (in press).
Davenport, C. B. and M. P. Ekas.
    1936. ''Statistical Methods in Biology, Medicine and Psychology.'' Pp. 216.
Ezekial, M.
    1941. ''Methods of Correlation Analysis.'' 2nd edition. New York: J. Wiley and Sons. Pp. 531.
Ginsburg, B. and W. C. Allee.
    1942. *Physiol. Zool.*, 15: 485–506.

Hoskins, R. G., H. M. Levine and S. Bevin.
    1939. *Endocrinology,* 25 : 143.
Jenkins, D.
    1943. ''Territory as a Result of Despotism and Social Organization in Geese.'' *Auk.* (In press.)
Kelley, T. L.
    1923. ''Statistical Method.'' New York: Macmillan Co. Pp. 390.
Lorenz, K.
    1931. *Jour. f. Ornith.,* 79 : 67–127.
Nice, M. M.
    1929. *Oologist,* 46 : 161–163.
Palmer, R. S.
    1941. *Proc. Boston Soc. Nat. Hist.,* 42 : 1–119.
Payne, F.
    1941. *Anat. Rec.,* 81 (Suppl.) :51.
Schelderup-Ebbe, T.
    1922. *Zeitschr. Psychol.,* 88 : 225–252.
    1935. ''Social behavior of birds.'' Chapter 20 in Murchison's ''Handbook of Social Psychology.'' Worcester, Mass.: Clark Univ. Press.
Shoemaker, H. H.
    1939a. *Proc. Soc. Exper. Biol. Med.,* 41 : 299–302.
    1939b. *Auk,* 56 : 381–406.
Wright, S.
    1921. *Jour. Agr. Res.,* 20 : 557–585.
    1934a. *Genetics,* 19 : 506–536.
    1934b. *Ann. Math. Stat.,* 5 : 161–215.

17

Reprinted from *Physiological Zoöl.*, **30**(3), 255–268 (1957)

SOME RELATIONS BETWEEN FLOCK SIZE AND AGONISTIC BEHAVIOR IN DOMESTIC HENS[1]

EDWIN M. BANKS[2] AND W. C. ALLEE

Department of Biology, University of Florida, Gainesville, Florida

INVESTIGATIONS concerned with a number of different problems involving the social hierarchy in flocks of the common domestic hen have been reported during the course of the last fif-

teen years. The general practice in these studies has been to limit the size of the flocks undergoing investigation to small numbers. This has been done to facilitate

[1] This work has been supported in part by a grant from the Rockefeller Foundation to the late Dr. W. C. Allee.

[2] Present address: Division of General Studies, University of Illinois, Urbana, Illinois. The senior author accepts full responsibility for this report. We are indebted to Dr. A. M. Guhl for critical reading of the manuscript.

the finding of dominance-subordinance relationships among flock members, the foundation upon which any study in this area is based. Since the number of unidirectional dominance-subordinance relationships in a flock may be determined by the formula $N[(N-1)/2]$, where N is the number of individuals (Carpenter, 1942), it can readily be appreciated that the 300 relationships existing in a flock of 25 members would require long hours of tedious observation before any kind of analytical program could be instituted. In general, most studies attempting to analyze various aspects of agonistic behavior[3] and social organization have been carried out on flocks ranging roughly from 6 to 15 individuals. However, Guhl (1953) reported a study of a flock containing 96 pullets. Despite the recording of over 17,000 pecks, "there were a number of unknown dominance relationships among the 4,460 possible pair combinations." That a hierarchy existed, in fact, in this large flock was clearly indicated in this report. The question as to whether the results of investigations using small numbers apply equally well to larger flocks remains open. As a first step toward an answer, it was thought worthwhile to attack the problem from one basic aspect, that is, Does flock size per se influence the course of agonistic behavior? To phrase the question in another manner, Is there any quantitative relationship between flock size and agonistic behavior in the domestic hen?

In designing the experiment, it was recognized that flock size, pen size, and the dimensions of the centrally placed grain hopper used only during observation periods were all amenable to controlled variation. Considerations of expediency

[3] The term "agonistic," introduced by Scott and Fredericson (1951), includes both aggressive and submissive phases of interindividual contact; in the present context this term includes pecks, threats, and avoids.

prompted the use of pens and hoppers of identical size for three different flock sizes as the first of three sets of observations. This was to be followed by a second project in which flock size and pen size were to be varied directly and, finally, a third set of observations in which flock, pen, and hopper size were to undergo similar variation. The present report is based on the first set of observations wherein only flock size was varied.

ANIMALS AND THEIR TREATMENT

The Single Comb White Leghorn hens used in this study were purchased from the Department of Poultry Husbandry of the University of Florida. This strain represents approximately 20 years of closed-flock mating, selection being based primarily on agreement with strain standards of physical conformation. The average age of these hens, at the beginning of the investigation, was two and one-half years, ranging from one and one-half to three years. Each bird was made individually recognizable by attaching a pair of yellow plastic wing badges with large black numerals.

The birds were placed in two sets of seven assembled groups, each set with four wooden chicken houses. The two sets were separated into a north and south block by an alley approximately 12 feet wide. Each of the houses was divided into two pens; all pens were 8×12 feet in width and length, respectively. Each had a dropping board 2 feet above the floor that was 8 feet in length and about $3\frac{1}{2}$ feet wide. Six inches above the latter were two or three 2×4-inch perches running the length of the dropping board. At its apex, the roof was $7\frac{1}{2}$ feet from the floor. The dropping boards and floors were covered with a 3-inch layer of wood shavings. Each pen was furnished with a water pail, a single metal mash hopper, nest boxes, and a grit and oyster-shell hopper; each also opened by a small door into a separate yard. The L-shaped yard was 22 feet long and 20 feet wide at its widest point. The yards of all pens were of the same area except that those of the pens to the south of the separating alley were larger than the north yards, being 38 feet in length. The area of the yards was of no importance in this study because all observations were made in the pens.

As noted previously, each chicken house was divided into two pens. The partition dividing each house was constructed in part of a wooden wall at the bottom and a wire-mesh wall at the top. A frame door with wire mesh covered with burlap completed the division. A hole, approximately 1 foot square, was cut out of the burlap to afford a clear view of the pen floor; only the head of the observer was clearly visible to members of the flock undergoing study. Visual contact among individuals from two adjacent pens was eliminated in the separate pen yards by lining all yard fences with a 3-foot-high strip of black-felt roofing paper.

Preceding the collection of data on which this report is based, a preliminary trial was performed in order to initiate both the birds and the observer into the experimental conditions to be presented. In seven of the north pens the following flocks were randomly assembled: one with 24, two with 12, and four with 6 hens. Observations of agonistic behavior of the north set of flocks began on September 16, 1952. A duplicate set of flocks was arranged in the south pens, and these were similarly subjected to three hours of study, extending from October 6 to November 12, 1952. All observation periods were restricted to early-morning or late-afternoon hours and were conducted in the following manner.

The flocks from each pen of a house were gently stimulated out into their respective yards, and the small doors through which they passed were closed. A specially constructed feed hopper for grains (mixed whole grains of corn, wheat, and oats) was then placed centrally on the floor of the pen to be observed. This hopper was a metal box 10 × 12 inches in width and length, and 6 inches deep, this size being arbitrarily adopted. It was fastened securely to a 2-foot-square, $\frac{3}{4}$-inch plywood base. In the adjacent pen, a tape recorder was set up. When completed, these preparations allowed the observer to watch the activity in the adjoining pen and quietly to record the interactions. The low speaking voice required to dictate the observations in this manner appeared in no way to distract the hens. The advantage of using a tape recorder over writing down observations is quite real, as the observer could continuously direct his attention to the activity in the flock before him. Just prior to starting the observation period, the hopper was half-filled with grains, the small yard door was opened, and, after the birds were gently moved into the pen, the door was closed.

Following the observation period, which was of 15 minutes' duration, the flock which had been studied was again moved out to its yard. The tape recorder was transferred to the now empty pen, and the grain hopper moved to the adjacent pen that had just previously been used by the observer. At this point, the second flock was moved from its yard into its pen, where the grain hopper had been placed centrally and refilled. The observer retired to the adjoining pen, where, with the tape recorder, he proceeded to make note of agonistic interactions for the 15-minute period.

Commercial laying mash, water, grit, and crushed oyster shell were always available to the birds. Grains were offered during observation periods only. When flocks in both north and south pens were undergoing study, each group was observed on alternate days, and the order in which the flocks of various size were studied was rotated, i.e., on one day the order was 6- to 12- to 24-hen flocks, whereas on the next round the order was 24- to 12- to 6-hen flocks. With a view toward maintaining a uniform state of motivation, the grains were broadcast on the pen floor on those days when no observations were being recorded.

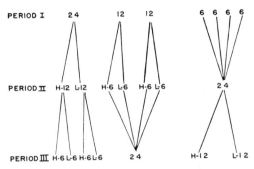

Fig. 1.—Regrouping scheme for north set of flocks.

In addition to manifestations of agonistic behavior such as pecks, threats, and avoids (Guhl, 1953), related phenomena were also recorded. Particular attention was paid to instances in which the behavior of given individuals was contrary to previously determined dominance-subordinance relationships. These phenomena, termed "peck-order violations" by Holabird (1955), will be discussed later.

Following the preliminary 3-hour training period, data were collected on the same flocks from October 23, 1952, to March 23, 1953, in the north pens, and from November 17, 1952, to March 24, 1953, in the south pens. In both instances the total observation time was 6 hours on each of the fourteen flocks. These records will be referred to as those from

Period I. The observation technique followed closely that employed during the training period, except that the birds became so habituated to the situation that the following innovation was introduced. Each of the small yard doors was fitted with a spring hinge, and a line attached to the door was so placed that the observer in the adjacent pen could open the door, record the order of entrance into the pen of the birds to be observed, and then, by releasing the line, close the yard door. It proved practical to limit this recording to the first three hens to enter the pen. In some instances it was necessary quietly to drive a few lagging flock members into the pen, but, in general, all individuals readily entered the door when it was pulled open by the observer. After all hens were in the pen, the door was closed, and the regular 15-minute observation period proceeded. The order-of-entrance record thus obtained began during the final hour of observations for Period I and was continued throughout the collection of the data for Periods II and III.

On completion of a total of 6 hours of observation per flock, both north and south sets of flocks were regrouped in the manner illustrated for the north set in Figure 1. This division separated the 12 highest ranking (H) of the 24 individuals from the 12 lowest ranking (L). Period I 12-hen flocks were similarly divided into eight 6-hen groupings, four of which were derived from the upper halves of their respective organizations and four from the lower. The eight 6-hen flocks of Period I were combined to form the two 24-hen arrays of Period II. By dividing and recombining the birds in this manner, it was hoped that the resulting data on frequency of display of agonistic interactions would aid further in the solution of the main problem of this study. The divisions and combinations so described were all made at the same time. In an attempt to upset the hens as little as possible, the regrouping was done at night when the birds were roosting.

Observations for Period II started on April 1, 1953, for the north pens. Collection of data from the south pens for the second period commenced on April 2, 1953. The method of study was the same as that employed on Period I flocks, and the recording of a total of 6 hours of observations per flock was completed on July 14 and 15, 1953, for the north and south pens, respectively. At this time the flock compositions were once more changed in the manner indicated above, and Period III was constituted. During the interval from July 23 to November 20, 1953, 6 hours of observations were made on the north pens. Beginning on July 24, the same period of time was used to study the south pens, the concluding 15 minutes being recorded on November 23, 1953.

In summary, as a result of regrouping, agonistic behavior was studied in a grand total of six flocks of 24, twelve flocks of 12, and twenty-four 6-hen flocks. Each flock was subjected to a total of 6 hours of observation.

Every hen from the original group which survived the 13 months during which this study was carried on was a member of each of the three sizes of flocks utilized. Some birds died in every flock grouping. These were immediately replaced by hens from a surplus group that had not been under observation, in order to keep the number of individuals in each flock at the desired level. The introduction of replacement birds into 12- and 24-hen flocks was made at night in an effort to inhibit excess pecking of the strange individuals. Introductions into 6-hen flocks was accomplished by arranging pair-contests (Collias, 1943) between the resident birds and the replace-

ment hen. This procedure was followed with the expectation that undue aggression directed to the new hen would be curtailed, an expectation borne out by subsequent observation, with one exceptional case to be mentioned later. Mortality and replacements will be discussed for each group in the following section.

OBSERVATIONS

a) PECKING ACTIVITY

In order to obtain a general inference conclusion regarding the effect of flock size on agonistic interactions, the data were analyzed as follows. For each 6-hour

TABLE 1

EFFECT OF FLOCK SIZE ON PECKING ACTIVITY

ANALYSIS OF VARIANCE

Source of Variation	D.F.	S.S.	M.S.
Major groups	17	294.4	.....
North versus south pens	1	14.2	14.2
Replications	2	1.4	0.7
Flock sizes	2	100.2	50.1*
Residual	12	133.6	11.13

* Significant at the 0.05–0.01 level.

period, the number of pecks delivered by every hen in a flock of a given size was determined. This total was found to range from 0 to approximately 400. The statistical analysis[4] to follow is based on the total number of pecks for each hen, in each size category, corrected according to Snedecor (1946).[5]

A summary of the analysis of variance performed on numbers of pecks delivered by hens in the three different-sized flocks is presented in Table 1. It can be seen that no difference in pecking activity was obtained between the north and south groups of flocks. Furthermore, there was no difference between Periods I, II, and III in terms of pecks per hen per period. Thus there was no detectable reflection

[4] Advice on the statistical techniques employed in this study was obtained from Dr. W. D. Hanson, Department of Agronomy, University of Florida.

[5] $\sqrt{N + 0.5}$.

on pecking activity of seasonal weather changes and the associated variations in the physiological condition of the birds. These findings indicate homogeneity in the pecking activity of our hens, irrespective of annual climatic variations and pen locations. The final source of

TABLE 2

COMPARISON OF SPECIFIC FLOCK SIZES

ANALYSIS OF VARIANCE

Source of Variation	D.F.	S.S.	M.S.
Sizes	2	100.2	.....
24 versus 6	1	0.8	0.8
24, 6, versus 12	1	99.4	99.4*
Residual	12	133.6	11.13

* Significant at the 0.05–0.01 level.

TABLE 3

PECKING ACTIVITY OF "HIGH" VERSUS "LOW" 12-HEN FLOCKS OF PERIODS II AND III

ANALYSIS OF VARIANCE

Source of Variation	D.F.	S.S.	M.S.
Within 12-hen flocks	6	40.86	.....
High versus low	1	24.00	24.00*
Residual	5	16.86	3.37

* Significant at the 0.05–0.01 level.

variation in the analysis was flock size, and here a significant difference was obtained. Table 2 presents a refinement of the size comparison, in which it is found that 12-hen flocks were observed to peck significantly more than did 6- and 24-bird flocks, whereas no difference was found between these two latter groups.

As described previously, the regrouping of birds to form the 6- and 12-hen flocks of Periods II and III was so devised that half the flocks of these sizes were composed of hens from the upper halves of their former social organizations and half from the lower. Flocks so constituted are hereafter referred to as being "high" and "low" flocks, respectively. An analysis of variance was performed to determine whether the flocks so separated differed in pecking activity. The results for 12-hen groups are summarized in Table 3, indicating that

flocks composed of formerly high-ranking hens pecked significantly more than did those composed of low-ranking individuals. Although a trend in the direction of greater pecking activity by high 6-hen flocks was apparent, the differences proved not to be statistically significant.

Table 4 presents the mortality record throughout the course of the study. Postmortem examinations were made on all birds to detect possible contagious diseases and prevent their spread through the colony;[6] none were encountered. Except in the one case described later, these

TABLE 4

MORTALITY RECORD

FLOCK SIZE	PERIOD I	II	III
N 24	1	4	0
S 24	0	0	1
N {12	0	0	1
{12	0	2	2
S {12	1	1	0
{12	1	2	1
N {6	1	1	1
{6	1	0	1
{6	0	0	2*
{6	0	0	4†
S {6	1	0	0
{6	1	0	0
{6	0	0	0
{6	0	0	0
Total	7	10	13

* Two hens disappeared.
† Three hens disappeared; one died.

deaths and the subsequent introduction of replacement birds resulted in no noticeable changes in pecking activity. During the fourth hour of Period III, 5 hens mysteriously disappeared from two 6-hen flocks, 2 from one and 3 from an adjoining flock. The flock in which 3 replacement birds were introduced exhibited a marked increase in rate of pecking. The number of pecks corrected for in all other 6-hen flocks ranged from 28.5 to

[6] The examinations were performed by the Poultry Pathology Laboratory of the University of Florida.

63.5 per hen per 6 hours. The number of pecks observed in the flock with 3 replacement hens was 79.4. In addition, this 6-hen flock for Period III had been derived from the lower portion of a low 12-hen group. It will be recalled that no difference was noted in pecking activity between high and low 6-hen flocks. An analysis made on all 6-hen groups except the two under discussion still failed to yield significance. Data from these two flocks were included in the analysis summarized in Table 2. The 6-hen flock into which 2 replacement hens were introduced exhibited pecking activity that fell within the range referred to.

b) THREATS AND AVOIDS

Threats have been defined as "undelivered pecks or instances in which the dominant bird raised its hackles or in an otherwise threatening manner caused the inferior bird to which this behavior was directed to avoid the threatening bird" (Guhl, 1953). In the present study, careful note was taken of such manifestations of agonistic behavior. These data were subjected to statistical analysis, summarized in Table 5. No differences were found between north and south flocks, nor did the replications differ with respect to threat behavior. Differences in threat activity related to flock size were found to be significant.

As indicated in Table 6, 24-hen flocks exhibited significantly fewer observed threats than did either 12- or 6-hen groups. The two latter size groupings did not differ in this regard.

Carrying the analysis further, it was found that high 12 hen flocks, when compared with their low counterparts in Periods II and III, exhibited a significantly higher frequency of threat reactions (Table 7). A similar type of analysis performed on data from high and low 6-hen flocks of Periods II and III re-

vealed no difference between these two categories with respect to threat frequency.

The most common response of a subordinate hen to the threat of a dominant was an avoidance reaction, in which the

TABLE 5

EFFECT OF FLOCK SIZE ON THREAT ACTIVITY

ANALYSIS OF VARIANCE

Source of Variation	D.F.	S.S.	M.S.
Major groups	17	490.62	......
North versus south pens	1	7.73	7.73
Replications	2	89.00	16.48
Sizes	2	361.42	180.71*
Residual	12	32.47	2.7

* Significant at the 0.01 level.

TABLE 6

COMPARISON OF DIFFERENT-SIZED FLOCKS WITH RESPECT TO NUMBER OF THREATS PER HEN PER 6-HOUR PERIOD

Size of flock....	24	versus	6	$P=0.001$
No. of threats..	665		972.1	
Size of flock....	24	versus	12	$P=0.001$
No. of threats..	665		912.1	
Size of flock....	6	versus	12	Not significant
No. of threats..	972.1		912.1	

TABLE 7

THREAT ACTIVITY OF HIGH VERSUS LOW 12-HEN FLOCKS OF PERIODS II AND III

ANALYSIS OF VARIANCE

Source of Variation	D.F.	S.S.	M.S.
Within 12-hen flocks	6	49.82	......
High versus low	1	29.37	29.37*
Residual	5	20.45	4.09

* Significant at the 0.05–0.01 level.

subordinate tended to increase the distance between itself and the dominant. Note was taken of these avoidance reactions, and, except under certain circumstances, avoids balanced threats in terms of the number of each type of activity observed in a given pen over a given period of time. Exceptions to this general situation were rare; on some occasions a subordinate hen exhibited a peck-order violation by threatening one of its dominants. In these instances the response to the threat would not be an avoid but either a peck or a threat by the dominant bird directed to the abnormally behaving subordinate. Cases of peck-order violations will be discussed further, the point at present being that the same conclusions may be drawn regarding the effect of flock size on avoiding reactions as applied to threat behavior; 24-hen flocks gave significantly fewer avoids than did either 12- or 6-hen groups. The two latter groupings did not differ in this regard.

c) EFFECT OF TIME ON RATE OF AGONISTIC ACTIVITY

Guhl and Allee (1944) found that in small organized flocks of domestic hens, social tensions, as reflected in pecking activity, tended to diminish with time when compared with small flocks undergoing continuous reorganization. The reduction in pecking activity was manifest after the flocks had been intact for 11–38 weeks. In the present experiment the duration of a given constitution of the flocks ranged from 15 weeks for Period II to 17 weeks for Period III and 19 weeks for Period I. Flock membership was held constant in the present study, except for the replacement of dead birds, and thus simulated some of the conditions which prevailed in the organized flocks of Guhl and Allee (1944). However, the feeding technique used here differs significantly from that used by Guhl and Allee except for the final phase of their experiment.

In an effort to determine the effect of time on agonistic activity, the following analysis was conducted. The number of pecks and threats for each size category was totaled for each of the twenty-four 15-minute sessions comprising the three 6-hour periods. A pooled mean was then calculated for each observation period, and correlation coefficient analyses were made between the number of 15-minute

observation sessions, i.e., the length of time during which the flocks were held intact, and mean numbers of pecks and threats for each 15-minute observation period for the three sets of data.

Treated so, no significant correlations existed between the duration of time in which the flocks were held intact and the numbers of pecks or threats given per observation period. Such correlations were lacking in all flock sizes tried except that the rate of pecking activity in 24-hen flocks increased with time, $r = +0.61$. This relationship is highly significant. The trend line shown in Figure 2 was established by the method of least squares. Social tensions as measured by pecks and threats did not diminish with time under the conditions of the present study, and in the six largest flocks social tensions increased.

d) PECK-ORDER VIOLATIONS

As defined by Holabird (1955), a peck-order violation is "an interaction in which a subordinate pecked or threatened a dominant or in which a dominant avoided a subordinate . . . provided that the peck order remained subsequently unchanged." Small flocks of the domestic hen exhibit a remarkably stable peck-right type of social organization in which dominance by one individual over another persists. The stability of this inter-individual relationship may prevail, in the absence of abnormal situations, as long as the two hens are kept together. It is therefore of some interest to note that such a phenomenon as the peck-order violation is not uncommon. The role of peck-order violations in the dynamics of the social hierarchy in domestic hens is incompletely understood.

Table 8 presents a summary for the entire study of the mean number of peck-order violations observed in each of the three sizes of flocks. Testing each mean against the other two revealed that the 24-hen flocks exhibited significantly more violations than did either the 12- or the 6-bird group. The difference between the two latter groups was also found to be significant, 12-hen flocks showing more violations than 6-hen groups. In all three comparisons the *P*-value was less than 0.001.

Examination of these data from another point of view reveals, as illustrated in Table 9, that 25 per cent of the total

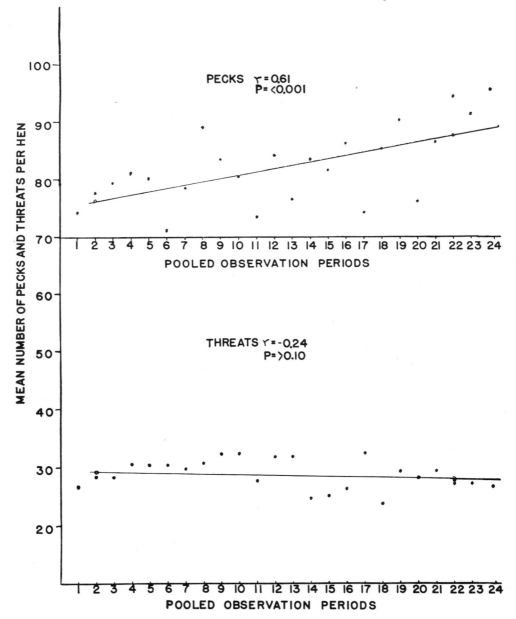

FIG. 2.—Correlation between cumulative length of organization and amount of pecking and threatening in six 24-hen flocks.

number of hens (452) used in this study were observed to exhibit peck-order violations. In the majority of these cases, each violator reacted against peck-order expectation between one and five times. Of particular interest is the record of hen No. 172, which gave 42 violations during the three periods (Table 10). While this

TABLE 8

PECK-ORDER VIOLATIONS

Flock Size	Mean No. Violations per 18-Hour Observation
6	4.8
12	13.7
24	20.3

TABLE 9

BREAKDOWN OF PECK-ORDER VIOLATIONS

	Number	Per Cent
Hens showing violations	114	25
Hens showing no violations	338	75

No. of Violations	No. of Hens
1– 5	97
6–10	11
11–20	5
Over 20 (42)	1

individual was a member of a 6-hen flock, only one violation was recorded for her. She committed 21 violations in a 24-hen flock, and 20 when a member of a 12-bird group. All 42 violations were committed against 2 dominant hens, Nos. 174 and 103—24 against the former, 18 against the latter. Both Nos. 174 and 103 maintained peck-right over No. 172 throughout the course of the study.

e) ORDER-OF-ENTRANCE OBSERVATIONS

As indicated earlier, it was possible to record the order of entrance of the hens into their respective pens after the yard doors had been opened by the observer. These data were collected, starting with the final hour of observations for Period I and including all of Periods II and III. Our interest in making these records was to ascertain whether the hens tended to enter the pen in any consistent order and also to determine the presence or absence of association between rank in the hierarchy and order of entrance into the pen.

An analysis of these data failed to reveal a consistent pattern in the order in which the flock members entered the pen. This lack of consistency prevented an analysis of correlation between social status and order of entrance into the pen. Fischel (1927) found that flocks of hens, when under range conditions, tended to move away from their range shelters in loose groups and that these groups were led by any hen. Leadership in the domestic hen does not appear to reside in the alpha individual of a flock.

TABLE 10

VIOLATION RECORD OF HEN No. 172

Period	No. of Violations	Size of Flock
I	1	6
II	21	24
III	20	12

DISCUSSION

There are many factors which must be considered in an interpretation of these data. One of these is obviously flock size per se. We found that, as the number of hens increased from 6 to 12, there followed a significant increase in the number of pecks per hen per unit time. A further increase in individuals from 12 to 24, however, did not follow this trend. In fact, 12-hen flocks pecked significantly more than did 24-hen groups. Furthermore, there was no essential difference in pecking activity per bird per unit time between the two extremes of flock size; 6- and 24-hen flocks exhibited about the same number of pecks

We look for an explanation of these findings in the following set of considerations. All pens were of the same size and, therefore, area per hen decreased as the flock size increased, i.e., in 6-hen flocks there were 16 sq. ft. of space per hen; in 12-hen flocks this area decreased to 8 sq. ft. per hen; and in 24-hen flocks it was 4 sq. ft. per hen. The expectation might be that, as a given area became more densely populated, contact between individuals would occur more frequently and thus the number of pecks per hen per unit time would increase. This appeared to be the case as flock size was increased from 6 to 12, but the expected increase in pecking of 24 over 12 did not occur. One reason for this may be because during observation periods the birds were purposely concentrated about a grain hopper that was constant in size, irrespective of the number of hens in the flock. Hence, as the number of birds per flock increased, the amount of feeding space decreased. The question then arises as to whether this reduced feeding space would not serve to accentuate the already reduced areal space per bird and thus to enhance the possibility of interindividual contacts.

There are two reasons which would lead to a negative reply to this question. First, the stability of dominance-subordinance relationships among hens is evidently based on the ability of the individual birds to distinguish or recognize one another and to behave accordingly when contact is made. It is further known that the most important visual cues guiding the recognition of one hen by another are the conformation and color of the head and face region (Schjelderup-Ebbe, 1923; Guhl and Ortman, 1953). All metal grain hoppers used in this study were 12 × 10 × 6 inches in length, width, and depth, respectively. After entering the pen, the hens proceeded rapidly to the hopper. Six birds were easily accommodated, with surplus unoccupied space and room for free move-

ment about the hopper. Competition for space at the hopper in 12-hen flocks was more severe; there was much moving and pushing about. A few of the birds holding lowest ranks in the 12-hen hierarchies rarely succeeded in reaching the hopper.

In 24-hen flocks, this situation was still more pronounced. The added population held movements about the hopper to a minimum. Only the most aggressive, high-ranking individuals actually fed during observation periods. The lowest-ranking birds stayed out of the center of activity and obtained no grain at all. Thus the reduced visual contact among aggressive individuals crowded around the hopper in 24-hen flocks, associated with the large number of subordinate birds evading the area of interactions, tended to reduce, on the average, the rate of pecking in this flock size, as compared with that observed in 12-hen flocks.

Another factor which must be considered is the possible failure of the observer actually to see all the interactions that occurred in the 24-hen flocks. All observations were made by one individual, and every effort was made to hold this type of error to a reasonably small percentage. It may well be that some interactions occurring during the 15-minute periods in all three flock sizes were not recorded. However, the fact that the average corrected number of pecks per hen per 6 hours of observation (7.69) in the 6-hen flocks was actually slightly higher than that obtained in the 24-hen groups (7.59) would lend weight to a factor or factors apart from observational error in an explanation of the results.

Consideration of two other aspects of agonistic behavior—threatening and avoiding activity—revealed a significant difference between 24-hen flocks and both 12- and 6-hen flocks. The two latter categories were observed to exhibit threatening and avoiding behavior more frequently than did 24-hen flocks. In this regard, 6- and 12-hen groups did not differ significantly.

The circumstances in which threatening behavior can be observed are essentially the same as those in which pecking occurs, with the additional factor of distance. Clear-cut threats have been observed between a dominant and a subordinate hen that were as much as 6 feet apart. On the other hand, a dominant hen may threaten a subordinate standing adjacent to her. In both situations the ability of the interacting individuals to recognize one another is undoubtedly of primary importance.

In 24-hen flocks, as has been pointed out, this ability was interfered with by the crowding in the vicinity of the centrally placed grain hopper. Our records indicate that the bulk of threatening behavior tabulated for 24-hen flocks took place among high-ranking individuals that remained at the grain hopper.

Turning to the effect of time on the rate of agonistic activity, we found that social tensions as measured by pecking and threatening activity did not diminish with time, and in the six largest flocks such behavior actually exhibited a significant increase. One suggestion which might explain the increased pecking rate in 24-hen flocks stems from the fact that in Periods II and III these groups were formed by the combination of four sets of strange hens. Guhl has observed (personal communication) that under similar conditions the individuals from the same group tend initially to stay together for some days, avoiding interactions with strangers. Such inertia would eventually break down, with a resultant increase in pecking activity with time. The observations of the present study, however, indicate that the most aggressive hens from the 6-hen flocks attained high status in

the 24-hen groups very rapidly. These high-ranking hens contributed the majority of pecks and threats in the 24-hen flocks from the onset of assemblage of the birds. The increase in the average number of pecks was due largely to an increase in the pecking activity of the high-ranking individuals and not to a general increase in pecking activity of all hens. Moreover, the lowest-ranking birds in all 24-hen flocks rarely approached the grain hopper and received few pecks from individuals higher in the social organization. This held true throughout the duration of constitution of the flocks.

A direct comparison of the present study with that of Guhl and Allee (1944) is not feasible because of a difference in feeding technique and the consequent difference in hunger motivation of the hens. Other factors which make such a comparison unprofitable are the difference in flock size and in strain of the White Leghorn breed used in the two investigations. That important strain differences exist in the display of agonistic behavior of the domestic hen was pointed out by the study of Allee, Foreman, Banks, and Holabird (1955).

A consideration of the peck-order violation data may be pertinent at this point. It was found that 6-, 12-, and 24-hen flocks all differed significantly from one another in this regard and that the differences in numbers of violations varied directly with flock size. In 24-hen groupings, which showed the largest number of peck-order violations, lack of recognition by the violator may have been the most important factor. However, Douglis (1948) in experiments with simultaneous membership in several flocks has shown that a hen can recognize and react in a predictive manner to a total of 27 other hens. Guhl (1953) alludes to the determination of the peck order of a flock of 96 pullets. Although full details of the study are not presented, he found few reversals in dominance. This may be a further indication that hens are capable of responding "properly" to the individual differences of a considerable number of flockmates.

Another possibility is that peck-order violations are manifestations of social tension between two hens, preceding attempts at reversal of dominance. At our present state of knowledge concerning this aberrant form of hierarchical behavior, however, it is not possible to distinguish between lack of recognition and attempted reversal as a causal basis for the peck-order violation.

The behavior of hen No. 172, related previously, demonstrates that, as this hen was shifted from 6- to 12- to 24-hen flocks, she committed an increasing number of violations. That all such violations were against only two flockmates that were moved with her is also impressive. Were these persistent, but unsuccessful, attempts by No. 172 at reversing her subordinate status, or were they simply lapses of memory? More information and analysis are required before a definitive answer can be given.

Crew and Mirskaia (1931) studied the effects of density on adult line-bred albino mice in the following manner. Varying numbers of mice (2, 4, 8, 16, and 24) were place in cages of identical size, and the relationship between population size and such factors as death rate, reproductive rate, and fecundity were observed. With regard to death rate, cages with 2 mice ranked lowest, whereas those with 8 ranked highest. Cages with 24 mice exhibited a somewhat lower death rate than did those with 8 individuals. The authors concluded that death, reproductive, and fecundity rates were all affected by population density, but they suggested that certain individuals were unable to adapt to conditions of crowding.

Scott and Fredericson (1951) have interpreted these results as follows: crowding tends to increase the probability of fighting up to a certain point; beyond that point, crowding exerts a favorable effect on mortality by allowing subordinate mice to avoid notice.

Translating the density-dependent measurement from mortality rate to phases of agonistic behavior such as those observed in the present study, the results of the two sets of observations offer an interesting parallel.

It is clear that the problem of the relationship between flock size and agonistic behavior in the domestic hen has not been completely solved by the present study. Variables not manipulated, such as feeding space and pen size, must be considered before a more complete answer to this problem can be given.

SUMMARY

1. To determine the effects of flock size on the rate of agonistic activity in the domestic hen, three different flock sizes were assembled as follows: six flocks of 24, twelve flocks of 12, and twenty-four flocks of 6 hens.

2. The flocks were subjected to careful observation for 15-minute periods until each had been studied for a total of 6 hours, distributed over 15–19 weeks.

3. An analysis of variance was performed on three aspects of agonistic behavior, i.e., pecks, threats, and avoids, with the following results.

4. Hens in groups of 12 pecked significantly more than did those in groups of 6 and 24 (variance ratio significant at the 0.05–0.01 level). No significant difference in rate of pecking was found between 6- and 24-hen groups.

5. With regard to threat activity, 6- and 12-hen groups exhibited significantly higher scores than did 24-hen flocks

(variance ratio significant at the 0.01 level). No significant difference in threat activity was found between 6- and 12-hen flocks.

6. Hens in groups of 6 and 12 exhibited significantly more avoiding reactions than did those in flocks of 24 (variance ratio significant at the 0.01 level).

7. No correlation was found between the duration of time in which the flocks were held intact and the number of pecks or threats given per observation period, except that the rate of pecking activity in 24-hen flocks increased significantly with time ($r = +0.61$).

8. Peck-order violations were committed by 25 per cent of the hens used in the study ($N = 452$). It was found that 6-, 12-, and 24-hen flocks all differed significantly from one another in this respect and that the differences in numbers of violations varied directly with flock size.

9. A brief discussion is presented of the interpretation of these data, with suggestions for further study.

LITERATURE CITED

ALLEE, W. C., FOREMAN, D., BANKS, E. M., and HOLABIRD, C. 1955. Effects of an androgen on dominance and subordinance in six common breeds of *Gallus gallus*. Physiol. Zoöl., **28**:89–115.

CARPENTER, C. R. 1942. Societies of monkeys and apes. Biol. Symp., **8**:177–204.

COLLIAS, N. E. 1943. Statistical analysis of factors which make for success in initial encounters between hens. Amer. Naturalist, **77**:519–38.

CREW, F. A. E., and MIRSKAIA, L. 1931. The effects of density on an adult mouse population. Biol. gen., **7**:239–50.

DOUGLIS, M. B. 1948. Social factors influencing hierarchies of small flocks of the domestic hen; interactions between resident and part-time members of organized flocks. Physiol. Zoöl., **21**:147–82.

FISCHEL, W. 1927. Beiträge zum Sociologie des Haushühns. Biol. Zentralbl., **47**:678–95.

GUHL, A. M. 1953. Social behavior of the domestic fowl. Kansas Agr. Exper. Stat. Tech. Bull., **73**:1–48.

GUHL, A. M., and ALLEE, W. C. 1944. Some measurable effects of social organization in flocks of hens. Physiol. Zoöl., **17**:320–47.

GUHL, A. M., and ORTMAN, L. L. 1953. Visual patterns in the recognition of individuals among chickens. Condor, **55**:287–98.

HOLABIRD, C. 1955. Social organization in flocks of Light Brahma hens as compared with other breeds. Physiol. Zoöl., **28**:239–55.

SCHJELDERUP-EBBE, T. 1923. Weitere Beiträge zur social und individual Psychologie des Haushühns. Zeitschr. f. Psychol., **92**:60–87.

SCOTT, J. P., and FREDERICSON, E. 1951. The causes of fighting in mice and rats. Physiol. Zoöl., **24**:273–309.

SNEDECOR, G. W. 1946. Statistical methods. Ames, Iowa: Iowa State College Press.

IV
Applications

Editor's Comments on Papers 18 Through 20

Many of the findings from biosociological studies had tremendous implications with respect to agricultural practices and the production of food and fiber. It seemed reasonable to assume that production would be more efficient, and therefore economically more sound, if the animal's normal behaviors were taken into account in devising management routines. Much of the responsibility for the applied emphasis rests with A. M. Guhl, who started as a student in Allee's laboratory in Chicago and subsequently developed his own strongly applied program at Kansas State University. Guhl's 1953 classic (Paper 9) was followed by many others alone and in conjunction with colleagues at Kansas State. At the present time, a considerable amount of time and effort is devoted to biobehavioral research at many agricultural research institutions and wildlife stations throughout the world. A new journal, *Applied Animal Ethology,* is testimony to the steady growth of the applied emphasis.

Space limitations preclude the reprinting of more than just a mere sample of applied articles, but it seems only natural that the 1945 Guhl, Collias, and Allee article (Paper 18) should lead off this section. The other two articles, one by Guhl, Craig, and Mueller (Paper 19) and the other by Craig and Tóth (Paper 20), trace the development of the basic idea presented by Guhl in Paper 18 through to the practical applications shown by Craig and Tóth in Paper 20. The report of a negative correlation between a hen's social status and her mating performance (Paper 18) suggested that very high ranking females would be economically undesirable. That the social status of a hen could easily be manipulated by selective breeding was clearly demonstrated in Paper 19. These ideas came together in Paper 20, which showed a lower rate of egg production in one variety of chickens selected for high-social-dominance abilities.

Many similar examples could have been drawn from the pages of the specialized agricultural and wildlife journals, and there is little doubt that the future will bring even more. Practical "fallout" from biobehavioral studies is inevitable, just as social dominance is a phenomenon that must be considered in dealing with many of our domestic animals.

18

Reprinted from *Physiological Zoöl.*, **18**(4), 365–390 (1945)

MATING BEHAVIOR AND THE SOCIAL HIERARCHY IN SMALL FLOCKS OF WHITE LEGHORNS[1]

A. M. GUHL, N. E. COLLIAS,[2] AND W. C. ALLEE

Whitman Laboratory of Experimental Zoölogy, University of Chicago

OUR original interest in this study stems from the reports on social organization in the flocks of common domestic fowl by Schjelderup-Ebbe (1922, 1935). The avian hierarchy in flocks of hens and of other birds has received much analytical attention by Allee and his associates (see Literature Cited). These studies have been concerned with endocrine, physiological, and psychological aspects of group behavior and have recently been summarized by Allee (1942, 1943, 1945) and by Collias (1944). It early became apparent that certain differences were associated with position in the so-called "peck-order" of the flock. High-ranking hens possess a greater freedom of activity as compared with those of low social status; for example, they have precedence to food (Masure and Allee, 1934). Many aspects of observed relations to social status are summed up in the statement that fewer eggs are laid by hens in the lower half of the social order than by those composing the more aggressive half (Sanctuary, 1932). We have here attempted to determine a part of the relation of the social order to the mating behavior in such flocks. In many ways the present study is a continuation and extension of the work of Skard (1937). A preliminary abstract of some of the results has been published by one of us (Guhl, 1941).

ANIMALS AND THEIR TREATMENT

All animals used in these studies were White Leghorns; we have reason to believe that all were pure-bred representatives of that breed. Observations by the first author on two flocks of cocks and hens provided the main body of data to be analyzed here; data have also been considered that were obtained during previous observations on some of these birds, as well as on others. These flocks were housed in the poultry-house at Whitman Laboratory in pens 14.5 feet long, 4.5–5.0 feet wide, and 8 feet high. They were continuously illuminated (between 6:00 A.M. and 6:00 P.M.) by a 60-watt ceiling lamp near the center of each pen in addition to the natural light entering from south windows. During the cold months the pens were heated by steam radiators and were roughly controlled at a temperature of $10°–15°$ C. A more complete description of the general situation may be found elsewhere (Allee, Collias, and Lutherman, 1939). Mash and water were available at all times, and a mixture of grain was fed during the afternoons. Feeding and general care, including cleaning of the pens, were in the hands of the observer, to whom the birds became accustomed and whom they appeared to ignore during observations as he watched them from the outside of the pen.

Flock D.—This flock contained 7 hens,

---

[1] We are indebted to Professor L. F. Payne and Dr. D. C. Warren for counsel given to the first author concerning the interpretation of the data, to Professor Sewall Wright for advice concerning statistical methods, and to Mr. Peter Frank for extensive statistical calculations.

[2] A. M. Guhl is now located at Kansas State College, and N. E. Collias has entered the United States Army.

which were mature when obtained in October, 1938. The 4 cocks were received as newly hatched chicks in August, 1939. They were reared with females of their own age and were separated from them about 3 weeks after treading (copulation) had become frequent. These males were then kept in a pen by themselves from early January, 1940, to April and had not been exposed to this flock of hens before the present experiments. Straight-line peck-orders prevailed among both hens and cocks; a change in order occurred among the cocks during the course of the observations and will be noted later.

Beginning April 13, 1940, the males were placed singly and successively into the pen of the hens of flock D on alternate days. This procedure, which left the hens without males every other day, was taken as a precaution against a probable carry-over of certain reactions of the hens toward a given cock that might influence their behavior toward his successor. The sequence in which the males were introduced remained the same. Each cock was placed in the pen during the preceding night or early morning of its scheduled day of exposure to the hens. The recorded observations of 2 hours' duration (1 hour on the initial day) were made between 3:00 and 6:00 P.M. until each male was exposed ten times and for a total of 19 hours of observation.

On June 22, all 4 cocks were introduced together into the pen, where they remained with the hens. No records were made until July 4, to allow for adjustment. The initially intense interaction caused the hens to remain on the roost. Cock IV died on July 24; and on September 8 the dominant cock, I, was removed for a period of 30 days, for reasons to be discussed later; cocks II and III were left with the hens throughout this period of observation.

Flock F.—These 9 hens and 4 cocks were received in February, 1941, when nearly 1 year old. All of the birds but cock VIII came from the same flock. Pecking triangles occurred among the hens for 10 days; hen GB developed a paralysis and was removed on May 27; and on the same day YG gained dominance over PB and so initiated the only triangle that persisted through most of the experiment with this flock.

The males of this group, especially VI, were not socially aggressive and did relatively little pecking. Threats and avoiding reactions were used to determine their peck-order; back-pecks and brief fights, more threatening than final, were common throughout much of the experiment. However, the order given in Tables 1–4 was definite for the period covering the time during which treading data were taken. When all the males were placed in the pen of these hens, pecking by the males occurred more frequently and the same order prevailed. Cock *I* of flock D was included in part of the observations of flock F for general comparative purposes.

Each day for a total of 60 days, beginning April 26, 1941 (exclusive of a 5-day interval at the end of the fourth round for the males), one cock in a wire cage was placed within the pen of the hens. He was released during the observation period of 30 minutes and was then replaced by another caged cock in preparation for the observations of the following day. This method greatly increased the number of matings and the amount of other intersexual behavior that was exhibited during the observation period (cf. Skard, 1937). All 4 cocks of flock F were introduced into the pen for permanent residence on July 3 and observed for 60 minutes daily for 48 days.

Flock G.—This group of 6 hens was

used as an accessory flock in certain tests of the males of flock D. It was composed of hens from several sources, of various ages, and also included a poulard.

Preliminary observations were made using the following additional flocks:

Flock A6.—The 6 birds composing this flock had lived together for over 4 months when observations started early in February, 1939. The flock included 4 moderately inbred young White Leghorn hens, received as pullets from Iowa State College, and 2 old hens from flock D.

Flock E.—Flock E was a small lot of hens from an Indiana Farm Bureau strain that had been intraflock bred for 6 years. These yearling hens were culled from a large flock and were received by us late in April, 1939; observations on mating frequency were begun after the hens had been under our care for a few weeks.

Two cocks, other than those mentioned for flocks D and F, were used in certain earlier tests. These were obtained from a local poultry farm when about 7 months old. One of these was used with flocks A6 and E; the other was with the flock-D hens in 1939 when the data for these three flocks (Table 9) were obtained. Both these males were very active sexually.

To recapitulate: The cocks and hens were kept in unisexual flocks to determine their respective peck-orders. The males were then placed singly with the hens of their respective flocks every other day in flock D and daily in flock F to study the sexual behavior of the hens in association with each individual cock. The order of introduction of the cocks into the pen of the hens on successive days remained unchanged. After several weeks of such observations the cocks of each flock were placed simultaneously into the pen of their respective female flockmates, where they remained for another series of observations on sexual behavior under these more complicated conditions. At certain times the same birds were used to study social discrimination, the results of which have been reported separately (Guhl, 1942). These flocks were under observation during the latter part of the afternoon, when matings occur most frequently (Heuser, 1916; Philips, 1919; Upp, 1928; Skard, 1937; Parker, McKenzie, and Kempster, 1940). The peck-orders given in Tables 1 and 2 were based on more than 9,000 pecks or threats; nearly 1,000 matings were observed during this particular series of observations and experiments.

OBSERVATIONS

Tables 1 and 2 present many of the basic data obtained in flock D and flock F in 1940–41. Table 1 summarizes the data obtained when the cocks were introduced singly and successively into the pens of their respective flocks of hens. Table 2 gives a similar summary for the observations when all 4 cocks of each flock were continuously in the pens of the hens. The results for each flock are separated in the body of the tables. The hens are designated by letters and are listed according to their position in the peck-order, beginning with the alpha hen. The cocks are indicated by Roman numerals, and their peck-order reads from left to right. It will be noted that the male order shown in Table 2 differs from that in Table 1, since in the meantime cock IV had defeated cock III.

The column headings marked "Treadings," "Courtings," "Sex Invitations," and "Avoidances" (described below) indicate the types of behavior tabulated; no avoidances were recorded in Table 1 for flock D, as the value of such data was not appreciated when this study began.

Early in the observations of flock D it was noted that hens would fly to the roost or the top of the nest boxes, where they remained for varying lengths of the time and where they were usually out of contact with the cocks. No matings occurred on these perches; and, since sustained avoidance would become a major factor in computing the frequency of sexual behavior between hens and cocks, it was found necessary to record, in minutes, the time each hen and each cock spent on the floor, where it was exposed on the floor, where it was exposed to the opposite sex. From such data each bird has been ranked, within its flock and sex, according to its mean rate in mating, in courting, and in giving or receiving "sex

TABLE 1

HETEROSEXUAL BEHAVIOR WHEN COCKS WERE INTRODUCED SINGLY
AND SUCCESSIVELY INTO THE PEN OF THE HENS

Results are given by individuals and by totals for each hen and each cock. (See Table 3)

Flock D (40 Observations between April 13 and June 20, 1940)

♀♀	♂ I				♂ II				♂ III				♂ IV				Totals			
PECK-ORDER	Treadings	Courtings	"Sex Invitations"	Avoidances	Treadings	Courtings	"Sex Invitations"	Avoidances	Treadings	Courtings	"Sex Invitations"	Avoidances	Treadings	Courtings	"Sex Invitations"	Avoidances	Treadings	Courtings	"Sex Invitations"	Avoidances
GY....	2	10	0	...	6	10	0	...	6	10	0	...	1	4	0	...	15	34	0	...
BB....	2	20	0	...	7	9	0	...	5	9	0	...	0	1	0	...	14	39	0	...
GG....	6	16	0	...	2	14	0	...	4	11	0	...	3	11	0	...	15	52	0	...
BR....	10	11	1	...	13	8	0	...	13	26	0	...	8	17	3	...	44	62	4	...
RY....	9	15	5	...	6	10	0	...	6	24	2	...	8	20	4	...	29	69	11	...
YY....	11	24	6	...	7	23	0	...	7	31	2	...	13	37	1	...	38	115	9	...
RW...	5	25	1	...	6	10	0	...	6	13	0	...	5	21	0	...	22	69	1	...
Totals	45	121	13	...	47	84	0	...	47	124	4	...	38	111	8	...	177	440	25	...

Flock F (60 Observations between April 26 and June 30, 1941)

♀♀	♂ I*				♂ V				♂ VI				♂ VII				♂ VIII				Totals			
PECK-ORDER	T	C	SI	A	T	C	SI	A	T	C	SI	A	T	C	SI	A	T	C	SI	A	T	C	SI	A
WG...	4	29	3	22	9	20	11	12	1	56	3	47	4	21	3	16	5	27	3	36	23	153	23	132
RG....	8	25	6	18	5	7	7	6	3	17	9	9	2	3	2	6	8	24	6	32	26	76	30	71
PB....	5	22	2	18	8	18	3	15	7	32	6	27	6	16	12	16	5	23	2	35	31	111	25	111
YB....	1	32	0	31	2	14	4	11	3	22	6	16	7	24	6	9	2	22	6	21	15	114	22	88
RB....	0	14	0	20	1	15	2	26	1	29	0	30	2	12	2	13	3	18	3	35	7	88	7	124
YG†...	9	46	2	30	6	18	12	12	14	34	5	18	10	15	12	6	5	30	5	34	44	143	36	100
GB‡...	2	9	0	9	2	11	1	16	0	6	1	6	0	4	0	6	4	7	4	15	8	37	6	52
YR....	7	25	3	13	18	28	4	7	11	22	9	8	6	11	3	6	15	55	5	46	57	141	24	80
WR...	7	35	3	23	7	17	6	7	11	39	5	17	6	23	2	15	9	46	11	50	40	160	27	112
Totals	43	237	19	184	58	148	50	112	51	257	44	177	43	129	42	93	56	252	45	304	251	1,023	200	870

* Cock *I* was not housed with the F-cocks.

† Hen YG gained dominance over PB on May 27.

‡ Hen GB died May 30.

invitations." The ranking was possible (body text continues from columns — note: the second column bottom text reads:)

spent on the floor, where it was exposed to the opposite sex. From such data each bird has been ranked, within its flock and sex, according to its mean rate in mating, in courting, and in giving or receiving "sex

TABLE 2

HETEROSEXUAL BEHAVIOR WHEN ALL 4 COCKS WERE CONTINUOUSLY IN THE PEN OF THE HENS
Results are given by individuals and by totals for each hen and each cock. (See Table 4)

PECK-ORDER	TREADINGS	COURTINGS	"SEX INVITATIONS"	AVOIDANCES	TREADINGS	COURTINGS	"SEX INVITATIONS"	AVOIDANCES	TREADINGS	COURTINGS	"SEX INVITATIONS"	AVOIDANCES	TREADINGS	COURTINGS	"SEX INVITATIONS"	AVOIDANCES	TREADINGS	COURTINGS	"SEX INVITATIONS"	AVOIDANCES
♀♀	♂ I				♂ II				♂ IV				♂ III				Totals			

Flock D: Period A (29 Observations between July 4 and August 4, 1940)

PECK-ORDER	TREADINGS	COURTINGS	"SEX INVITATIONS"	AVOIDANCES	TREADINGS	COURTINGS	"SEX INVITATIONS"	AVOIDANCES	TREADINGS	COURTINGS	"SEX INVITATIONS"	AVOIDANCES	TREADINGS	COURTINGS	"SEX INVITATIONS"	AVOIDANCES	TREADINGS	COURTINGS	"SEX INVITATIONS"	AVOIDANCES
GY........	7	31	2	28	2	3	0	19	0	0	0	0	0	0	0	6	9	34	2	53
BB........	5	44	0	38	2	15	0	49	1	8	0	7	0	1	0	8	8	68	0	102
GG........	9	27	0	23	0	10	0	27	4	14	0	11	0	1	0	1	13	52	0	62
BR........	17	52	2	32	4	16	0	31	11	34	0	26	0	0	0	6	32	102	2	95
RY........	22	23	10	19	3	11	0	43	4	11	0	28	0	0	0	5	29	45	10	95
YY........	26	55	7	33	1	13	0	25	12	12	0	10	0	1	0	2	39	81	7	70
RW.......	2	10	0	9	1	1	0	7	1	4	0	7	0	0	0	0	4	15	0	23
Totals...	88	242	21	182	13	69	0	201	33	83	0	89	0	3	0	28	134	397	21	500

Flock D: Period B (34 Observations between August 5 and September 8, 1940)

PECK-ORDER	TREADINGS	COURTINGS	"SEX INVITATIONS"	AVOIDANCES	TREADINGS	COURTINGS	"SEX INVITATIONS"	AVOIDANCES			AVOIDANCES	TREADINGS	COURTINGS	"SEX INVITATIONS"	AVOIDANCES
Totals...	69	198	28	174	1	45	0	258	Died		48	70	243	28	480

Flock D: Period C (31 Observations between September 9 and October 11, 1940)

PECK-ORDER	TREADINGS	COURTINGS	"SEX INVITATIONS"	AVOIDANCES	TREADINGS	COURTINGS	"SEX INVITATIONS"	AVOIDANCES	TREADINGS	COURTINGS	"SEX INVITATIONS"	AVOIDANCES	TREADINGS	COURTINGS	"SEX INVITATIONS"	AVOIDANCES
Totals...	58*	125*	22*	181*	1	17	0	89	Died		4	5	2	116	5	22	2	205

Flock F (48 Observations between July 3 and August 20, 1941)

PECK-ORDER	TREADINGS	COURTINGS	"SEX INVITATIONS"	AVOIDANCES	TREADINGS	COURTINGS	"SEX INVITATIONS"	AVOIDANCES	TREADINGS	COURTINGS	"SEX INVITATIONS"	AVOIDANCES	TREADINGS	COURTINGS	"SEX INVITATIONS"	AVOIDANCES	TREADINGS	COURTINGS	"SEX INVITATIONS"	AVOIDANCES
♀♀	♂ V				♂ VI				♂ VII				♂ VIII				Totals			
WG.......	16	84	4	59	2	23	1	18	1	32	0	25	17	97	0	159	36	236	5	261
RG†......	5	41	0	27	10	47	11	31	0	8	0	15	6	38	0	65	21	134	11	138
PB........	16	96	0	72	8	76	7	58	2	25	0	26	10	66	1	133	36	263	8	289
YB........	0	29	0	26	7	47	5	29	1	26	1	28	0	16	0	55	8	118	6	138
RB........	0	14	0	13	3	23	1	22	0	5	0	11	0	26	0	60	3	68	1	106
YG‡......	8	78	2	54	6	51	3	38	2	15	1	17	6	63	0	139	22	207	6	248
YR........	21	111	1	79	10	76	3	56	0	14	1	14	11	79	1	119	42	280	6	268
WR.......	7	61	0	38	20	94	7	69	1	19	1	18	2	71	0	107	30	245	8	232
Totals...	73	514	7	368	66	437	38	321	7	144	4	154	52	456	2	837	198	1,551	51	1,680

* Cock *I* with flock G.

† Hen RG died August 2.

‡ Hen YG pecked PB.

invitations" and to the mean rate at which the hens avoided the cocks. These rates and ranks are given in Tables 3 and 4, in terms of (a) the frequency of each type of behavior as the mean number per hour and (b) the relative rank for each bird among the members of its flock and sex in its participation in these types of behavior. Equal ranking is indicated by half-steps. The same method of ranking was used with flock F, although the number of minutes each bird spent on the floor was the same for all individuals, since the roost and nest boxes were screened off during observation time. Mean rates are also given for each hen and each cock and for the sexes as groups. The latter rates were computed from total observations (Tables 1 and 2) and aggregate time rather than as the means of the rates given in the body of these tables.

OBSERVATIONS ON GENERAL MATING
BEHAVIOR WITHIN FLOCKS

The initial sexual reactions of the cocks toward a hen were recorded as courtings and involved such behavior as raising the hackle while pointing the head toward a female; catching a hen by the comb or neck; placing a foot on the saddle of a prospective mate; or pursuing her on the run. The wing-flutter or so-called "waltz," also often included as a form of courting, was recorded separately in the protocol, since this behavior seems to be characteristically somewhat different from other courtship patterns. In the waltz the cock lowers and flutters the wing on the opposite side from the female and makes an arc about the hen with a series of short, quick steps.

Often the hens seem to be sexually indifferent to the presence of the cock, but they may squat or crouch before him; such behavior by the hen was recorded as a "sex invitation." A hen may respond to active courtship by squatting, in which case copulation usually follows; she may be indifferent; or she may avoid the male. Her avoidance may take various forms with varying intensities, from merely stepping aside to running; or if the male mounts her, she may struggle to get away; or she may go to the roost and remain there, thus giving a sustained avoidance.

The movement of a hen was not considered as avoidance unless some overt behavior by the cock toward her could be associated with her reaction. Most of the recorded avoidances were responses to courting or to attempts to tread. No records of avoidance were taken when the flock-D males were introduced singly into the pen of the D-hens. Avoiding reactions may be accompanied by vocalizations ranging from a faint cry to a loud squawk. The types of sexual behavior we have observed are similar to those outlined by Skard (1937).

It is well known that to copulate the male mounts the squatting hen, grasps her comb or hackle with his beak, and balances himself on her back with a treading motion. The hen crouches low and moves her tail to one side, while the cock rears up and, spreading his tail, the vents meet. An attempt was made to determine and record whether the copulations were initiated by the male or by the female, or simultaneously by both.

In some instances the cock mounted a struggling hen and such matings appeared to be forced. As compared with the obviously male-initiated and female-initiated ("invited") matings, there is a third category, composed of incidences in which mounting by the cock and crouching by the hen occurred simultaneously. In the latter the initiator could not be determined, since the crouch may have

TABLE 3

HETEROSEXUAL BEHAVIOR PATTERNS WHEN THE COCKS WERE INTRODUCED SINGLY AND SUCCESSIVELY INTO THE PEN OF THE HENS*

(See Table 1 and p. 370)

Flock D (40 Observations between April 13 and June 20, 1940)

♀♀ (Peck-Order)	♂ *I* Treadings Fr.	Rk.	Courtings Fr.	Rk.	"Sex Invitations" Fr.	Rk.	Avoidances Fr.	Rk.
GY	0.17	7	0.89	6.5	0.00	6	....	....
BB	0.21	6	2.10	1.0	0.00	6	....	....
GG	0.62	4	1.68	4.0	0.00	6	....	....
BR	0.81	1	0.89	6.5	0.08	3	....	....
RY	0.72	3	1.20	5.0	0.40	2	....	....
YY	0.79	2	1.73	3.0	0.43	1	....	....
RW	0.38	5	1.91	2.0	0.07	4	....	....
Mean rates	0.54	3	1.47	3.0	0.15	1	....	....
Rates of ♂♂	3.05	3	8.23	2.0	0.88	1	....	....

♀♀ (Peck-Order)	♂ II Treadings Fr.	Rk.	Courtings Fr.	Rk.	"Sex Invitations" Fr.	Rk.	Avoidances Fr.	Rk.
GY	0.91	6	1.53	5	0.00	....	....	....
BB	1.16	4	1.50	6	0.00	....	....	....
GG	0.34	7	2.41	3	0.00	....	....	....
BR	2.20	2	1.36	7	0.00	....	....	....
RY	1.25	3	2.08	4	0.00	....	....	....
YY	1.00	5	3.28	2	0.00	....	....	....
RW	2.40	1	4.00	1	0.00	....	....	....
Mean rates	1.21	1	2.17	1	0:00	4	....	....
Rates of ♂♂	4.56	1	8.15	3	0.00	4	....	....

♀♀ (Peck-Order)	♂ III Treadings Fr.	Rk.	Courtings Fr.	Rk.	"Sex Invitations" Fr.	Rk.	Avoidances Fr.	Rk.
GY	0.57	4.5	0.95	7	0.00	....	....	....
BB	0.61	2.0	1.09	6	0.00	....	....	....
GG	0.48	7.0	1.32	4	0.00	....	....	....
BR	1.12	1.0	2.24	3	0.00	....	....	....
RY	0.59	3.0	2.38	2	0.19	....	....	....
YY	0.55	6.0	2.46	1	0.16	....	....	....
RW	0.57	4.5	1.23	5	0.00	....	....	....
Mean rates	0.65	2.0	1.72	2	0.05	3	....	....
Rates of ♂♂	3.31	2.0	8.73	1	0.28	3	....	....

♀♀ (Peck-Order)	♂ IV Treadings Fr.	Rk.	Courtings Fr.	Rk.	"Sex Invitations" Fr.	Rk.	Avoidances Fr.	Rk.
GY	0.09	6	0.37	6	0.00	5.5	....	....
BB	0.00	7	0.10	7	0.00	5.5	....	....
GG	0.31	5	1.15	5	0.00	5.5	....	....
BR	0.65	3	1.38	4	0.24	2.0	....	....
RY	0.72	2	1.80	2	0.36	1.0	....	....
YY	0.94	1	2.68	1	0.07	3.0	....	....
RW	0.41	4	1.72	3	0.00	5.5	....	....
Mean rates	0.47	4	1.39	4	0.10	2.0	....	....
Rates of ♂♂	2.46	4	7.20	4	0.52	2.0	....	....

♀♀ (Peck-Order)	(no cock) Treadings Fr.	Rk.	Courtings Fr.	Rk.	"Sex Invitations" Fr.	Rk.	Avoidances Fr.	Rk.
GY	....	....	....	....	....	....	....	....
BB	....	....	....	....	....	....	....	....
GG	....	....	....	....	....	....	....	....
BR	....	....	....	....	....	....	....	....
RY	....	....	....	....	....	....	....	....
YY	....	....	....	....	....	....	....	....
RW	....	....	....	....	....	....	....	....
Mean rates	....	....	....	....	....	....	....	....
Rates of ♂♂	....	....	....	....	....	....	....	....

♀♀ (Peck-Order)	Mean Rates Treadings Fr.	Rk.	Courtings Fr.	Rk.	"Sex Invitations" Fr.	Rk.	Avoidances Fr.	Rk.
GY	0.38	7	0.87	7	0.00	6	....	....
BB	0.41	6	1.16	6	0.00	6	....	....
GG	0.45	5	1.56	4	0.00	6	....	....
BR	1.04	1	1.47	5	0.09	3	....	....
RY	0.75	3	1.78	3	0.28	1	....	....
YY	0.80	2	2.43	1	0.19	2	....	....
RW	0.57	4	1.79	2	0.03	4	....	....
Mean rates	0.65	....	1.61	....	0.09	....	....	....
Rates of ♂♂	3.23	....	8.02	....	0.45	....	....	....

Flock F (60 Observations between April 26 and June 30, 1941)

♀♀ (Peck-Order)	♂ *I*† Treadings Fr.	Rk.	Courtings Fr.	Rk.	"Sex Invitations" Fr.	Rk.	Avoidances Fr.	Rk.
WG	0.65	7	4.75	4	0.49	3.5	3.61	4.0
RG	1.31	2	4.09	6	0.98	1.0	2.95	7.5
PB	0.82	5	3.61	7	0.32	5.5	2.95	7.5
YB	0.16	8	5.23	3	0.00	8.0	5.08	1.0
RB	0.00	9	2.29	9	0.00	8.0	3.28	6.0
YG‡	1.47	1	7.54	1	0.32	5.5	4.91	2.0
GB§	0.80	6	3.60	8	0.00	8.0	3.60	5.0
YR	1.25	3	4.46	5	0.53	2.0	2.32	9.0
WR	1.15	4	5.73	2	0.49	3.5	3.77	3.0
Mean rates	0.85	....	4.68	....	0.37	....	3.65	....
Rates of ♂♂	7.05	....	38.85	....	3.11	....	30.16	....

♀♀ (Peck-Order)	♂ V Treadings Fr.	Rk.	Courtings Fr.	Rk.	"Sex Invitations" Fr.	Rk.	Avoidances Fr.	Rk.
WG	1.47	2	3.28	3.0	1.80	2	1.97	4.5
RG	0.82	6	1.15	9.0	1.15	3	0.98	9.0
PB	1.31	3	2.95	4.5	0.49	7	2.46	3.0
YB	0.32	8	2.29	8.0	0.65	6	1.80	6.0
RB	0.16	9	2.46	7.0	0.32	9	4.26	2.0
YG‡	0.98	5	2.95	4.5	1.97	1	1.97	4.5
GB§	0.80	7	4.40	2.0	0.40	8	6.40	1.0
YR	3.21	1	5.00	1.0	0.71	5	1.25	7.0
WR	1.15	4	2.79	6.0	0.98	4	1.15	8.0
Mean rates	1.14	1	2.92	3.0	0.98	1	2.21	3.0
Rates of ♂♂	9.50	1	24.26	3.0	8.19	1	18.36	3.0

♀♀ (Peck-Order)	♂ VI Treadings Fr.	Rk.	Courtings Fr.	Rk.	"Sex Invitations" Fr.	Rk.	Avoidances Fr.	Rk.
WG	0.16	7.5	9.18	1	0.49	7.0	7.70	1
RG	0.49	5.5	2.79	8	1.47	2.0	1.47	8
PB	1.15	4.0	5.24	4	0.98	3.5	4.43	3
YB	0.49	5.5	3.61	7	0.98	3.5	2.62	6
RB	0.16	7.5	4.75	5	0.00	9.0	4.91	2
YG‡	2.29	1.0	5.57	3	0.82	5.5	2.95	4
GB§	0.00	9.0	2.40	9	0.40	8.0	2.40	7
YR	1.96	2.0	3.93	6	1.61	1.0	1.43	9
WR	1.80	3.0	6.39	2	0.82	5.5	2.79	5
Mean rates	1.01	3.0	5.12	1	0.87	3.0	3.50	2
Rates of ♂♂	8.36	3.0	42.13	1	7.21	3.0	29.01	2

♀♀ (Peck-Order)	♂ VII Treadings Fr.	Rk.	Courtings Fr.	Rk.	"Sex Invitations" Fr.	Rk.	Avoidances Fr.	Rk.
WG	0.65	6.0	3.44	3	0.49	5.0	2.62	1.5
RG	0.32	7.5	0.49	9	0.32	7.0	0.98	8.5
PB	0.98	4.5	2.62	4	1.97	1.5	2.62	1.5
YB	1.15	3.0	3.93	1	0.98	3.0	1.47	6.0
RB	0.32	7.5	1.97	7	0.32	7.0	2.13	5.0
YG‡	1.64	1.0	2.46	5	1.97	1.5	0.98	8.5
GB§	0.00	9.0	1.60	8	0.00	9.0	2.40	4.0
YR	1.17	2.0	2.16	6	0.58	4.0	1.17	7.0
WR	0.98	4.5	3.77	2	0.32	7.0	2.46	3.0
Mean rates	0.86	4.0	2.57	4	0.85	4.0	1.85	4.0
Rates of ♂♂	7.05	4.0	21.14	4	6.88	4.0	15.24	4.0

♀♀ (Peck-Order)	♂ VIII Treadings Fr.	Rk.	Courtings Fr.	Rk.	"Sex Invitations" Fr.	Rk.	Avoidances Fr.	Rk.
WG	0.82	6	4.43	4	0.49	7.5	5.90	4.0
RG	1.31	4	3.93	5	0.98	4.5	5.24	8.0
PB	0.82	6	3.77	6	0.32	9.0	5.73	5.5
YB	0.32	9	3.61	7	0.98	4.5	3.44	9.0
RB	0.49	8	2.95	8	0.49	7.5	5.73	5.5
YG‡	0.82	6	4.91	3	0.82	6.0	5.57	7.0
GB§	1.60	2	2.80	9	1.60	2.0	6.00	3.0
YR	3.00	1	11.00	1	1.00	3.0	9.20	1.0
WR	1.47	3	7.54	2	1.80	1.0	8.19	2.0
Mean rates	1.11	2	5.03	2	0.90	2.0	6.06	1.0
Rates of ♂♂	9.18	2	41.31	2	7.37	2.0	49.83	1.0

♀♀ (Peck-Order)	Mean Rates Treadings Fr.	Rk.	Courtings Fr.	Rk.	"Sex Invitations" Fr.	Rk.	Avoidances Fr.	Rk.
WG	0.75	6	5.03	3.0	0.75	6	4.34	1
RG	0.85	5	2.50	9.0	0.98	2	2.33	9
PB	1.02	4	3.65	6.0	0.82	5	3.65	5
YB	0.49	8	3.75	5.0	0.72	7	2.89	8
RB	0.23	9	2.89	8.0	0.23	9	4.07	3
YG‡	1.44	2	4.70	4.0	1.18	1	3.29	6
GB§	0.64	7	2.96	7.0	0.48	8	4.16	2
YR	2.12	1	5.26	1.5	0.89	3	2.98	7
WR	1.31	3	5.26	1.5	0.88	4	3.68	4
Mean rates	0.99	....	4.05	....	0.79	....	3.45	....
Rates of ♂♂	8.26	....	33.65	....	6.58	....	28.62	....

* Fr. = frequency; Rk. = rank. Italic figures indicate rank.

† Cock *I* was hot housed with the F-cocks.

‡ Hen YG gained dominance over PB on May 27.

§ Hen GB died May 30.

TABLE 4

HETEROSEXUAL BEHAVIOR PATTERNS WHEN ALL 4 COCKS WERE CONTINUOUSLY IN THE PEN OF THE HENS*

(See Table 2 and p. 370)

Peck-Order	Treadings Fr.	Treadings Rk.	Courtings Fr.	Courtings Rk.	"Sex Invitations" Fr.	"Sex Invitations" Rk.	Avoidances Fr.	Avoidances Rk.
Flock D: Period A (29 Observations between July 4 and August 4, 1940)								
♀♀	♂ *I*							
GY.....	0.38	*5*	1.72	*4*	0.11	*3*	1.55	*4*
BB.....	0.23	*7*	2.07	*3*	0.00	*6*	1.78	*1*
GG.....	0.50	*4*	1.50	*5*	0.00	*6*	1.28	*6*
BR.....	0.88	*3*	2.69	*2*	0.10	*4*	1.65	*3*
RY.....	1.08	*2*	1.13	*7*	0.48	*1*	0.94	*7*
YY.....	1.36	*1*	2.87	*1*	0.36	*2*	1.72	*2*
RW.....	0.29	*6*	1.49	*6*	0.00	*6*	1.33	*5*
Mean rates..	0.71	*1*	1.97	*3*	0.17	*1*	1.48	*4*
Rates of ♂♂..	3.18	*1*	8.76	*1*	0.76	*1*	6.59	*3*
Flock D: Period B (34 Observations between August 5 and September 8, 1940)								
Mean rates..	0.81	*1*	2.35	*1*	0.33	*1*	2.06	*3*
Rates of ♂♂..	3.41	*1*	9.80	*1*	1.38	*1*	8.61	*2*
Flock D: Period C (31 Observations between September 9 and October 11, 1940)								
Mean rates..	1.15	†	2.48	†	0.44	†	3.60	†
Rates of ♂♂..	6.04	...	13.02	...	2.29	...	18.85	...
Flock F (48 Observations between July 3 and August 20, 1941)								
♀♀	♂ V							
WG.....	0.35	*2.5*	1.84	*3*	0.08	...	1.29	*3*
RG‡....	0.17	*4.5*	1.45	*5*	0.00	...	0.97	*5*
PB.....	0.35	*2.5*	2.10	*2*	0.00	...	1.57	*2*
YB.....	0.00	*7.5*	0.63	*7*	0.00	...	0.57	*7*
RB.....	0.00	*7.5*	0.31	*8*	0.00	...	0.28	*8*
YG§....	0.17	*4.5*	1.71	*4*	0.04	...	1.18	*4*
YR.....	0.46	*1.0*	2.43	*1*	0.02	...	1.73	*1*
WR.....	0.15	*6.0*	1.33	*6*	0.00	...	0.83	*6*
Mean rates..	0.21	*1.0*	1.48	*1*	0.02	*2*	1.06	*2*
Rates of ♂♂..	1.60	*1.0*	11.27	*1*	0.15	*2*	8.07	*2*

Peck-Order	Treadings Fr.	Treadings Rk.	Courtings Fr.	Courtings Rk.	"Sex Invitations" Fr.	"Sex Invitations" Rk.	Avoidances Fr.	Avoidances Rk.
Flock D: Period A								
♀♀	♂ II							
GY.....	0.35	*3.5*	0.53	*6*	0.00	...	3.39	*6*
BB.....	0.35	*3.5*	2.67	*2*	0.00	...	8.75	*1*
GG.....	0.00	*7.0*	2.32	*4*	0.00	...	6.27	*3*
BR.....	0.78	*1.0*	3.13	*1*	0.00	...	6.07	*4*
RY.....	0.60	*2.0*	2.20	*5*	0.00	...	8.60	*2*
YY.....	0.18	*6.0*	2.45	*3*	0.00	...	4.71	*5*
RW.....	0.34	*5.0*	0.34	*7*	0.00	...	2.41	*7*
Mean rates..	0.38	*3.0*	2.02	*2*	0.00	...	5.89	*2*
Rates of ♂♂..	1.34	*3.0*	7.11	*2*	0.00	...	20.72	*2*
Flock D: Period B								
Mean rates..	0.04	*2*	1.78	*2*	0.00	...	1.20	*1*
Rates of ♂♂..	0.10	*2*	4.45	*2*	0.00	...	25.54	*1*
Flock D: Period C								
Mean rates..	0.02	*2*	0.36	*1*	0.00	*2*	1.86	*2*
Rates of ♂♂..	0.12	*2*	2.15	*1*	0.00	*2*	11.26	*2*
Flock F								
♀♀	♂ VI							
WG.....	0.04	*8*	0.50	*7.5*	0.02	*7.5*	0.39	*8*
RG‡....	0.36	*2*	1.67	*2.0*	0.39	*1.0*	1.10	*4*
PB.....	0.17	*4*	1.66	*3.5*	0.15	*2.5*	1.27	*2*
YB.....	0.15	*5*	1.03	*6.0*	0.11	*4.0*	0.63	*6*
RB.....	0.06	*7*	0.50	*7.5*	0.02	*7.5*	0.48	*7*
YG§....	0.13	*6*	1.12	*5.0*	0.06	*5.5*	0.83	*5*
YR.....	0.22	*3*	1.66	*3.5*	0.06	*5.5*	1.22	*3*
WR.....	0.43	*1*	2.06	*1.0*	0.15	*2.5*	1.51	*1*
Mean rates..	0.19	*2*	1.25	*3.0*	0.11	*1.0*	0.92	*3*
Rates of ♂♂..	1.44	*2*	9.58	*3.0*	0.83	*1.0*	7.03	*3*

Peck-Order	Treadings Fr.	Treadings Rk.	Courtings Fr.	Courtings Rk.	"Sex Invitations" Fr.	"Sex Invitations" Rk.	Avoidances Fr.	Avoidances Rk.
Flock D: Period A								
♀♀	♂ IV							
GY.....	0.00	*7*	0.00	*7*	0.00	...	0.00	*7*
BB.....	0.10	*6*	0.82	*6*	0.00	...	0.72	*6*
GG.....	0.46	*3*	1.60	*2*	0.00	...	1.26	*4*
BR.....	1.14	*2*	3.54	*1*	0.00	...	2.70	*2*
RY.....	0.45	*4*	1.25	*4*	0.00	...	3.18	*1*
YY.....	1.31	*1*	1.31	*3*	0.00	...	1.09	*5*
RW.....	0.30	*5*	1.21	*5*	0.00	...	2.12	*3*
Mean rates..	0.58	*2*	1.46	*4*	0.00	...	1.56	*3*
Rates of ♂♂..	2.53	*2*	6.38	*3*	0.00	...	6.84	*4*
Flock D: Period B								
Mean rates..	Died							
Rates of ♂♂..								
Flock D: Period C								
Mean rates..	Died							
Rates of ♂♂..								
Flock F								
♀♀	♂ VII							
WG.....	0.02	*4.0*	0.70	*1*	0.00	...	0.55	*3*
RG‡....	0.00	*7.0*	0.28	*7*	0.00	...	0.53	*4*
PB.....	0.04	*1.5*	0.55	*3*	0.00	...	0.57	*2*
YB.....	0.02	*4.0*	0.57	*2*	0.02	...	0.61	*1*
RB.....	0.00	*7.0*	0.11	*8*	0.00	...	0.24	*8*
YG§....	0.04	*1.5*	0.33	*5*	0.02	...	0.37	*6*
YR.....	0.00	*7.0*	0.31	*6*	0.02	...	0.31	*7*
WR.....	0.02	*4.0*	0.41	*4*	0.02	...	0.39	*5*
Mean rates..	0.02	*4.0*	0.41	*4*	0.01	*3*	0.44	*4*
Rates of ♂♂..	0.15	*4.0*	3.15	*4*	0.08	*3*	3.37	*4*

Peck-Order	Treadings Fr.	Treadings Rk.	Courtings Fr.	Courtings Rk.	"Sex Invitations" Fr.	"Sex Invitations" Rk.	Avoidances Fr.	Avoidances Rk.
Flock D: Period A								
♀♀	♂ III							
GY.....	0.00	...	0.00	...	0.00	...	33.33	*3*
BB.....	0.00	...	6.25	...	0.00	...	50.00	*1*
GG.....	0.00	...	9.09	...	0.00	...	9.09	*6*
BR.....	0.00	...	0.00	...	0.00	...	37.50	*2*
RY.....	0.00	...	0.00	...	0.00	...	31.25	*4*
YY.....	0.00	...	6.25	...	0.00	...	12.50	*5*
RW.....	0.00	...	0.00	...	0.00	...	0.00	*7*
Mean rates..	0.00	*4*	3.00	*1*	0.00	...	28.00	*1*
Rates of ♂♂..	0.00	*4*	2.50	*4*	0.00	...	23.33	*1*
Flock D: Period B								
Mean rates..	0.00	*3*	0.00	*3*	0.00	*3*	3.31	*2*
Rates of ♂♂..	0.00	*3*	0.00	*3*	0.00	*3*	8.57	*3*
Flock D: Period C								
Mean rates..	0.08	*1*	0.10	*2*	0.04	*1*	2.44	*1*
Rates of ♂♂..	0.48	*1*	0.60	*2*	0.24	*1*	13.97	*1*
Flock F								
♀♀	♂ VIII							
WG.....	0.37	*1.0*	2.12	*1*	0.00	...	3.48	*1*
RG‡....	0.21	*4.0*	1.35	*6*	0.00	...	2.31	*6*
PB.....	0.22	*3.0*	1.44	*4*	0.02	...	2.91	*3*
YB.....	0.00	*7.5*	0.35	*8*	0.00	...	1.20	*8*
RB.....	0.00	*7.5*	0.57	*7*	0.00	...	1.31	*7*
YG§....	0.13	*5.0*	1.38	*5*	0.00	...	3.04	*2*
YR.....	0.24	*2.0*	1.73	*2*	0.02	...	2.62	*4*
WR.....	0.04	*6.0*	1.55	*3*	0.00	...	2.34	*5*
Mean rates..	0.15	*3.0*	1.31	*2*	0.005	*4*	2.40	*1*
Rates of ♂♂..	1.14	*3.0*	10.00	*2*	0.04	*4*	18.35	*1*

Peck-Order	Treadings Fr.	Treadings Rk.	Courtings Fr.	Courtings Rk.	"Sex Invitations" Fr.	"Sex Invitations" Rk.	Avoidances Fr.	Avoidances Rk.
Flock D: Period A								
♀♀	Mean Rates							
GY.....	0.28	*6*	1.08	*7*	0.06	*3*	1.69	*7*
BB.....	0.21	*7*	1.84	*3*	0.00	*6*	2.77	*2*
GG.....	0.41	*4*	1.67	*4*	0.00	*6*	1.99	*5*
BR.....	0.93	*2*	2.97	*1*	0.05	*4*	2.76	*3*
RY.....	0.84	*3*	1.31	*5*	0.29	*1*	2.78	*1*
YY.....	1.15	*1*	2.40	*2*	0.20	*2*	2.07	*4*
RW.....	0.31	*5*	1.16	*6*	0.00	*6*	1.77	*6*
Mean rates..	0.62	...	1.85	...	0.09	...	2.33	...
Rates of ♂♂..	2.54	...	7.54	...	0.40	...	9.50	...
Flock D: Period B								
Mean rates..	0.56	...	1.96	...	0.22	...	3.86	...
Rates of ♂♂..	1.95	...	6.76	...	0.78	...	13.37	...
Flock D: Period C								
Mean rates..	0.05	...	0.23	...	0.02	...	2.11	...
Rates of ♂♂..	0.30	...	1.35	...	0.12	...	12.57	...
Flock F								
♀♀	Mean Rates							
WG.....	0.19	*2.5*	1.29	*4*	0.02	*7.0*	1.42	*3*
RG‡....	0.18	*4.0*	1.19	*5*	0.09	*1.0*	1.22	*6*
PB.....	0.19	*2.5*	1.44	*2*	0.04	*2.5*	1.58	*1*
YB.....	0.04	*7.0*	0.64	*7*	0.03	*5.0*	0.75	*7*
RB.....	0.01	*8.0*	0.37	*8*	0.005	*8.0*	0.57	*8*
YG§....	0.12	*6.0*	1.13	*6*	0.03	*5.0*	1.35	*4*
YR.....	0.23	*1.0*	1.53	*1*	0.03	*5.0*	1.46	*2*
WR.....	0.16	*5.0*	1.34	*3*	0.04	*2.5*	1.27	*5*
Mean rates..	0.14	...	1.11	...	0.03	...	1.20	...
Rates of ♂♂..	1.08	...	8.50	...	0.28	...	9.20	...

* Fr. = frequency; Rk. = rank. Italic figures indicate rank.

† Cock *I* with flock G.

‡ Hen RG died August 2.

§ Hen YG pecked PB.

resulted either from the weight of the cock or coincidentally from a spontaneous reaction by the hen; crouches occurring under these conditions were not considered as "sex invitations." Matings were recorded as incomplete whenever it was obvious that the vents failed to meet.

Although not systematically studied, it was observed in the earlier tests with flock D and other flocks that thirst, food, and roosting drives, especially thirst, seemed readily to take precedence over the sex drive. A cock not watered on a given day was likely to spend much of the observation period drinking, in place of paying attention to the hens. If the observation periods were late in the afternoon, the cock was likely to go to roost with the hens instead of attempting to mate with them. An effort was made to avoid these complicating factors.

COMPARISONS AMONG THE HENS

The analysis in the present section is limited to the relatively simple situation that obtained when the cocks were introduced singly and successively. There was some daily variation in mating frequency by the individual hens. Heuser (1916) made a similar observation. To test for the consistency of matings, the records from April 13 to June 20 of flock D and from April 26 to June 30 of flock F were selected. These data were not complicated by the effect of the presence of more than one cock or by the molting season and were all the more significant, since the time of the year includes the peak of the normal laying cycle. The number of treadings for odd-numbered and even-numbered days were tabulated for each hen, and paired comparisons for all the hens of each flock were then made by the method of "Student" and found to be statistically insignificant (P-values > 0.57),[3] which means that there was no real difference between the number of matings by these hens for the alternate days during the season of the year sampled.

The hens usually paid no attention when one of them was being trod, unless the one being mated cried or struggled vigorously to get free. On rare occasions a hen tried to peck an inferior flockmate while she was being held down by the cock. This was more likely to happen if there was an apparent antipathy between them.

Hen BR of flock D began to molt in July, and her rank in matings dropped from first place (Table 3, mean rate) to second place (Table 4, period A); she continued to mate less frequently as her molt became heavy, and consequently ranked sixth for period B. As a result of BR's decrease in sexual activity and YY's cessation of mating (see below), RY was mated more frequently and rose in mating rank from third place (Table 3) to top rank during period B. About 3 weeks after the close of period C, RY also entered a heavy molt, ceased to mate, and avoided all the males. The molt of the remaining D-hens was light and extended over a longer period of time. The observations of flock F did not extend through the molting season and offered no evidence of its effect on their sexual activity.

The periodicities of the molt and of reproduction appear to be independent (van der Meulen, 1939) but may affect each other secondarily, and their respective cycles may show more or less overlapping. The hens of the flocks that molted heavily were very sensitive to handling, and we received the impression

[3] P-values of 0.05 are taken as the upper limit of statistical significance, the smaller the fraction the higher the statistical probability.

that this hypersensitive condition of the skin contributed much to the tendency of molting individuals to avoid close physical contact with the males.

Hen YY of flock D was originally second in mating rank (Table 3, mean rates) and was first during period A (Table 4), when BR molted. On the first day of period B, YY received a pellet of testosterone propionate, implanted subdermally. Within 3 days she ceased to mate; and after the three treadings that occurred during this transitional period she became lowest in mating rank, remaining so for the remainder of period B.

Inspection of Tables 3 and 4 shows a suggestion of a negative correlation between social ranks and ranking based on the frequency at which the hens were trod and at which they gave the so-called "mating invitation" to the cocks. These and other possible correlations will be considered further on page 380, after other pertinent observations have been presented.

COMPARISONS AMONG THE COCKS

Cocks introduced singly and successively.—As with the hens, the cocks of both flocks showed a day-by-day consistency in their treading frequencies for the period during which this matter was tested with the hens. The *P*-values for the odd-numbered versus the even-numbered days of both groups were greater than 0.58. There was no evidence that molting, which was light in all of the D-cocks, influenced their observed behavior.

A comparison of the cocks is more complicated than that of the hens, as the positive and negative reactions of the hens toward the cocks as individuals may influence, to a greater or less degree, the frequency at which a rooster may have the opportunity to mate, court, receive "sex invitations," or be avoided.

The differential behavior of hens toward different cocks was more marked than was the comparable reaction of the cocks to the hens as individuals. Differences between the cocks may be considered in several ways: (*a*) by comparing the total number of matings and other behavior patterns for each cock under like conditions and during equal periods of observations; (*b*) by the examination of the mean frequencies of the matings and courtings of each cock; (*c*) by studying the actions and reactions directed toward each cock by other cocks; (*d*) by a statistical comparison of the frequencies of the female reactions toward each of the cocks; and (*e*) by descriptive remarks on the observed behavior of the cocks.

The recorded treadings by the D-cocks when introduced singly and successively (Table 1), with the possible exception of cock II, showed no very marked differences between them in their success at mating with the hens as a group. Cock II courted the least number of times and was never "invited" to tread. Since some of the D-hens spent much time on the roost, out of contact with certain males, the individuality of the cocks becomes clearer when the rates of male activities are considered as calculated from the time the given cock was closely associated with hens on the floor of the pen. The frequency rates of the males and the mean rates of the females as a group, given in Table 3, suggest greater differences between these cocks than those shown by the raw data.

To test for the statistical significance of the differences between the cocks in the rates presented in Table 3, paired comparisons were made of the frequencies of the behavior patterns of the individual hens when in association with each cock. The mean mating frequency of all the hens when in association with cock II, of 1.21 matings per hour, is significantly greater than the mean rates of the other 3 cocks (P-values: 0.03–0.02). The mean rate (2.17) at which the hens were courted by II was decidedly greater than either of these frequencies with I or IV (P-values: 0.04 and 0.01). When considered in the different possible pair-combinations, there is no statistical difference between the rates for these types of behavior among the other cocks; differences in the rates at which each of the 4 cocks received "sex invitations" failed to show any statistical significance, in part because the 3 top-ranking hens gave no "sex invitations" during this period. The analysis points to cock II as the most active sexually. His comparative libido contrasts with the fact that he was the only male that all the hens failed to "invite" by crouching.

Statistical tabulations often do not give a complete picture of behavior. This is particularly true if intensities of reactions are considered; hence descriptive remarks have value in the interpretation of the results. Cock II approached the hens quickly or chased them from one end of the pen to the other; this caused them to fly to the roost, where they tended to remain while he was in the pen. With the same total elapsed observation time the females spent a total of 2,322 minutes on the floor with cock II, as compared with 4,926 minutes with cock I. Of II's matings, 76 per cent were forced, and none of the hens "invited" him by squatting. Cock I approached the hens slowly from the rear and held his head over a female as though ready to grasp her comb, or he raised his hackle; if the hen moved away, he diverted his attention to another; if a hen squatted, he trod. Only 15 per cent of I's matings appeared to be forced; 29 per cent were "invited"; and in the remaining 55 per

cent courting and squatting appeared to be simultaneous.

Cocks III and IV appeared to grade between I and II as to sex drive but closer to I than to II, with III probably more sexually active than IV; both forced 21 per cent of their copulations. In general, this comparison agrees with the preceding analysis.

The negative reactions of the hens toward each of the cocks were not tabulated for flock D. But this factor was considered by an analysis of sustained avoidance made by paired comparisons of the total time these hens spent on the floor with each cock, the P-values of which are given in Table 5.

These statistics support direct observations and show that the hens gave significantly more sustained avoidance toward II than toward any other rooster, and the least toward cock I. These individual differences also became significant when these same birds were tested in a discrimination cage (Guhl, 1942), where it was found that discriminations were notably influenced by individuality and that the birds acted and reacted according to former pair-contact experience. These observations are in agreement with a conclusion drawn by Tinbergen (1931) that "the members of a flock in which a peck-order occurs know each other individually revealing not only an amazing power of discrimination, but at the same time the working of rather intricate conditioning processes by which an animal forms separately special habits in relation to every individual."

The raw data concerning the mating of the F-males (V, VI, VII, and VIII) given in Table 1 show no marked differences between individuals in total numbers of matings or "sex invitations" received. Cocks VI and VIII courted more and were avoided more than were V and

VII. Some of these differences become more evident when activity rates are compared in a hen-by-hen analysis of the frequency of the female reactions (Table 3) toward each of the cocks.

The following differences have statistical probabilities (*P*-values) that lie between 0.05 and 0.0004. Other possible comparisons are less valid statistically.

TABLE 5

A COMPARISON OF THE TIME WHICH THE D-HENS, AS A GROUP, SPENT ON THE FLOOR IN THE PRESENCE OF EACH OF THE D-COCKS WHEN THE LATTER WERE INTRODUCED SINGLY AND SUCCESSIVELY

COMPARISON OF COCKS	TIME (MINUTES)*		*P*-VALUE
	Total	Mean Difference	
I............ II..........	4,926⎫ 2,322⎭	372	0.006
I............ III.........	4,926⎫ 4,325⎭	86	0.003
I............ IV.........	4,926⎫ 4,781⎭	20	†
II.......... III.........	2,322⎫ 4,325⎭	286	0.001
II.......... IV.........	2,322⎫ 4,781⎭	351	0.004
III......... IV.........	4,325⎫ 4,781⎭	65	†

* The data in the column "Total Time" give the sum of the number of minutes each hen was on the floor with each cock during observation periods. The difference, in minutes, between the cocks is given as the mean difference.

† Denotes a *P*-value of 0.10 or greater.

(It is well to remember that the peck-order of the F-cocks is indicated by the ordinal rank of the numerals and that cock *I* did not meet the F-cocks, and hence his relative social status is not known.) Cocks *I*, VI, and VIII courted more than did V or VII. The hens squatted less frequently for cock *I* than for V, VI, or VIII. They avoided *I* and VI more than VII, and they avoided VIII more than any of the other four males.

Cock VIII had slightly more forced treadings and was less skilful than the others. Cock VII would stand idly, as though oblivious of his female penmates; and VI courted very mildly when a hen came near him. None of these cocks displayed much intensity in sex behavior, and they seemed to be somewhat abnormal, with VIII more nearly approaching the normal pattern in sex behavior. A possible explanation of their apparent abnormality will be suggested later.

As with cock II and the hens of flock D, the cocks that the F-hens avoided at the higher rates (*I*, VI, and VIII) were those that courted with a significantly higher frequency.

Although the F-cocks were relatively weak in sexual initiative, the rates of courting and mating were higher than those found in flock D (Table 3). Each of the following modified factors probably helped produce this result: (*a*) For most of the day preceding his period of free association with the hens each cock was placed in a wire cage in the pen near the hens, to augment sexual readiness. (*b*) The roosts and other possible retreats were blocked off, and the available floor space was approximately halved. (*c*) There was a ninefold increase in the rate of "sex invitations" by the F-hens as compared with those of the D-flock. Confronted by sexually unaggressive males, such reactions by the hens may be classed as compensatory behavior.

Cock *I* of the D-flock was included in tests with F-hens in an attempt to secure a better basis for cross-comparisons of the mating behavior of the two sets of hens. The utility of cock *I* for comparative purposes is marred by complications. In addition to the differences just outlined, this cock was now a year older; he was not penned with the F-cocks but had a pen to himself; he was now placed with young, rather than aged, hens; and these younger birds, between days with cock I, associated with the relatively unaggressive F-cocks. Contrariwise, between periods of association with cock I, the D-hens had experience with the notably aggressive cock II. Thus, several elements of both the physical and the social environment differed in the two sets of tests.

We received the impression that cock *I* was relatively less acceptable to the F-hens than was to be expected from his comparative showing with the D-cocks.

There was no discernible modification of the earlier sex technique shown by *I* in the later situation. He courted and trod the F-hens at a heightened rate, indicating that the generally increased frequencies of these behavior patterns among the F-hens resulted from altered conditions or modified traits of the hens.

With both flocks D and F the observations made when cocks are introduced singly and successively into the pens with hens indicate that the social position and the sexual drive of the cocks were not necessarily related and that the reproductive efficiency in the flock was only partially determined by the relative libido of the cocks. However, when 4 cocks were placed with a small group of hens, the interrelationship of their relative social aggressiveness and their sexual drive became apparent.

Four males in the pen of the hens.— When all 4 D-cocks were placed in the pen of the D-hens, the male interaction became intense, and for 11 days the hens spent most of their time on the roost. The observations were divided into three periods (Tables 2 and 4). Period A, of 29 observations, may be considered as transitional; period B, of 34 observations, as a climax; and period C, of 31 observations, as a test. Cock IV died accidentally

on July 24, shortly before the end of period A.

The previous male peck-order, with the cocks arranged in the straight line of *I*, II, III, and IV, changed on June 28, when IV defeated III. Cock II pecked III in the eye; and as a result of this temporary visual handicap, cock III lost a contest with IV and consequently his social position. He remained on the roost and lost 13 ounces in weight; by the very end of period A he ventured to the floor to feed, but then only when *I* and IV went to the roost. His hunger dominated his sex drive almost to the exclusion of the latter. Cocks II and III developed a toleration for each other on the roost, where they spent much time together. The sexually active II became conditioned to ignore the hens by a slower process. The alpha cock, *I*, charged at him when he approached a hen and drove him to the roost. This caused all the hens to scatter hurriedly and fly to any perch. Cock II soon learned to pass more time on the roost, and the hens learned to run whenever he came to the floor to feed. He lost 6 ounces in weight, and his hunger drive approximated that of III. Of II's 13 matings, 12 were incomplete (the vents did not meet), as a result of the resistance of the hens or the interference of *I* and/or IV.

The intense antagonism of *I* toward II contrasted with *I*'s curious and extreme toleration of IV; the latter would interfere even when *I* was in the act of treading a hen; IV did so 30 times, while *I* interfered with 20 of IV's matings. The hens learned to avoid IV, and it appeared as though they might nullify his sexual activity by means of their negative discriminations; he succeeded in obtaining 15 of his 33 copulations by pushing *I* from treading position and mounting immediately. At first *I* would peck his usurper, but later he showed no agitation. This unusual toleration probably facilitated IV's successful revolt over II, which occurred the day before IV died.

During period B both II and III spent more time on the floor, but they were conditioned to remain at a distance from the alpha cock and the hens. Cock II still courted at intervals, and his single copulation occurred early in this period. Cock III was psychologically castrated, i.e., completely suppressed sexually; and II nearly so. Cock II crowed at times, but III was not heard to crow after he lost his social position. The size of comb did not fluctuate to any marked degree in any of the males during the time when some were sexually suppressed. A simple test was made to determine whether the frustrated cocks II and III were capable of sexual behavior. Cocks I, II, and III were each placed singly on two different days with a strange flock for 30-minute periods. Cock II trod 7 times and courted 27 times; III trod 5 times and courted 40 times, which compared favorably with I's 6 matings and 41 courtings under the same conditions. However, when II and III were then exposed singly for 30 minutes with the D-hens, neither of them trod, and courtings were limited to 2 wing-flutters by III. This indicated that their sexual inhibition depended on their associates and was definitely related to the hens of the home pen.

The dominant cock, I, was removed during period C to note whether the sexual activity of the inferiors would become re-established. Cock I was placed with flock G. Over an interval of 33 days cock II was seen to tread 3 times (2 were out of the formal observation time), and III mated 11 times (7 out of the formal observation period). All of II's copulations were forced, but III trod only on invitation. Cock II pecked III when the latter trod, and there was some indication that II might in time suppress III completely. Both cocks would at times give the food-call; but when a hen responded, they would peck her. When these results are considered with those of the partial control situation summarized in Table 1, the slight difference (in period C over B) in sexual behavior appeared to be negligible and suggests either that the conditioning effects associated with psychological castration tend to persist or that the continuing conditioned avoidance displayed by the hens toward these roosters may have been a factor in the failure of cocks II and III to copulate at more nearly a normal rate.

Another test was made to determine the nature, if any, of heterosexual conditioning involving the frustrated males. The alpha cock, I, was used as a control to evaluate, in a general way, the relative sexual behavior of the hens. The cocks were exposed to flocks D and G singly on alternate days; each day one cock was placed with the D-hens and another with the G-hens for 20 minutes. This rotation continued until each rooster had 8 exposures to each group of hens; between tests these males were housed together in a separate pen. The results are given in Table 6, from which it is evident that the inhibited sexual activity of cocks II and III when with the D-females affected their behavior when with the hens of flock G. Their courting increased notably, although they were not very successful in mating. The evidence supports the previous test, which showed that these two roosters were conditioned not to approach the D-hens. An analysis of sustained avoidance indicates that the D-hens were also conditioned to avoid these particular cocks.

The observer gained the impression

that the G-hens also learned to spend more time on the roost when cocks II and III were introduced—an impression that was confirmed in part by statistical analysis, since hens of flock G spent 37 per cent less time on the floor with II than with *I* (*P*-value, 0.03) and 25 per cent less time with III than with *I* (*P*-value, 0.07). The D-hens spent 34 per cent less time with II than with the alpha cock (*P*-value, 0.005) and 40 per cent less with III than with *I* (*P*-value, 0.006). Here, as in the discrimination tests (Guhl, 1942), the hens as a statistical group showed differential behavior toward the cocks as individuals.

We turn now to consider the F-males, which, as has been previously stated, were rather weak, so far as intensities of reactions were concerned, in both their social and sexual behavior. When all 4 were placed in the pen of the F-hens, there was an increase in pecking frequency among these cocks; but the suppressive influence of low status in their peck-order was less pronounced than

TABLE 6

SUMMARY OF THE SEXUAL BEHAVIOR OF THE D-COCKS WHEN TESTED SINGLY AND ALTERNATELY WITH THE HENS OF FLOCKS D AND G

PECK-ORDER	WITH FLOCK D		WITH FLOCK G	
	Tread-ings	Court-ings	Tread-ings	Court-ings
I.........	9	54	21	93
II.........	0	19	3	126
III........	0	2	3	95
Total....	9	75	27	314

with cocks II and III in flock D. Cock VII approached the psychological castration stage, as he trod only 7 times (Table 2), 4 of these times resulting in incomplete copulations. The differences

between the frequencies of VII's matings and courtings and those of the other three F-cocks were statistically significant (*P*-values ranged from 0.03 to 0.01). There were no significant differences between these records of the other three males.

TABLE 7

SUMMARY OF THE INTERFERENCES WITH MATING OR ATTEMPTS TO MATE AMONG THE F-COCKS

The figures present the number of interferences by each of the cocks listed on the left with attempts to copulate by each of the other cocks.

Peck-Order	V	VI	VII	VIII	Total
V............	.....	55	7	57	119
VI.........	17	.....	2	9	28
VII........	0	0	.....	0	0
VIII.......	13	1	0	.....	14
Total.....	30	56	9	66	161

Interferences with mating or attempts to mate are summarized in Table 7. Only the frustrated cock VII was completely inhibited from taking an aggressive role in competition for the possession of a receptive hen. These attacks were not limited to cocks of a particular status, although the alpha cock interfered in about 74 per cent of the observed instances. During the early part of this 48-day period these obstructions to copulation were caused by pecks delivered on the treading male; by the fourth week the pecks became very light, or the interfering male would merely approach the mating pair; and by the fortieth day there was a tendency for the interloper to ignore the copulating male and to grasp the comb of the hen being trod.

When an inferior rooster crowed or gave the food-call, he was usually challenged by the alpha V, with the result that such vocalizations disappeared during most of these weeks. The male interaction lessened in intensity after the twenty-fifth day, and both these calls then reappeared to some extent. The hens showed a tendency toward sustained avoidance of the partially frustrated males by moving to neutral corners. This female reaction could not be qualified to the same extent as with the D-hens, since the roost was screened off and the floor space restricted during observations to one-half that of the normal pen. With such closer contact one might have anticipated greater interaction between the cocks and more evidence of psychological castration among normal young cocks.

At the end of the observations correspondence with the breeder who furnished the birds revealed some interesting information on the early history of these cocks. Cocks V, VI, and VII were part of a group of 24 that had been penned with about 200 hens and were shipped to us as culls, which suggests that they were probably inferiors in this group of cocks. The indications are that they were conditioned when received, as 4 of the original 24 cocks had been killed by their superiors. Little was known about cock VIII; he came from a different source. If this assumption is correct, one might conclude, as with the D-cocks, that conditioning of mating behavior patterns in roosters, once formed, tends to persist.

The data for the cocks do not show any strong correlations between the social position per se of the individual males and the frequency of their mating behavior when introduced singly and successively into a pen of hens ($r = 0.4$; P-value > 0.1). Cocks tended to be consistent in the frequency of matings when not inhibited by the presence of other males. However, as has just been stated,

when 4 cocks were placed together in a pen containing a small number of hens, a form of suppression developed among the cocks; and the dominant male almost completely castrated psychologically some of the low-ranking roosters. This shows that, as a result of their greater freedom of the pen, the dominant cocks possessed precedence to mating in addition to precedence to food. There is some evidence that the activities leading to the frustration associated with sexual suppression, or the suppression itself, although it does not incapacitate a cock sexually, conditions the hens to avoid a suppressed cock or modifies his behavior in a manner which makes him less acceptable to the hens.

SEX DIFFERENCES IN BEHAVIOR

The males are more aggressive than the females and tend to avoid other males strongly or to fight back rather than to submit to them, whereas the hens with less aggressiveness than the males submit more readily to cocks and to aggressive hens by lowering the head or by squatting. Like the cocks, they also frequently avoid social contacts. The peck-orders of cocks are somewhat unstable, and those of hens are relatively stable (Sanctuary, 1932; Masure and Allee, 1934). An alpha hen, "antipathies" aside, becomes tolerant of low-ranking individuals. This reduction of social interaction between individuals of widely separated ranks becomes greater between the males and the females, with the consequence that males normally do not peck females but tolerate them; and the females show no marked avoidance of normal males, so that typically two essentially unisexual peck-orders occur in a flock composed of several cocks and several hens. Carpenter (1942) has described a similar situation in his hordes of rhesus monkeys on Santiago Island off the coast of Puerto Rico. In a flock of intersexual Brown Leghorns, Domm and Davis (1941) found that, in general, the most masculine birds were at the top, and the most feminine were at the bottom, of the ranking among the males. The hens appear to learn readily that cocks will not peck them, although they also learn to avoid a rough and sexually active cock such as II.

INTERSEXUAL BEHAVIOR

Hens secrete androgens, as well as estrogens, as shown by comb regression to the capon level in bilaterally ovariotomized hens (Domm, 1927). Koch (1937) has demonstrated by results obtained by applying the hormone to the surface of the comb that comb size can be used as an indication of the amount of androgen present. A marked correlation exists between aggressiveness and comb size in normal or in testosterone-injected hens (Collias, 1943). We have records of pronounced masculine behavior in normal hens that merit brief attention. In a previous report (Guhl, 1942) mention was made of 3 hens that wing-fluttered to other hens, and this reaction has since been observed under somewhat similar conditions in other flocks. In a more recent experiment the food-call appeared to be rather common among some hens, whether in the pen, out of doors, or in isolation. The crowing of hens is a well-known phenomenon. In addition to GG of flock D, 5 normal hens of a different flock than those mentioned here have been heard to crow; one of these (RG) crowed quite regularly, whether in isolation or in the pen with others, but in the latter case only when she was the alpha bird.

We have a record of 20 unisexual matings in which 3 normal hens (none of which were in the flocks discussed in this paper) took the male role with 6 different hens. None of these had received experimental treatment with hormones. The vents are known to have met in 10 of these cases. Observations were facilitated by the fact that the feathers in the region of the vents were lacking. Treading movements with the feet were noted in 9 cases. In nearly all these instances, the hen that took the normal role of the female squatted submissively before the threatening superior before the latter mounted. When one hen (GB) trod, she everted her cloaca; and on two occasions when the vents failed to meet, she discharged a stream of rather clear fluid. None of the treading hens had been heard to crow. All 3 hens assuming the male position were either laying or laid within 4 days of these pseudo-matings, and all were socially dominant over the hen that was trod. No cocks were kept with these flocks.

Heuser (1916) reported a case of a hen that attempted to mate 35 times, "in as perfect a manner as the best male," with 13 different individuals. This, he stated, occurred at times of egg production and nonproduction, mating and nonmating, and while the males were present or absent. Heuser kept no records of social position; hence we do not know the relative social dominance of his malelike hen. More observations are needed before a complete behavioristic interpretation can be attempted, but these cases do raise the question of how widely the capacity for male copulatory behavior is distributed among otherwise normal hens and also what social and other conditions are required to elicit this response in a female. Whitman (Carr, 1919) noted that the potentialities of male and female sexual behavior patterns in pigeons were almost on a par in the two sexes, and he dis-

357

cusses some of the conditions under which either males or females may exhibit activities peculiar to the opposite sex.

Androgen-treated hens became highly aggressive, and their female penmates learned to avoid them rather than to submit. Since the submissive squat appears to be essential to copulation, the treated hens in our flocks did not secure the opportunity to display the male copulatory pattern if it was induced. In the cases in which we saw unisexual mating, the treading hen, although the dominant of the pair, did not peck hard or often; instead she frequently elicited submission rather than avoidance. Since individuals develop special habits in relation to every individual (Tinbergen, 1931; Guhl, 1942) and since, as discussed by Allee, Collias, and Lutherman (1939), social inertia may enforce or replace aggressive behavior patterns, it appears logical to assume that habit may be important in the patterns of sex behavior shown by hormonally treated fowl. For example, when YY received a pellet of testosterone propionate, her aggressive behavior was noted in the discrimination pen (Guhl, 1942) long before it became evident in her home pen; and when she was placed in the pen of flock G for a test, she drove all these hens about their pen, but on return to her own flock she assumed pair-behavior patterns in accord with her social status of that day (cf. Douglis, 1944). The same dominance of learned behavior over physiological state was shown by cocks II and III, which, although capable of copulation, did not mate with the D-hens even when the suppressive influence of cock *I* was removed. Allee, Collias, and Lutherman (1939) found that low-ranking hens treated with androgen not only rose in the social order but maintained their new social status long after

injections had ceased and the recipient hens had lost their androgen-mediated aggressiveness.

CORRELATIONS BETWEEN OBSERVED RANKINGS AMONG THE HENS

Before summarizing observed correlations, it is necessary to pause briefly to introduce pertinent background material about flocks A6, D_{1939}, and E. The records of these last three flocks were obtained during a total of 32 hours of observations on flock A6, 35 hours on flock D_{39}, and 8.5 hours on flock E, all during the first 8 months of 1939. Observation periods were 10 or 15 minutes in length and usually occurred late in the afternoon. Flocks A6 and E were composed of 6 hens; D_{39}, of 7 hens. The basic relations are summarized in Table 8.

There was a pecking triangle in flock A6 at this time; a linear pecking order existed in flock E and during about half of this period for flock D_{39}. Two reversals of social position occurred in flock D in the second half of the period, but these did not concern the two top birds; the social order indicated in Table 8 represents the earlier situation.

The data from the 1939 observations are somewhat complicated as a result of hormone treatments given in connection with other experiments. The treatments consisted of injections of estradiol benzoate or of thyroxin into birds selected from various social levels and, except for the higher doses of thyroxin, produced little or no effect on the rate of mating. All data are omitted from Table 8 that were recorded at the time when some of the top birds (2 hens in flock D and 1 in flock E) received doses of thyroxin sufficient markedly to depress their frequency of mating.

Mention was made previously of a tendency toward a negative correlation between the social position of the hens in flocks D_{1940} and F and the rates at which they were trod or gave the sexual squat. In the more careful statistical study of these matters it was decided to concentrate on data secured when each cock was with his respective flock of hens singly and successively. The presence of 3 or 4 cocks in the pen at the same time introduced complications that we have been at some pains to describe. So far as

We are interested in the correlations within each flock considered as a unit and within various flock combinations. The combined correlations given in Table 9 are for the 1940 observations on D_{40} and F; on the 1939 observations on A6, E, and D_{39}; and on all the flocks considered together. In this last instance the D-flock for both years was considered as one unit and given the rank-order based on mean rate of performance. The com-

TABLE 8

SOCIAL POSITIONS OF THE HENS AS RELATED TO THEIR FREQUENCY OF
MATING AND "SEX INVITATIONS" (SEE P. 370)*

Observations made on 3 flocks during 1939

SOCIAL POSITION	NO. OF TREADINGS			NO. OF "SEX INVITATIONS"		
	A6	D_{1939}	E	A6	D_{1939}	E
1...............	5	30	0	0	14	0
2...............		38	0		18	0
3...............	9, 77, 40†	39	32	0, 27, 6†	10	27
4...............		46	31		32	10
5...............		93	13		43	10
5.5.............	30, 37‡			5, 0‡		
6...............		66	4		43	1
7...............		82			51	

* Blank spaces indicate that there were no hens at these levels in the social order. Half-steps in social position mark deviations from a straight-line peck-order.
† Members of a pecking triangle.
‡ Two hens tied for the omega position.

flocks D_{40} and F are concerned, the first correlation studies are based on data given in Table 3; the relations in the other flocks have just been given in Table 8. The correlation coefficient (r) was calculated from the regression formula

$$r_R = \frac{\epsilon v_1 v_2 - \overline{V} \epsilon v_2}{\epsilon v_2^2 - \overline{V} \epsilon v_2}.$$

Further, since social status is given here summarily in terms of rank achieved rather than by listing in detail the pecks delivered and received, it is probably best to examine the correlation of social ranks with rankings based on the rates for other observed activities.

parison of correlations between flocks that differ in size was accomplished by the simplest-possible treatment. The top-ranking hens were regarded as equivalent; so were those in bottom rank; the intermediate ranks were adjusted by simple "stretching" of the smaller number of ranks to give them their appropriate place in the larger series. The correlations are listed in Table 9, with the statistically significant values shown in italic type.

Negative correlations.—Although not always statistically significant in all flocks, there were negative correlations between ranks based on frequency of mating in relation to position in the peck-order: the top-ranking hens were trod less than their subordinates. These results are statistically significant

TABLE 9

CORRELATION BETWEEN RANKINGS AMONG THE HENS

Statistically significant correlations are given in italics. (D_{39} and D_{40} signify the D-flock for 1939 and 1940, respectively.)

FLOCK	No. OF PAIRS	CORRELATION	P-VALUE*
Social Rank : Mating Rank			
A6........	6	−0.2857	>0.1
E.........	6	−0.3642	>0.1
D_{39}.......	7	−0.8929	<0.01
D_{40}.......	7	−0.6785	<0.1
F.........	9	−0.4000	>0.1
D_{40}, F.....	16	−0.5264	<0.05
A6, E, D_{39}.	19	−0.5427	<0.02
All........	28	−0.6407	<0.01
Social Rank : Rank in "Sex Invitations"			
A6........	6	−0.1613	>0.1
E.........	6	−0.5757	>0.1
D_{39}.......	7	−0.8909	<0.01
D_{40}.......	7	−0.7600	<0.05
F.........	9	−0.0833	>0.1
D_{40}, F.....	16	−0.2677	>0.1
A6, E, D_{39}.	19	−0.6033	<0.01
All........	28	−0.3210	<0.1
Rank in "Sex Invitations" : Mating Rank			
A6........	6	+0.7714	>0.1
E.........	6	+0.9740	<0.01
D_{39}.......	7	+0.8214	>0.02
D_{40}.......	7	+0.8909	<0.01
F.........	9	+0.8666	<0.01
D_{40}, F.....	16	+0.8213	<0.01
A6, E, D_{39}.	19	+0.5185	>0.02
All........	28	+0.5640	<0.01
Social Rank : Courting Rank			
D_{40}.......	7	−0.9286	<0.01
F.........	9	−0.4622	>0.1
D_{40}, F.....	16	−0.6794	<0.01

* The P-values are based on Fisher (1936, p. 202).

TABLE 9—Continued

FLOCK	No. OF PAIRS	CORRELATION	P-VALUE*
Courting Rank : Mating Rank			
D_{40}........	7	+0.6071	>0.1
F.........	9	+0.7500	0.02
D_{40}, F.....	16	+0.5465	>0.02
Rank in "Sex Invitations" : Courting Rank			
D_{40}........	7	+0.6429	>0.1
F.........	9	+0.3109	>0.1
D_{40}, F.....	16	+0.4629	>0.05
Courting Rank : Avoidance Rank			
F.........	9	+0.1167	>0.1
Mating Rank : Avoidance Rank			
F.........	9	−0.3000	>0.1
Rank in "Sex Invitations" : Avoidance Rank			
F.........	9	−0.5667	>0.1
Social Rank : Avoidance Rank			
F.........	9	−0.0333	>0.1

for the three flock combinations and for D_{39}.

The correlations between social rank and the frequency of giving the sexual crouch (so-called "sex invitations") were also steadily negative. They were statistically significant with the D-flocks for both years and when all observations for 1939 (A6, F, D_{39}) are considered together. Although still negative, the value of *r* for flock F is very low. The D_{40} hens gave no "sex invitations" to cock II and very few to cock III.

The correlations between social rank and rank based on the frequency at which the hens were courted by the cocks are also steadily negative, significantly so for D_{40} and for D_{40} plus F. While statistically not significant, the negative correlations between mating and avoidance ranks and between those of "sex invitations" and avoidance are in accordance with expectation. The insignificant correlation between social and avoidance ranks shows that in flock F, the only one for which data are available, the social independence of the top-ranking hens included a strong tendency not to avoid the cocks. These observations do not give a basis for predicting what the behavior of the F-hens might have been in the presence of cocks with more highly developed sexual initiative.

Positive correlations.—The correlations between "sex invitations" and matings were uniformly high and were statistically significant except for flock A6. The correlations between courting and mating ranks were also all positive and were statistically significant for the F-hens and for F and D_{40} flocks considered together. The similarly positive correlations between "sex invitations" and courting lack statistical significance, as does also the low correlation between courting and avoidance.

Probably the most interesting of these correlations is the inverse relationship between high social rank and frequency of being trod by the cock. Our data on this point were obtained independently by two observers and are supported by certain observations in the mating pens. As Tables 3 and 8 show, in flocks A6, D (1939 and 1940), and E the alpha hens were trod at the lowest rate. The bird in

flock A6 for which only 9 treadings were observed was formerly the top hen in this flock, although at the time of this experiment she was tied for third place with two other hens.

The behavior of GG, in 1939 the alpha hen of flock D, is also illuminating. This hen was at that time the least trod of her flock, "invited" the cock to tread the least, laid well in comparison with the other hens, was a good fighter in staged encounters with strange hens, and the cock frequently courted and waltzed around her. Hen GG also displayed some cocklike behavior: when the cock wing-fluttered to her, she would reciprocate by waltzing around him, giving a rudimentary male-behavior pattern; at times she pecked at him. Only GG among the normal hens of flock D was heard to crow or seen to wing-flutter in the pen. When GY and BB were transferred to flock D in 1940 from flock A6, where they were at the bottom of the peck-order, these hens moved to the top of the new hierarchy and replaced GG at the bottom of the treading order (Table 3).

Although we have not attempted to make a similarly elaborate statistical analysis of the observed correlations when all cocks were placed together with their respective flocks of hens, we have compared the ranks based on the behavior when each cock was introduced singly (Table 3) with that shown when all were present together with the hens (Table 4). The social ranks remained constant in both flocks D_{40} and F—the two for which data are available—during the entire period under discussion. Hen GB, of flock F, died before the second set of observations (those recorded in Table 4) were begun; but her death merely caused YR and WR to become Nos. 7 and 8, respectively, in the peck-order, rather than Nos. 8 and 9.

All the calculated correlations between ranks based on mating, courting, so-called "sex invitations," and avoidances, with 1 cock present, as compared with the same reactions with 3 or 4 present, are positive. The only correlation that is statistically significant is that regarding treading ranks under the two sets of conditions. Here $r = +0.8055$ with a P-value of <0.01. The ranks based on courting show that $r = +0.3711$ ($P > 0.1$), and for "sex invitations" $r = +0.4343$ ($P > 0.1$). Data on avoidances are available for flock F only, and there $r = +0.3095$ ($P > 0.1$).

It is possible to subject the available data to various other statistical analyses. These are all that have seemed profitable for us to discuss. We trust that the data are given in sufficient detail so that anyone interested may make such statistical and other tests as meet his needs.

DISCUSSION

In order to discuss the detailed observations more intelligently, it may be well to pass in review, in summarized form, the correlations that we consider to be fairly valid, or at least suggestive, when all our data and experience are considered. For the hens these are presented in the following informal tabulation.

I. NEGATIVE CORRELATIONS

1. Social position : frequency of mating
2. Social position : frequency of being courted
3. Social position : frequency of "inviting" the cock to mate

II. POSITIVE CORRELATIONS

1. Frequency of giving the sexual crouch : frequency of mating
2. Frequency of being courted : frequency of mating

In addition, there were suggestions of a positive correlation in each test made between frequency of being courted and the frequency of giving "sex invitations." All these correlations are consistent wherever cross-checking is possible.

The interesting negative correlation between ranks based on frequency of mating and social rank requires further discussion. Our results at this point are to be contrasted with those of Skard (1937), who obtained a positive correlation of 0.53–0.55 (P-value about 0.05) between mating "preference" and the peck-order with one cock and a statistically insignificant positive value of only 0.24 with the same cock in another flock of hens. The performance of a second cock, Skard says, "does not show any conformity with the first cock's preferences."

Skard obtained her "mating-preference" order (a) by summing all the approaches of her cocks to the different hens or (b) by considering weighted values for different recognizable aspects of male-initiated sex behavior, with a value of 1 for each "weak impulse" and of 6 for each copulation. Skard's recorded data do not indicate either the sign or the degree of correlation that she found for either sex between rank in the peck-order and the number of actual copulations; hence they are not directly comparable with our results.

Murchison (1935b) reports a positive correlation between the number of times a hen was trod and the distance she traveled toward another hen in his "social reflex runway." The hen that traveled farthest had the highest frequency in mating. He found also (1935a) that the hen traveling the greater distance in his special runway dominated the other of the contact-pair in physical combats. Hence, in our terminology,

Murchison found a positive correlation between rank based on the number of times a hen is mated and her rank in the peck-order of the female flock. He observed 292 copulations between 5 pullets and 3 cocks and gives no statistical analysis. An examination of his published records, using the statistical methods we have applied to our own data (in which we based mating rank on the total observed treadings for each hen) indicates a positive coefficient of correlation (r) of 1.0 when 0.9587 would have a statistical probability (P-value) of 0.01. Pertinent differences of Murchison's (1935b) methods of observation from ours are as follows: (a) Observations on mating were made when his birds were 16–48 weeks old. (Sixteen weeks is an unusually early age at which to expect sexual maturity.) (b) The dominance relationships among the pullets were determined in the "social-reflex runway" when they were about 32 weeks old, or 16 weeks after observations on matings began. (c) The mating data obtained when cockerels were used singly and successively were pooled, without distinction, with those recorded when all the males were penned with the pullets. (d) Murchison does not describe the cage in which the pullets were housed with the cockerels. We have, therefore, no way of knowing whether the matings Murchison observed are comparable with ours, made under other conditions and on a different breed.

In our flocks the over-all correlation between social rank and ranked frequency in mating is regularly negative rather than positive—and significantly so, according to accepted statistical standards, when larger numbers of comparisons are made. It is interesting to note that Schjelderup-Ebbe (1935), whose observations we have often verified, also reported that his top-ranking hens mated less frequently than those with low social status. Schjelderup-Ebbe presented naturalistic observations only without supporting statistical analysis.

We can make some tentative suggestions concerning the important discrepancy between our observations, together with those of Schjelderup-Ebbe, and the findings of Murchison and probably of Skard. We know that differences in ancestry, in previous history, in housing, and in treatment often produce important differences in behavior. Hens high in the social order have greater freedom of movement. Such freedom can be important in allowing more frequent meeting of dominant hens with somewhat sluggish cocks; and Murchison's cocks may all have been sluggish, since 3 of his 6 cocks did not mate. Contrariwise, a negative correlation between mating and social rank may well be the normal relationship when cocks show the usual amount of sexual initiative. Under such conditions the greater tendency of hens low in the social order to give the submissive sexual crouch did much to determine mating frequency. Skard's experience (1937), like ours, shows that there are differences between flocks even under similar experimental conditions; and doubtless there are still greater differences under the divergent conditions used by different investigators. Perhaps there are differences between breeds.

The situation is not simple and becomes less so when other social relations are considered. Despite the advantage of a high social level of aggressiveness in competitions for food and space, beyond some undefined point, female aggressiveness appears to interfere with the mating process. Also, beyond a considerably higher level, aggressiveness of a cock toward associated hens—cock II, for example—lessens his chances for mating.

When different cocks were introduced singly and successively into the pen of the hens, individual differences were revealed in courting and mating habits and in the relative frequency of these activities. The individual differences, though sometimes statistically valid, were not startling until after the cocks had been placed together simultaneously with the hens. Then followed, especially with flock D, a severe conditioning of certain subordinate cocks. This was continued until the individuals were sexually suppressed so far as these hens were concerned, even in the absence of the alpha cock. The suppressed males were less effective even when they were introduced singly into a flock of strange hens (Table 6).

Heuser (1916), who did not describe peck-orders among his birds, found evidence of sexual suppression among males that he related to constitutional vigor. He stated: "It is generally believed that males of high constitutional vigor will overcome the low vigor birds and prevent their mating. However this is not always true, as shown by the behavior of 'black,' one of the low vigor birds, that had third highest number of matings although chased and bullied by all the other males in the pen." This suggests a peck-order relation to what he calls "constitutional vigor." If one could assume that rank in vigor is correlated with social position, one could find parallels between his results and those presented in this paper, such as: (a) "high-vigor" males mated more frequently than medium or "low-vigor" males; (b) "low-vigor" females mated, on the average, more frequently than hens with high "constitutional vigor"; and (c) the correlation between "constitutional vigor" of the females and mating power is weaker than with the males; it is also negative with hens and positive with cocks.

Murchison (1935b) found that 3 of his 6 cocks did not mate with the hens with which all of them were associated. This may have been an unrecognized instance of sexual suppression similar to that shown diagrammatically in our 1940 D-flock. In his flock, the alpha, beta, and omega cocks did all the treading. In our flock D, males with intermediate ranks also were suppressed. We found evidence of so-called "favoritism" between cocks I and IV. There is no recorded reason why Murchison's omega cock trod hens when those in third, fourth, and fifth ranks did not.

Our observations on domestic fowl under laboratory conditions may be compared to some others made on another gallinaceous species in nature. Simon (1940) and Scott (1942) studied the mating behavior of the sage grouse (Centrocercus urophasianus) and found basic behavior phenomena that bear rather close resemblance to those of Gallus domesticus.

During the breeding season the cocks of the sage grouse develop a dominance organization composed of a master-cock, a subcock, 3–6 guard cocks, and a variable number of "outside" or isolated cocks. A formalized "wing-beating" constitutes the chief manner of fighting between the males. The hens are more passive but may peck other females. Strutting increases in the presence of females, and mating by inferior males appears to stimulate an attack on the copulating cock. The hens initiate coition by a squatting reaction in front of a cock. Satiation on the part of the master-cock may cause receptive hens to crouch before a subcock or guard cock (Scott, 1944). At daybreak and again during the late afternoon the cocks gather at their strutting grounds and arrange themselves in a number of dominance orders. The hens enter the assembly later and appear to avoid a poorly integrated group of cocks engaged in much fighting. In these particular observations the master-cocks were involved in 74 per cent of the matings, the subcocks obtained 13 per cent of the copulations, and the guard cocks 3 per cent. The more numerous isolated cocks at the edge of the breeding area secured less than 10 per cent of the total matings observed. It should probably be noted that neither Simon nor Scott stated that the observed cocks were marked and recognized as individuals. These conditions strongly resemble the situation that we have described for flock D_{40} when all cocks were present together.

Skard (1937) was apparently impressed by the so-called mating preferences in her flocks. Warren and Kilpatrick (1929) found evidence of "preferential mating," since three of their hens did not mate with one of the cocks although they did with others introduced alone into the pen. Almost all students of the social behavior of the domestic fowl, from Schjelderup-Ebbe (1922) on, have seen reactions that can most briefly be characterized by the anthropomorphic terms of preference and its opposite, antipathy. We have not yet had better illustrations of such reactions than the antagonistic behavior of cock I of flock D_{40} toward cock II or by his weak reaction ("preferential behavior") to the aggression of the otherwise well-subordinated cock IV.

We found little evidence that the males which we observed gave marked positive or negative attention to particular hens. As a rule, the gradation was fairly gradual from the most- to the least-trod hen. Exceptions did occur. For example, when introduced singly, cock II mated with RW at the rate of 2.40 times per hour and with BR 2.20 per hour, but with the hen third in mating frequency,

at the rate of only 1.25 per hour. Although cock II courted GG at a fairly high rate, he mated with her infrequently. Cock III copulated much more frequently with BR, cock V (flock F) with YR, and cock VIII with YR, than was to be expected on the basis of their treading of other hens. Cock I, when with the F-hens, may have avoided mating with YB and RB. When 4 cocks were together in the flock at one time, I and IV mated more with one individual (YY) than with any of the others.

These particular instances of mating "preferences" or "antipathies" may have been the result of the small sample rather than a reality; it would not be surprising if they were real. On the other hand, these differences in the rates at which a given male mates with several hens may be as much a matter of the reactions of the individual hens toward the male as of the behavior of the cock toward the different hens. Some evidence in this regard is fairly apparent in Table 8 and in the avoidance shown by the hens of flock D_{40} to cocks II and III in their later association with those sexually suppressed individuals. However, an adequate discussion of this phase of mating behavior would lead us beyond the scope of the present report.

Finally, what is the evidence that the social position of an individual or its sexual-behavior complex may affect its reproductive success? It is obvious that a socially suppressed cock loses much of his opportunity for mating. It is probable that sexual suppression of a cock need not be carried to the point of psychological castration to produce a reduction in the number of his progeny. Crew (1926) and Warren and Kilpatrick (1929) report that the competitive efficiency of sperm may be of relatively short duration (about 3 days) and that sperm of a new male may supplant those of a preceding mate. Malbandov and Card (1943) found that eggs fertilized by stale (senescent) sperm show an increase in failure to hatch in proportion to the age of the sperm, although no significant differences in embryo mortality occurred until the sperm were more than 10 days old. They concluded that "it is possible that unpopularity of individual hens with males and insufficient number of males may cause some hens to lay fertile eggs that are incapable of hatching because they were fertilized by stale sperm." Fertilization by artificial insemination depends, among other factors, upon the concentration of sperm (Munro, 1938; Hartman, 1939). Kosin (1944) demonstrated that the numerical relationship between functionally active germ cells and those functionally inactivated by X-rays determine the proportion of eggs fertilized by the specimen of semen treated. Parker and Spadden (1943) found that inanition decreased the semen volume, the number of spermatozoa per collection, and the fertilizing capacity of the males.

Lamoreux (1940) in a study of infertility of fowl lists, among other factors, some causes of infertility that may be influenced by heterosexual behavior, such as "preferential mating," number of matings, and the time elapsed since last mating. All the evidence at hand indicates that these factors may be influenced by the individuality of the cocks and, to a lesser degree, by that of the hens. They are also affected by the learned pair-reaction patterns of the hens and the cocks.

The already complicated problem is rendered still more complex by indications that in the domestic fowl the competition between cocks for mates does not end when a given cock is successful in matings. The sperm from successive matings by two individuals may continue to compete in the reproductive tract of the hens both by numerical advantage and by relative physiological virility. This problem is being directly attacked by one of us with attention to all the known variables (cf. Guhl, Warren, and Payne, 1944).

There can be no doubt but that social dominance among males, whether of *G. domesticus*, the sage grouse (Simon, 1940; Scott, 1942), or rhesus monkeys (Carpenter, 1942), allows the dominant animal an opportunity to sire more offspring. The relation between high social status and reproductive success of common hens is not so obvious. Sanctuary (1932) found that hens from the lower half of the peck-order lay fewer eggs than do their more aggressive flockmates. The high-ranking hens obtain more food and otherwise lead more secure lives. Opposed to these trends, there is the tendency for socially high-ranking hens to mate less frequently than those lower in the social scale. In our small flocks each hen mated frequently enough to insure fertile eggs and viable offspring. A preliminary report of a direct test (Guhl, Warren, and Payne, 1944) of the subject indicates that, despite the negative correlation with mating frequency, high-ranking hens, as well as cocks, produce more offspring than their low-ranking flockmates.

SUMMARY

1. Approximately 1,630 treadings of domestic fowl were observed and studied in relation to associated behavior and social status.

2. Statistically significant negative correlations were found between the social position of the hens and the frequencies at which they were courted by the cocks.

3. Although rarely statistically significant in any given flock, there was a consistent tendency toward a negative correlation between the frequency at which the hens were mated and their position in the social order that became statistically significant when results from different flocks were considered together.

4. In 4 out of 5 flocks the hens highest in the social order either failed to "invite" the cock to mate or crouched less frequently than most of their penmates. The consistent negative correlation between social rank and the frequency of giving the sexual crouch was suggested but not proved by our data.

5. There was a high correlation in all observed flocks between the number of times the individual hens were mated and their frequencies in giving "sex invitations" to the males.

6. There was no statistically significant relationship between the social status of the cock in his unisexual group and his sexual activities when placed alone in a flock of hens. However, when 4 cocks were introduced together into a relatively small but uncrowded pen containing hens, a form of suppression developed which practically inhibited mating behavior by some of the low-ranking cocks and demonstrated that the dominant males possess a greater freedom to mate and so sire more offspring than their socially inferior penmates.

LITERATURE CITED

ALLEE, W. C. 1936. Analytical studies of group behavior in birds. Wilson Bull., 48:145–51.
———. 1938. The social life of animals. New York: Norton. Pp. 283.
———. 1942. Social dominance and subordination among vertebrates. Biol. Symposia, 8:139–62.
———. 1943. Where angels fear to tread: a contribution from general sociology to human ethics. Science, 97:517–25.
———. 1945. Human conflict and co-operation: the biological background. Chap. xx, pp. 321–67, of Approaches to national unity: fifth symposium of the Conference on Science, Philosophy and Religion, ed. L. BRYSON, L. FINKELSTEIN, and R. M. MACIVER. New York: Harper & Bros. Pp. 1037.
ALLEE, W. C.; COLLIAS, N. E.; and LUTHERMAN, C. Z. 1939. Modification of the social order in flocks of hens by the injection of testosterone propionate. Physiol. Zoöl., 12:412–40.
CARPENTER, C. R. 1942. Sexual behavior of free ranging rhesus monkeys (Macaca mulatta). I. Specimens, procedures and behavioral characteristics of estrus. Jour. Comp. Psychol., 33:113–42.
CARR, H. A. 1919. The behavior of pigeons. Posthumous works of Charles Otis Whitman. Washington: Carnegie Institution of Washington. Pp. 161.
COLLIAS, N. E. 1943. Statistical analysis of factors which make for success in initial encounters between hens. Amer. Nat., 77:519–38.
———. 1944. Aggressive behavior among vertebrate animals. Physiol. Zoöl., 17:83–123.
CREW, F. A. E. 1926. On fertility in the domestic fowl. Proc. Roy. Soc., Edinburgh, 46:230–38.

DOMM, L. V. 1927. New experiments on ovariotomy and the problem of sex inversion in the fowl. Jour. Exper. Zoöl., 48:31–150.
DOMM, L. V., and DAVIS, D. E. 1941. Sexual behavior of intersexual fowl. Proc. Soc. Exper. Biol. and Med., 48:665–67.
DOUGLIS, MARJORIE. 1944. Hens that are members of as many as four flocks may maintain a different social status in each. Anat. Rec., 89:23 (abstract).
FISHER, R. A. 1936. Statistical methods for research workers. London: Oliver & Boyd. Pp. 339.
GUHL, A. M. 1941. The frequency of mating in relation to social position in small flocks of White Leghorns. Anat. Rec., Suppl., 81:113 (abstract).
———. 1942. Social discrimination in small flocks of the common domestic fowl. Jour. Comp. Psychol., 34:127–48.
GUHL, A. M.; WARREN, D. C.; and PAYNE, L. F. 1944. Number of progeny related to social position in a flock of chickens. Anat. Rec., Suppl., 89:31 (abstract).
HARTMAN, C. G. 1939. Ovulation, fertilization and the transport and viability of eggs and spermatozoa. Chap. iv in Sex and internal secretions (2d ed.; ed. E. ALLEN), pp. 630–719. Baltimore: Williams & Wilkins.
HEUSER, G. F. 1916. A study of the mating behavior of the domestic fowl. Thesis, Master of Science of Agriculture degree, Graduate School of Cornell University. (Unpublished.)
KOCH, F. C., 1937. The male sex hormones. Physiol. Rev., 17:153–238.
KOSIN, I. L. 1944. Some aspects of the biological ac-

tion of X-rays on cock spermatozoa. Physiol. Zoöl., **17**:289–319.

LAMOREUX, W. F. 1940. The influence of intensity of egg production upon infertility in the domestic fowl. Jour. Agric. Res., **61**:191–206.

MALBANDOV, A., and CARD, L. E. 1943. Effect of stale sperm on fertility and hatchability of chicken eggs. Poultry Sci., **22**:218–26.

MASURE, R. H., and ALLEE, W. C. 1934. The social order in flocks of the common chicken and the pigeon. Auk, **51**:306–27.

MEULEN, J. B. VAN DER. 1939. Hormonal relation of molt and ovulation. Seventh World's Poultry Cong., pp. 109–12.

MUNRO, S. S. 1938. The effect of dilution and density on the fertilizing capacity of fowl suspensions. Canadian Jour. Res., **16**:281–99.

MURCHISON, CARL. 1935a. The experimental measurement of a social hierarchy in *Gallus domesticus*. I. The direct identification and direct measurement of social reflex No. 1 and social reflex No. 2. Jour. Gen. Psychol., **12**:3–39.

————. 1935b. The experimental measurement of a social hierarchy in *Gallus domesticus*. III. The direct and inferential measurement of social reflex No. 3. Jour. Genet. Psychol., **46**:76–102.

PARKER, J. E.; McKENZIE, F. F.; and KEMPSTER, H. L. 1940. Observations on the sexual behavior of New Hampshire males. Poultry Sci., **19**:191–97.

PARKER, J. E., and SPADDEN, B. J. 1943. Influence of feed restriction on fertility in male domestic fowl. Poultry Sci., **22**:170–77.

PHILIPS, A. G. 1919. Preferential mating in fowls. Poultry Husbandry Jour., **5**:28–32.

SANCTUARY, W. C. 1932. A study in avian behavior to determine the nature and persistency of the order of dominance in the domestic fowl and to relate these to certain physiological reactions. Thesis for M.S. degree, Massachusetts State College, Amherst. (Unpublished.)

SCHJELDERUP-EBBE, T. 1922. Beiträge zur socialpsychologie des Haushuhns. Zeitschr. f. Psychol., **88**:225–52.

————. 1935. Social behavior in birds. Chap. xx in Murchison's Handbook of social psychology, pp. 947–72. Worcester, Mass.: Clark University Press.

SCOTT, J. W. 1942. Mating behavior of the sage grouse. Auk, **59**:477–98.

————. 1944. Additional observations on mating behavior of the sage grouse. Anat. Rec., Suppl., **89**:24 (abstract).

SIMON, J. R. 1940. Mating performance of the sage grouse. Auk, **57**:467–71.

SKARD, ASE G. 1937. Studies in the psychology of needs: observations and experiments on the sexual needs of hens. Acta psychologica, **2**:175–232.

SMITH, KIRSTINE. 1922. The standard deviations of fraternal and parental correlation coefficients. Biometrika, **14**:1–22.

TINBERGEN, N. 1931. On the analysis of social organization among vertebrates, with special reference to birds. Amer. Midl. Nat., **21**:210–34.

UPP, C. W. 1928. Preferential mating in fowls. Poultry Sci., **7**:225–32.

WARREN, D. C., and KILPATRICK, L. 1929. Fertilization in the domestic fowl. Poultry Sci., **8**:237–56.

Erratum

Page 390, line 6 should read: Nalbandov, A., and Card, L. E.

19

Reprinted from *Poultry Sci.*, **39**(4), 970–980 (1960)

Selective Breeding for Aggressiveness in Chickens[1]

A. M. GUHL, J. V. CRAIG AND C. D. MUELLER[2]

Department of Zoology and Department of Poultry Husbandry, Kansas State University, Manhattan

(Received for publication October 28, 1959)

EXCEPT for Game breeds, the domestic fowl has not been consciously selected for behavior traits. However, poultrymen have been aware of some breed differences in behavior, especially between the so-called light and heavy breeds. Some preliminary tests with two breeds and their reciprocal crosses by the senior author (unpublished) gave good evidence of a genetic background for aggressiveness. Komai, Craig and Wearden (1959) have estimated the heritability of social aggressiveness in the domestic fowl to average near 0.30, when data for six strains were analyzed. Breed differences were observed in social dominance relations (Potter, 1949; Holabird, 1955; Tindell and Craig, 1959) and in the effects of androgen on aggressiveness (Allee and Foreman, 1955). Breed differences in the fighting behavior of cocks treated with androgen were reported by Hale (1954), and Siegel (1959) obtained significant differences between aggressiveness and sex drive among cocks of inbred lines.

In the experiment reported here, an attempt was made to determine whether two strains differing in levels of aggressiveness could be obtained by selective breeding. Social agressiveness is considered here as a tendency to be self-assertive as evidenced by the ability to dominate others in a peck-order or in winning dominance status over strange individuals. An experiment (Guhl and Eaton, 1948; Eaton, 1949), which preceeded the one reported here, suggested some modification of techniques (Collias, 1943) for measuring aggressiveness. Because information on the development of aggressiveness in chicks was unavailable at that time, it was deemed essential to conduct, concurrently, a study on the establishment of social organization in chicks (Guhl, 1958).

PROCEDURE

Birds and Pens. The White Leghorn breed was used in this experiment because of its known social activity. Foundation males were of the Ghostley strain and the original females were from Kansas State University exhibition stock. Nothing was known about the relative aggressiveness of the two strains. By using two different strains for the parental generation, poten-

---

[1] Contribution No. 296, Department of Zoology, and No. 249, Department of Poultry Husbandry, Kansas Agricultural Experiment State, Kansas State University, Manhattan.

[2] Now at Glastonbury, Conn.

tial effects of inbreeding were expected to be minimized.

The birds of the parental generation were penned in unisexual groups, and those of each sex were placed into two pens in approximately equal numbers. The offspring of successive generations were reared with other birds on the poultry farm in the usual manner, which included a summer range. When the pedigreed birds of the filial generations were housed they were also divided into two groups for each sex. The membership in each group was assigned in a manner which placed equal, or approximately equal, numbers of each line of selection into each pen. As wing-bands showed parentage, pen assignments followed numerical order, but with about the same number of individuals from each family in each of the pens. The person making the pen assignment did not handle or see the birds when flock membership was determined. The individuals were identified during observations by marks made with colored dyes on wings, backs, and saddles.

Feeding, care, and egg collection were all managed by the employees on the poultry farm, and in the usual manner.

Measurement of Relative Aggressiveness. Two methods for determining levels of aggressiveness in chickens have been used; the ranks in a peck-order and the number of initial encounters won (Allee, Collias and Lutherman, 1939; Collias, 1943). Neither of these yields an absolute measure, but rather a relative one within the sample tested. Since several factors associated with age may alter a bird's ability to establish social dominance (Schjelderup-Ebbe, 1935; Allee, Collias and Lutherman, 1939) tests between generations were not made. The relative merits of the rank-order method and those of initial encounters have been discussed by Guhl (1953) with indications that the so-called paired en-

counters may be more reliable, because some of the environmental factors which make for dominance can be controlled. Both methods were used in the tests made during this experiment, but the number of encounters won received more weight in the selection of breeding stock.

Initial paired encounters, as the term implies, are conducted between two unacquainted birds, and in an unfamiliar (or neutral) area. Since chickens usually lose ability to recognize penmates after at least two weeks of separation (Schjelderup-Ebbe, 1935), the individuals to be tested by paired encounters may be separated by subflocking or by partial (contactual) isolation (laying cages may be satisfactory). Both of these methods have some disadvantages. In a previous selection experiment (Eaton, 1949) it was found that caged birds showed less indication toward aggression, probably due to changes in footing and some debility resulting from lack of activity. With subflocking, the habits of dominating, or of being dominated, might have a psychological influence on agonistic behavior in a test situation. Such effects seemed to be in evidence in certain flocks when marked social tensions were observed.

Peck-orders were determined in both flocks of each sex, and individuals were ranked according to the percentage of individuals each dominated. Initial encounters were not staged until several weeks after the flocks were assembled, when the peck-orders were well established, and inter-individual toleration became apparent (*c.f.*, Guhl and Allee, 1944).

The initial paired encounters were held in a strange pen. Two birds of the same sex were placed simultaneously into an 24″ × 24″ × 27″ exhibition cage (51″ × 24″ × 27″, with the partition removed for males) which was on the litter covered floor. An encounter was considered as won,

370

or dominance established, when one bird consistently avoided the other irrespective of the intensity of aggression (or the lack of it, *i.e.*, an escape reaction without any back threatening or pecking). Each bird in one flock met each in the other of its own sex and generation. All possible pair combinations between strange birds were tested. The person conducting the encounters did not know whether an individual was the progency of the high or low aggressive line of selection. There was a total of 3,889 initial paired encounters, including those of the parental generation and each succeeding generation through the F_4. The number of encounters between individuals representing each of the two lines totals 1,921, and 344 encounters were between crosses of the lines and the parental lines in the F_4.

Selection of Breeding Stock. Two or three males which rated high in the number of encounters won, and also in social rank, were selected for breeders with several females also having high ratings for aggressiveness. One of these males was penned with the females and the others held in reserve. The amount of inbreeding was reduced by using two males each generation. There was a time-lapse between males during which eggs were not held for incubation, to avoid questionable pedigrees. Selection for a line of low level aggressiveness was based on selection of males and females with the fewest encounters won and low rank in the social order, and approximately the same number of birds was retained for breeding. In the F_3 breeding pens, when enough eggs were collected, the males were removed for 10 days and the lines were then crossed.

The eggs were pedigreed and incubated along with others from other experiments. At hatching some small families were discarded. When the birds were housed in the fall, the number of families was reduced

further if more birds were available than could be housed and/or tested in the space allotted to this experiment.

Factors Associated with Dominance. Collias (1943) analyzed some factors for success in initial encounters between hens. He considered comb size the best convenient indicator of male hormone output. Hens injected with androgen showed comb growth and gained rank in the social order (Allee, Collias and Lutherman, 1939). Therefore data on comb size (length of blade plus height of point over the eyes) were obtained in this study. Our tests were made during early maturity; the exact age at which measurements were made varied somewhat from year to year. These data were taken when the encounters were completed.

RESULTS

Table 1 gives the number of surviving individuals, by generations and sex, tested by paired encounters. The parental genera-

TABLE 1.—*The number of individuals that survived the initial paired encounters in each generation of selection*

Generation and sex	Number in high line	Number in low line	Total
F_1 ♂♂	33	33	66
F_1 ♀♀	24	29	53
F_2 ♂♂	12	20	32
F_2 ♀♀	22	23	45
F_3 ♂♂	13	12	25
F_3 ♀♀	15	14	29
F_4 ♂♂	18	9	27
F_4 ♀♀	17	9	26
F_4 Crossed ♂♂			
F_3 high ♂ × F_3 low ♀	3		
F_3 low ♂ × F_3 high ♀	7		10
F_4 Crossed ♀♀			
F_3 high ♂ × F_3 low ♀	1		
F_3 low ♂ × F_3 high ♀	7		8
Totals	159	18 149	326

tion contained 36 males and 43 females on which there were complete data. These are small numbers for a selection experiment but still quite large when the number of paired encounters is considered.

Figure 1 presents, in graphic form, the results of initial paired encounters between individuals of the high and low lines. The points on the abscissa indicate generations. The top row of numbers under these points show the number of breeders whose offspring were used in successive generations (*i.e.*, breeders whose line was later discontinued are not included). The lower figures give the number of offspring surviving all tests. The ordinates indicate the percentage of encounters won by the birds selected for breeding and means of their offspring of the same sex. The points not connected by lines, in the F_4, give means for crosses of the "high" and "low" lines. In each graph the upper of the two points is the mean of a cross between a "low" sire and a "high"

dam, and the lower of two points is for the reverse cross.

Figure 1 shows progressive differences in aggressiveness between the two lines. The divergence is somewhat greater and more uniform among females than among males. When the lines were crossed, the means tended to become intermediate, but these results must be considered with some reservations. Unfortunately, circumstances beyond our control left only 10 males and eight females of the crosses available for testing. The upper point for males is the mean for seven cockerels, the lower for only three. For the females the numbers are seven and one, respectively.

Figure 2 is an abbreviated pedigree chart showing breeders only. The numbers indicate individual females (band numbers) and the letter, the identification colors of the males. Coefficients of inbreeding are given below the figure for the two lines of selection.

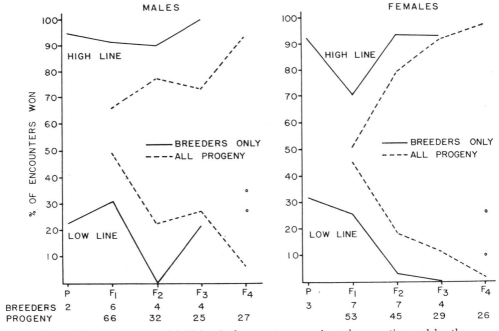

FIG. 1. Mean percentages of initial paired encounters won in each generation and by the individuals selected for breeding. (see text)

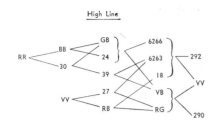

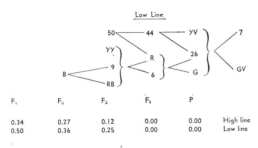

F₁	F₃	F₂	F₁	P	
0.34	0.27	0.12	0.00	0.00	High line
0.50	0.36	0.25	0.00	0.00	Low line

Males indicated by letters; Females by numbers.

FIG. 2. Abbreviated pedigree chart with coefficients of inbreeding.

Since each pen contained birds of both lines, and since paired encounters were within and between lines, chi-square analyses were made to compare the lines in aggressiveness as measured by both the number of birds dominated in the home pen and the number of encounters won with strangers in the other pen. The data in Table 2 are the number of birds in the low line dominated by those in the high line and the number of high line birds dominated by those in the low line, i.e., dominance between birds of the same line are not included. The data in Table 3 are the number of encounters won between individuals representing the two lines of selection, i.e., intra-line encounters were not included. Staged encounters which failed to establish dominance were excluded from the analyses, of which there were five, 13, and five in the $F_1(I)$ ♂ ♂, F_1 ♀ ♀ and F_2 ♀ ♀ respectively. In addition to probabilities, the ratios between the actual

numbers are included. Three assumptions were made, (1) that there was no difference between "high" and "low" lines; (2) that each individual had an equal chance to be in the middle of the peck-order; and (3) that each had an equal chance to win half of the initial paired encounters. Table 2 shows that the low line males in the F_1 dominated significantly more inter-line individuals than those in the high line. The F_1 females and all the succeeding generations for both sexes showed that the birds in the high line dominated significantly more individuals than did those in the low line. With the exception of the F_3 males, among all progeny in the F_2 and F_4 the ratios of the actual data deviate progressively more from 50-50, indicating separate populations. As measured by encounters won, Table 3 shows a progressive increase in the high line and decrease in the low line of mean levels of aggressiveness, which became significantly different by the F_2 generation. The ratios show a clear-cut separation of lines beginning with the F_2.

Heritability estimates are presented in Table 4 by generations and also on a cumulative basis. Data on males and females were pooled for these calculations. Since the differences between the lines became progressively greater, generation by generation, and were in general highly significant (see Tables 2 and 3), the unweighted mean heritabilities and the cumulative heritabilities of the second and later generations may also be considered significant. An examination of the heritability estimates obtained at the various generations suggests that selection over a period of four generations had produced about 20% of the differences selected for and no marked loss of additive genetic variation is apparent, i.e., selection continued to be effective. It is equally apparent, however, that further differentiation of the lines relative to each other would be next to impossible

TABLE 2.—*Chi-square analyses of the number of birds dominated in the peck-order between high and low lines*

Progeny		Expected	Actual	X^2	P	Ratio: No. dominated by lows
						No. dominated by highs
F_1 ♂♂ (I)	H	89	67	43.50	0.01	1.66*
	L	89	111			
F_1 ♂♂ (II)	H	44	38	6.54	0.02	1.32*
	L	44	50			
F_1 ♀♀	H	163.5	175	9.75	0.01	0.87
	L	163.5	152			
F_2 ♂♂	H	58	80	66.75·	0.01	0.45
	L	58	36			
F_2 ♀♀	H	126.5	208	419.11	0.01	0.22
	L	126.5	45			
F_3 ♂♂	H	39	51	24.82	0.01	0.53
	L	39	27			
F_3 ♀♀	H	52.5	78	99.08	0.01	0.35
	L	52.5	27			
F_4 ♂♂	H	41	82	328.00	0.01	0.00
	L	41	0			
F_4 ♀♀	H	38.5	77	312.05	0.01	0.00
	L	38.5	0			

* More individuals were dominated by the males of the low line. (I), (II) F_1 males were tested in two sets because of the large number.

(see Figure 1). Thus, a situation was obtained in the F_4 in which further testing on an inter-line basis would have been of no value, since most or all contests between high and low line individuals were won by the high line birds. Intra-line testing might have continued to be effective in producing genetic differences but measuring further differentiation between lines would require much ingenuity.

Using the method of path coefficients, Collias (1943) determined the relative importance of androgen output as indicated by comb size, thyroxin secretion as indicated by molting, social rank in the home flock, and body weight. These four factors gave a multiple correlation of 0.75, and he estimated that 44 percent of all factors were unknown or unmeasured. In the results reported here, molting was not considered because the birds were less than a year old and did not enter a seasonal molt. The other three factors were known, if percentage dominated is accepted as a measure of social status.

Since selection resulted in distinct aggressiveness lines, as measured by initial encounters and social levels, the difference in body weights and comb size were determined, Table 5, to note whether these factors were also affected by selection. Of the nine groups measured, five showed statistically significant differences between lines in body weight, with the high line birds heavier in four and the low line in one. Significantly larger combs in the more aggressive high line were found only in the F_4 females, and contrary to expectation, the low line birds in four groups tended to have larger combs.

A. M. Guhl, J. V. Craig and C. D. Mueller

Table 3.—*Chi-square analyses of the number of initial paired encounters won between high and low lines*

Progeny		Expected	Actual	X^2	P	Ratio: No. dominated by lows / No. dominated by highs
F_1 ♂♂ (I)	H	95.5	121			
	L	95.5	70	54.47	0.01	0.58
F_1 ♂♂ (II)	H	44	48			
	L	44	40	0.18	0.70	0.83
F_1 ♀♀	H	178	189			
	L	178	167	5.44	0.02	0.89
F_2 ♂♂	H	62	96			
	L	62	28	148.16	0.01	0.29
F_2 ♀♀	H	124	201			
	L	124	47	382.52	0.01	0.24
F_3 ♂♂	H	39	57			
	L	39	21	66.46	0.01	0.37
F_3 ♀♀	H	52.5	92			
	L	52.5	13	237.75	0.01	0.12
F_4 ♂♂	H	40	75			
	L	40	5	245.00	0.01	0.07
F_4 ♀♀	H	38	75			
	L	38	1	288.21	0.01	0.01

(I) (II) F_1 males were tested in two sets because of the large number.

Table 4.—*Heritability estimates of social aggressiveness*

Selected generation	Estimate by generations			Cumulative estimate		
	Differences (high-low)[1]		Heritability[3]	Differences (high-low)[2]		Heritability[3]
	Produced	Selected for		Produced	Selected for	
	A. Based on Percentage of Flock Dominated					
1	4.5	69.8	0.06	4.5	69.8	0.06
2	24.5	57.4	0.43	29.0	127.2	0.23
3	− 8.3	46.4	−0.18	20.7	173.6	0.12
4	27.6	66.1	0.42	48.3	239.7	0.20
Unweighted mean h^2	—	—	0.18	—	—	—
	B. Based on Percentage of Initial-Pair Encounters Won					
1	7.6	69.6	0.11	7.6	69.6	0.11
2	23.8	59.3	0.40	31.4	128.9	0.24
3	− 1.0	38.8	−0.03	30.4	167.7	0.18
4	14.5	38.4	0.38	44.9	206.1	0.22
Unweighted mean h^2	—	—	0.22	—	—	—

[1] Differences on a generation-by-generation basis calculated as: difference (high-low) of specified generation minus difference (high-low) of preceding generation, e.g. in generation 2 of Part B, difference produced $= 31.4 - 7.6 = 23.8$ and difference selected for $= 128.9 - 69.6 = 59.3$.

[2] Differences on a cumulative basis calculated as: (a) difference produced = actual difference (high-low) of specified generation, (b) difference selected for = difference (high-low) selected for in specified generation plus difference selected for previously, e.g. in generation 3 of Part B, difference selected for $= 38.8 + 128.9 = 167.7$.

[3] Heritability = (difference produced)/(difference selected for).

TABLE 5.—*Differences between lines of selection in the means of body weight and comb size*

Generations	n	"High" line	n	"Low" line	Difference
Mean body weights					
F_1 ♂♂ (I)	17	5.4	22	5.4	0.0
F_1 ♂♂ (II)	16	6.1	11	5.5	0.6*
F_1 ♀♀	24	4.9	29	4.5	0.4
F_2 ♂♂	12	6.0	20	5.4	0.6**
F_2 ♀♀	22	4.5	23	4.0	0.5
F_3 ♂♂	13	5.0	12	4.7	0.3
F_3 ♀♀	15	3.6	14	3.9	−0.3*
F_4 ♂♂	18	5.5	9	4.9	0.6**
F_4 ♀♀	17	4.6	9	3.9	0.7**
Mean comb size					
F_1 ♂♂ (I)	17	175.2	22	176.8	−1.6
F_1 ♂♂ (II)	16	183.0	11	177.5	5.5
F_1 ♀♀	23	126.5	29	125.1	1.4
F_2 ♂♂	12	170.0	20	173.0	−3.0
F_2 ♀♀	22	119.2	23	109.9	9.3
F_3 ♂♂	10	159.0	12	157.0	2.0
F_3 ♀♀	15	113.0	14	119.0	−6.0
F_4 ♂♂	18	168.0	9	173.7	−5.3
F_4 ♀♀	17	130.0	9	109.0	21.0**

* $P \leq 0.05$.
** $P \leq 0.01$.

Correlations, within lines, sex, and generation, were determined between the methods of measuring aggressiveness, and between each of the two methods of measuring aggressiveness with body weight and comb size, Table 6. Except for the low line males, the rank correlations between percentage of initial encounters won and percentage dominated are large and are significant for most flocks. Because there is a significant heterogeneity among these correlations, an unweighted correlation (0.59), is given for all flocks. The causes for the low and nonsignificant correlations for the $F_1(II)$, F_2 and especially the F_3 males are difficult to explain. Maintaining flocks of males in pens presents some management problems because the social tension causes low ranking cocks (usually the low line) to remain on roosts. Each pen contained approximately equal numbers of high and low males. Even though a feed trough was placed on the roost and water was nearby at about the same level to reduce the stress on the low ranking birds, it might be postulated that the social stress in the home pen may have confounded the potential ability of low line males to win initial paired contests.

The coefficients between comb size, body weight, and measures of aggressiveness lacked significance for most of the flocks. However, all weighted correlations for the complete experiment were significant, although low. It may be concluded that, within lines, comb size and body weight are significant though minor factors associated with attainment of social status and the ability to win initial encounters. This agrees in general with Allee, Collias and Lutherman (1939) and Collias (1943).

Although a number of factors are controlled in conducting initial paired encounters (Collias, 1943), and this technique is presumably the best now available to measure levels of aggressiveness, the intensity of stimuli which evoke aggression or submission are not under complete control. Occasionally neither bird stimulated the other, and dominance was not established. Dominance may be won by having one bird avoid the other (passive submission), by threatening and ensuing avoidance; by pecking with subsequent avoidance; or by fighting until one avoids the other (escape reaction). A record was made of the kind of activity which resulted in the attainment of dominance in a total of 2,842 encounters in which the establishment of dominance was observed.

Table 7 gives, for each generation and sex, the percentage of encounters won as a result of avoiding without aggression, threatening, pecking, or fighting. There appears to be no consistency from generation to generation within either sex, nor do the "high" and "low" lines show consistent differences from generation to generation. An examination of the mean percentages for all four generations does, however, indicate differences between the sexes. The males engaged in more fighting than did the females, but the "high" males did not appear to fight oftener than did the "low" males. The mean difference between lines of females is largely due to the low frequency of fighting by the F_4 "highs." The males tended to submit more by avoiding behavior, without aggression, than did the females. Threats and pecks were more decisive among females than among males.

DISCUSSION

The results show that strains differing in relative aggressiveness may be obtained by selective breeding. How readily this may be accomplished with other strains is still to be determined. We used two different strains in the parental generation, about which comparative data on aggres-

siveness were unavailable.

Although the initial paired encounters provide the best known method of measuring relative aggressiveness in chickens, there are a number of factors (probably minor, Collias, 1943) which cannot be easily controlled. One may question whether the method measures more than aggressiveness. For example, aggressiveness and submissiveness may well be independent factors, rather than extremes of a gradient. These are different behavior patterns; one is mediated by an androgen (c.f., Allee, Collias and Lutherman, 1939; Collias, 1943; Guhl, 1958) and the other

TABLE 6.—*Rank correlation coefficients among social aggressiveness measurements, comb size, and body weight*

Group	No. birds	% Won: % Dominated	% Won: Comb size	% Dominated: Comb size	% Won: Body wt.	% Dominated: Body wt.
High line males						
F_1 (I)	16	.777**	.508*	.397	.130	.068
F_1 (II)	17	.551*	.592*	.429	.675**	.676**
F_2	12	.726**	.357	.689*	.733**	.701*
F_3	13	.897**	.761*[1]	.800**[1]	.365	.495
F_4	18	.678**	.342	.085	.177	.236
Low line males						
F_1 (I)	11	.655*	.470	.248	.096	.325
F_1 (II)	22	.291	.105	.457*	.615**	.638**
F_2	20	.271	.180[1]	−.096[1]	.292	.329
F_3	12	−.386	−.012	.353	−.294	.715**
F_4	9	.729*	.296	.629	.175	.267
High line females						
F_1	24	.690**	−.056[1]	.182[1]	.337	.413*
F_2	22	.542**	.128	.462*	.062	.319
F_3	15	.728**	.711**	.483	.539*	.055
F_4	17	.895**	.588*	.574*	.304	.387
Low line females						
F_1	29	.857**	.059	.055	−.141	−.142
F_2	23	.757**	.364	.423*	.168	.269
F_3	14	.722**	.382	.486	.591*	.475
F_4	9	.392	.329	.467	.488	.625
Total	303					
Weighted correlations			.319**	.369**	.298**	.366**
Unweighted correlation		.598[2]				

* $P < 0.05$.
** $P < 0.01$.
[1] Because of comb removal or molt, the number of birds in groups Male F_3 High, Male F_2 Low and Female F_1 High were 10, 18 and 23, respectively.
[2] Because of significant heterogeneity ($P < 0.01$) among correlations for % won and % dominated an unweighted correlation coefficient is presented. Heterogeneity was lacking among correlations for the other variables.

TABLE 7.—*Percentages of dominance relationships won by either passive submission or by each of three intensities of aggression*

Generation	"High" line					"Low" line				
	Encounters n	Avoid. %	Threat. %	Peck %	Fight %	Encounters n	Avoid. %	Threat. %	Peck %	Fight %
Males										
F_1 (I)	194	17.0	36.1	30.9	16.0	185	24.3	35.1	30.3	10.3
F_1 (II)	110	19.1	28.2	36.4	16.3	74	17.6	32.4	36.5	13.5
F_2	126	37.3	19.0	17.5	26.2	124	20.2	19.4	29.8	31.5
F_3	114	49.1	21.9	15.8	13.2	66	57.6	7.5	18.2	16.7
F_4	239	49.8	24.6	15.0	10.4	51	49.0	11.7	21.5	17.6
Mean		34.46	25.96	23.12	16.42		33.74	21.22	27.26	17.92
Females										
F_1	281	7.8	36.3	46.6	9.3	302	10.3	42.0	41.0	6.7
F_2	311	22.2	43.1	26.7	8.0	177	12.4	31.1	46.9	9.6
F_3	147	22.4	43.5	22.4	11.6	62	35.5	35.5	16.1	12.9
F_4	236	37.2	43.6	16.5	2.5	43	20.9	34.8	25.5	18.6
Mean		22.40	41.63	28.05	7.85		19.78	35.85	32.38	11.95

by an estrogen (*c.f.*, Allee and Collias, 1940; Guhl, 1958). Gamecocks usually do not show submissiveness (observations at cockfights by two of us) whereas the Red Junglefowl appears to show high levels of both aggressiveness and submissiveness (general observations by A.M.G.). If this assumption is correct, then one may question whether the initial encounters measure both aggressiveness and submissiveness or whether one is detected more readily than the other.

The selection was from a small number of individuals in each generation, although the number was somewhat larger than some experiments which showed breed differences in aggressiveness (Potter, 1949; Allee and Foreman, 1955; Holabird, 1955).

The results raise some interesting questions. There is the matter of sex linkage, although the very limited data from reciprocal crosses of high and low lines do not suggest it. Siegel (1959) found no heterotic effect with regard to aggressiveness and sex drive. Do the genes influencing aggressiveness control hormonal output? which hormone(s)? and the neural patterns?

and/or their thresholds? Hale (1954) suggested that the genetic differences in fighting behavior of cocks depend on neural factors unrelated to endocrine activity. A number of reports by Young and his associates (*e.g.*, Valenstein, Riss and Young, 1954; Riss, Valenstein, Sinks and Young, 1955; Valenstein, Riss and Young, 1955) have shown that genetic differences in levels of sexual excitement in guinea pigs are not overcome by exogenous androgen. Wood-Gush (1956), Siegel (1959), and McDaniel and Craig (1959) found a correlation between levels of aggressiveness and courtship displays and between aggressiveness and completed matings in cocks. However, Guhl, Collias and Allee (1945) found no significant relationship between the social rank of a cock and his sexual activity when placed alone in a flock of hens. Guhl (1951) reported a case in which the least sexually active cock was socially dominant over two others when placed with hens. Possible correlations with other factors in reproduction, in both sexes, need to be explored. Although it may seem desirable, from a managerial viewpoint, to have an unaggressive strain, if reproductive behavior and gamete production are related to high aggressiveness, then the designs for genetic selection might give more attention to social behavior.

SUMMARY

Selection for high and low levels of aggressiveness was based on the results of initial paired encounters as a measure of relative aggressiveness. Ranks in the peck-order were used as supporting evidence for relative aggressiveness. Two different strains of White Leghorns were used in a one-way cross in the parental generation to reduce excessive inbreeding, and selection was carried to the F_4 generation. Beginning with the F_2 generation the two lines showed significant differences in the percentage of encounters won or lost as well as in high or low ranks in the peck-orders. Heritability estimates of 0.22 and 0.18 were obtained when based on the percentages of contests won and individuals dominated, respectively.

REFERENCES

Allee, W. C., and N. E. Collias, 1940. The influence of estradiol on the social organization of flocks of hens. Endocrinology, 27: 87–94.

Allee, W. C., N. E. Collias and C. Z. Lutherman, 1939. Modification of the social order in flocks of hens by the injection of testosterone propionate. Physiol. Zool. 12: 412–440.

Alee, W. C., and D. Foreman, 1955. Effects of an androgen on dominance and subordinance in six common breeds of *Gallus gallus*. Physiol. Zool. 28: 89–115.

Collias, N. E., 1943. Statistical analysis of factors which make for success in initial encounters between hens. Amer. Nat. 77: 519–538.

Eaton, R. C., 1949. Breeding for aggressiveness in the fowl. Master of Science Thesis. Department of Poultry Husbandry, Kansas State University, Manhattan.

Guhl, A. M., 1951. Measurable differences in mating behavior of cocks. Poultry Sci. 30: 687–693.

Guhl, A. M., 1953. Social behavior of the domestic fowl. Tech. Bull. 73, Kansas Agricultural Experiment Station, Manhattan.

Guhl, A. M., 1958. The development of social or-ganization in the domestic chick. Animal Behaviour, 6: 92–111.

Guhl, A. M., and W. C. Allee, 1944. Some measurable effects of social organization in flocks of hens. Physiol. Zool. 17: 320–347.

Guhl, A. M., N. E. Collias and W. C. Allee, 1945. Mating behavior and the social hierarchy in small flocks of White Leghorns. Physiol. Zool. 18: 365–390.

Guhl, A. M., and R. C. Eaton, 1948. Inheritance of aggressiveness in fowl. Poultry Sci. 27: 665.

Hale, E. B., 1954. Androgen levels and breed differences in the fighting behavior of cocks. Bull. Ecol. Soc. Amer. 35: 71–72.

Holabird, C., 1955. Social organization in flocks of Light Brahma hens as compared with other breeds. Physiol. Zool. 28: 239–255.

Komai, T., J. V. Craig and S. Wearden, 1959. Heritability and repeatability of social aggressiveness in the domestic chickens. Poultry Sci. 38: 356–359.

McDaniel, G. R., and J. V. Craig, 1959. Behavior traits, semen measurements and fertility of White Leghorn males. Poultry Sci. 38: 1005–1014.

Potter, J., 1949. Dominance relations between different breeds of domestic hens. Physiol Zool. 22: 261–280.

Riss, W., E. S. Valenstein, J. Sinks and W. C. Young, 1955. Development of sexual behavior in male guinea pigs from genetically different stocks under controlled conditions of androgen treatment and caging. Endocrinology, 57: 139–146.

Siegel, P. B., 1959. Evidence of a genetic basis for aggressiveness and sex drive in the White Plymouth Rock cock. Poultry Sci. 38: 115–118.

Schjelderup-Ebbe, T., 1935. Social behavior in birds. In Murchison's Handbook of Social Psychology, pp. 947–972. Clark Univ. Press, Worcester.

Tindell, D., and J. V. Craig, 1959. Effects of social competition on laying house performance in chickens. Poultry Sci., 38: 95–105.

Valenstein, E. S., W. Riss and W. C. Young, 1954. Sex drive in genetically heterogeneous and highly inbred strains of male guinea pigs. J. Comp. Physiol. Psychol. 47: 162–165.

Valenstein, E. S., W. Riss and W. C. Young, 1955. Experimental and genetic factors in the organization of sexual behavior in male guinea pigs. J. Comp. Physiol. Psychol. 48: 397–403,

Wood-Gush, D. G. M., 1956. The agonistic and courtship behaviour in the Brown Leghorn cock. Brit. J. Ani. Behav. 4: 133–142.

20

Reprinted from *Poultry Sci.*, **48**(5) 1729–1736 (1969)

Productivity of Pullets Influenced by Genetic Selection for Social Dominance Ability and by Stability of Flock Membership[1,2]

J. V. CRAIG AND A. TÓTH[3]

Kansas State University, Manhattan, Kansas 66502

(Received for publication May 19, 1969)

INTRODUCTION

SOCIAL stress is usually associated with lowered productivity in chickens. Strange hens introduced into organized flocks are initially harassed and their egg production is depressed (Sanctuary, 1932; Guhl and Allee, 1944; Guhl, 1953; and Morgan and Bonzer, 1959). Low peck-order status also leads to a stressful intra-flock environment for the individual hen and is associated with lower productivity (see Guhl, 1953).

Tindell and Craig (1959) confirmed the presence of intra-flock associations between social rank and productivity traits for the early part of the laying year, but found later egg production to be essentially uncorrelated with rank. They also compared 6 diverse genetic strains in intermingled and separate flocks. Inter-strain competition resulted in higher performance levels for the more aggressive strains; whereas, due to their lack of competitive ability, the less aggressive strains matured later, fed less often, had a lower rate of lay and poorer livability as compared to their performance when penned separately.

McBride (1958) suggested that a nonlinear association exists between peck-order status and egg mass produced per pullet. According to his hypothesis productivity drops logarithmically as peck-order position decreases below a critical level. He found that variance of egg mass for birds of low social status was significantly

[1] Contribution No. 735, Department of Dairy and Poultry Science, Kansas Agricultural Experiment Station, Manhattan, Kansas 66502.

[2] Supported in part by Grants G7069, G19853 and GB1720 from the National Science Foundation.

[3] Fellow of the Ford Foundation Hungarian Exchange Program. Present Address: Poultry Breeding Institute, Gödöllö, Hungary.

greater than that of high ranking birds. McBride (1960, 1962, 1968) has developed further hypotheses along the same lines. He pointed out that poor husbandry conditions, such as increased population density, inadequate feeding and watering space and social disturbances, probably increase the frequency of aggressive interactions. It was then predicted that under a series of different husbandry conditions means and variances would be inversely related and that management procedures causing increased social stress would produce a greater negative skew to the frequency distribution curve for productivity traits than ordinarily found. Noting the results of Tindell and Craig (1959), McBride (1962) postulated that high aggressiveness in a strain would have the same effect on productivity as poor husbandry.

Bidirectional selection for social dominance ability within each of two breeds by Craig *et al.* (1965) produced significant differences between strains in several components of social behavior. Preliminary comparisons of a few productivity traits (Craig, 1968) indicated that correlated responses had occurred. The present study compares female productivity of those selected strains under conditions of stable and unstable flock membership. The data were also analyzed to determine whether greater intra-strain variance was present within the more aggressive strains as predicted from McBride's hypothesis. Meaningful comparisons of the relative ability of the strains to withstand social stress were precluded due to lack of evidence of stress due to the unstable flock environment and because of heterogeneous variances within the high and low strains.

MATERIALS AND METHODS

Genetic Stocks. High and low social dominance strains produced by bidirectional selection within the White Leghorn

(W.L.) and Rhode Island Red (R.I.R.) breeds as described by Craig *et al.* (1965) were used. They found that high strain birds in pair contests were more likely to attack or threaten an opponent, their social interactions were physically more severe and contests between strains, within breeds, were predominantly won by highs. A subsequent study (Craig *et al.*, 1969) indicated that initial pair contests between high and low strain males of both breeds were still won mostly by highs even after 4 generations of relaxed selection. Furthermore, high strain W.L. pullets in floor flocks had greater frequencies of social interactions and variances of social tension index, indicating higher levels of social stress. R.I.R. high strain pullets had 13 percent more frequent interactions than lows in that study and 32 percent more in a later study (unpublished data), but these differences were statistically nonsignificant.

Chicks used in this study represented the second generation of relaxed selection. Hatching eggs were obtained from three multiple sire matings per strain, each consisting of seven males and from 17 to 34 females, representing all seven sire families of the preceding generation.

Management and Experimental Procedures. Chicks of all strains were hatched concurrently in the same incubator. They were sexed, wingbanded and vaccinated for bronchitis and Newcastle disease at hatching. Brooding and rearing were in two intra-strain flocks of 30 males and 95 to 100 females for each W.L. strain and three flocks of 20 males and 89 females for each R.I.R. strain in randomly assigned pens. Chicks were vaccinated for fowl pox and Newcastle disease and half of the upper beak was removed at eight weeks to inhibit cannibalism.

One-hundred and forty-four pullets were randomly selected per strain at 18 weeks of age and divided equally with 18 per flock into eight pens. Each pen was 1.5 by 2.3 m. and all were alike in feeding, watering, nesting and roosting facilities. Pens were arranged in four rows of eight. Two flocks of each strain were placed in each row, one designated as a stable and the other as an unstable flock. Position within rows was randomly determined.

Stable flocks were undisturbed in membership, whereas unstable flocks were changed in membership every seven days from 18 to 30 weeks of age. Membership in unstable flocks was determined weekly by a random scheme, except that essentially equal numbers were placed in the four flocks within each strain. All individuals were returned to their original flocks at 30 weeks of age and remained there until termination of the study at 42 weeks of age.

Individual body weights were recorded at 22 and 30 weeks of age and all flocks were trapnested three days a week. Age at first egg was recorded for each pullet and rate of lay was calculated from first egg through 30 weeks of age and from 32 through 41 weeks of age. Survival of strain-treatment groups was calculated for the 18 through 30 week period and by flocks for the 32 through 41 week period.

Statistical Analyses. Effects of strain differences on variability of productivity traits (other than percentage survival) were estimated by comparing pooled mean squares for individuals within high strain stable flocks with comparable mean squares for low strain stable flocks.

Heterogeneity of intra-flock variability was found for several traits in comparing high and low strains (see below). This heterogeneity, along with the procedure of changing flock memberships weekly in unstable flocks from 18 to 30 weeks of age, introduced difficulties into analyses of the data. Thus, there was no appropriate error term for testing the interaction of genetic groups by treatment (stable *vs.* unstable).

Main effects for most traits were tested for significance using analyses of variance of unweighted strain-treatment means and the interaction mean square as the error term.

Analyses for breed and strain effects on rate of lay from 32 through 41 weeks and percentage survival for the same period used flocks within strains mean squares as error terms.

RESULTS

Intra-flock Individual Variability. Pooled intra-flock variances for individuals were compared for high and low strains, within breeds, as shown in Table 1. High strain W.L. pullets were significantly more variable for 30-week body weight, age at first egg and rate of lay from 32 through 41 weeks of age. They also approached being significantly more variable for rate of lay from first egg through 30 weeks of age. Frequency distributions for the high and low W.L. strains for rate of lay in the second period are presented in Figure 1.

Effect of Unstable Flock Membership. The results shown in Table 2 fail to yield convincing evidence of any loss of productivity associated with unstable flock membership. There was even a suggestion (P = 0.06) of the opposite effect on age at first egg with earlier onset of sexual maturity in the unstable flocks.

Effect of Breed. Breed comparisons, though not of primary importance in this study, indicated heavier body weights, earlier sexual maturity and lower rate of lay from 32 through 41 weeks of age for the R.I.R. breed. It may be noted that the earlier sexual maturity of the R.I.R. is largely influenced by the high strain of that breed (Table 2).

Productivity of High and Low Social Dominance Strains. Body weight differences between high and low social dominance strains were found at 22 weeks of age within both breeds (P < 0.05).

382

Table 1.—*Variability in production traits for strains within each of two breeds*[1]

Trait	Breed	Strain	D.F.	M.S.	$F = \dfrac{\text{High M.S.}}{\text{Low M.S.}}$
Body weight, kg., 22 weeks of age	W.L.	High	67	0.020	1.25
		Low	66	0.016	
	R.I.R.	High	67	0.073	1.33
		Low	68	0.055	
Body weight, kg., 30 weeks of age	W.L.	High	61	0.052	1.73*
		Low	65	0.030	
	R.I.R.	High	65	0.089	1.10
		Low	68	0.081	
Age at first egg, days	W.L.	High	53	349	1.58*
		Low	61	220	
	R.I.R.	High	62	346	0.87
		Low	57	397	
Rate of lay, percent, first egg— 30 weeks of age	W.L.	High	38	245.7	1.53*
		Low	44	160.8	
	R.I.R.	High	53	237.3	1.10
		Low	40	216.7	
Rate of lay, percent, 32–41 weeks of age	W.L.	High	98	464.9	2.54***
		Low	124	183.2	
	R.I.R.	High	114	264.9	0.92
		Low	108	288.3	

[1] Comparisons are for individuals within stable flocks only.
* P<0.10, * P<0.05, *** P<0.005.

Though of similar sign and magnitude, differences at 30 weeks of age were significant within the R.I.R. breed only (Table 2).

The more aggressive strains within both breeds began laying at earlier ages. The W.L. high strain pullets also had lower rates of lay and higher mortality, particularly during the 32 through 41 weeks of age period (P < 0.01 for both traits, see Table 2).

DISCUSSION

Separation of chickens for two to three weeks results in failure of recognition of former pen mates (Schjelderup-Ebbe, 1935). Since strange birds commonly interact vigorously and with high frequency until a group becomes well integrated (see Guhl, 1968) it was expected that the unstable flocks of the present study would be under considerable social stress. Craig et al. (1969), in a study of the social behav-

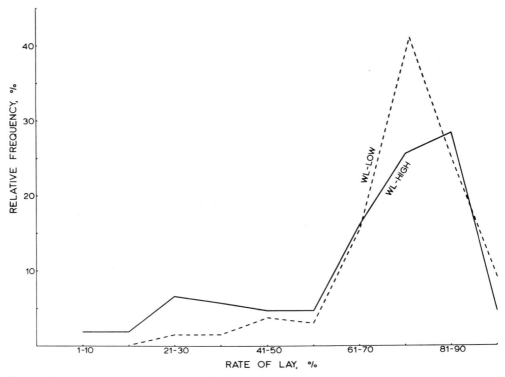

FIG. 1. Frequency distributions for rate of lay, for the period 32 through 41 weeks of age, within the W.L. high and low social dominance strains.

ior of these pullets, found the unstable flocks to have more agonistic interactions (including fights, peck-avoidances and threat-avoidances) than the stable flocks (P < 0.005).

In view of the higher frequency of interactions in the unstable flocks it might be assumed that greater social stress was present. As already noted, there was no convincing evidence of loss of productivity in the unstable flocks. These results imply that greater frequency of agonistic interactions *per se* does not necessarily indicate greater stress. We hypothesize from these results that frequent changes in group membership are beneficial to those individuals which would otherwise be at the bottom of the peck order in a stable group. Under a system of changing group membership

they would have the opportunity to rise in the hierarchy whenever a new group was formed.

Correlated responses in body weight and egg production traits associated with selection for social dominance in these strains, as reported by Craig (1968), are confirmed in the present study. Differences in age at first egg, suggested by results of that study, were more evident, and generally more significant in this experiment. Superior rate of egg production was again found for the W.L. low as compared to the W.L. high strain, but clear evidence of a difference for this trait between the R.I.R. strains was again lacking.

McBride (1962, 1968) predicted that higher levels of agonistic interactions within flocks of chickens, under high den-

TABLE 2.—*Number, means and treatment differences of behavior strain pullets of the W.L. and R.I.R. breeds in stable and unstable flocks for several traits*

Breed	Strain	Number of pullets		Treatment means		Treatment differences[1]		
		Stable	Unstable	Stable	Unstable	High-Low	WL-RIR	Stable-Unstable
				Body weight, kg. (22 weeks of age)				
W.L.	High	71	72	1.55	1.57	0.09*		
	Low	70	72	1.48	1.46		−0.65***	0.01
R.I.R.	High	71	72	2.15	2.12	−0.08*		
	Low	72	72	2.21	2.21			
				Body weight, kg. (30 weeks of age)				
W.L.	High	65	71	1.87	1.90	0.09		
	Low	69	71	1.82	1.76		−0.68***	0.01
R.I.R.	High	69	66	2.48	2.44	−0.12*		
	Low	72	70	2.56	2.60			
				Age at first egg, days				
W.L.	High	57	58	186.18	184.52	−5.15*		
	Low	65	68	191.85	189.16		6.80***	2.24(*)
R.I.R.	High	66	62	175.33	171.35	−15.58***		
	Low	61	61	189.23	188.62			
				Rate of lay, percent (first egg—30 weeks of age)				
W.L.	High	42	46	75.45	70.74	−4.42(*)		
	Low	48	52	79.58	75.46		0.11	2.88
R.I.R.	High	57	58	75.12	72.55	−2.72		
	Low	44	44	76.61	76.50			
				Rate of lay, percent (32–41 weeks of age)				
W.L.	High	106	—	68.09	—	−8.29**		
	Low	132	—	76.38	—		4.56*	—
R.I.R.	High	122	—	68.78	—	2.22		
	Low	116	—	66.56	—			
				Survival, percent[2] (18–30 weeks of age)				
W.L.	High	72	72	90.28	98.61	−2.78		
	Low	72	72	95.83	98.61		−0.35	−1.05
R.I.R.	High	72	72	95.83	91.67	−4.86		
	Low	72	72	100.00	97.22			
				Survival, percent[2] (32–41 weeks of age)				
W.L.	High	136	—	85.41	—	−10.24**		
	Low	140	—	95.65	—		−4.08	—
R.I.R.	High	135	—	93.45	—	2.33		
	Low	142	—	95.78	—			

[1] Differences tested as main effects in the analyses of variance. Significance levels indicated as:
(*) P=0.06, * P<0.05, ** P<0.01 and *** P<0.005.

[2] Percentages were transformed to arcsin $\sqrt{percentage}$ for analyses.

sity husbandry, would be associated with reduced mean levels of productivity and increased individual variability due to negative skew of the frequency distribution. We did not find clear evidence of reduced performance with the highly interactive unstable flocks, as previously discussed. Our results with the W.L. high and low strains for percentage survival and rate of egg production are, however, in reasonable agreement with McBride's predictions. The W.L. high strain pullets, which had significantly more social interactions per unit of time, had reduced survival and lower rates of lay (Table 2). They also showed greater individual variability for rate of lay, particularly during the second period (Table 1) and there was more negative skew associated with their frequency distribution (Figure 1).

Selection for high and low social dominance ability on the basis of pair contest results, though effective in both the W.L. and R.I.R., produced some inconsistent correlated responses. High strain W.L. and R.I.R. pullets were usually dominant to low strain pullets when mixed, within breeds, in floor flocks (Craig et al., 1965). However, though high strain W.L. pullets had significantly more agonistic interactions per unit time than low strain pullets in intra-strain flocks, the R.I.R. strains did not show differences of nearly the same magnitude (Craig et al., 1969). Changes in body weight were considerable and in opposite directions for the two breeds (Craig, 1968; and present study). Earlier sexual maturity was characteristic of high strains within both breeds, but was of greater magnitude in the R.I.R. Survival percentage and rate of lay, traits of major importance, did not differ much between the R.I.R. strains, but large and significant differences were observed between the W.L. strains in floor flocks. The latter results appear consistent with expectation on the basis of McBride's hypothesis (1962, 1968).

SUMMARY AND CONCLUSIONS

White Leghorn (W.L.) and Rhode Island Red (R.I.R.) pullets representing strains selected for high and low social dominance ability within both breeds were compared for productivity under conditions of stable and unstable flock membership. Although unstable flocks were undergoing weekly social reorganization, adverse effects on productivity were not detected.

Correlated responses associated with changes in social dominance produced by selection were consistent for sexual maturity. The more aggressive strains matured earlier in both breeds. Adult body weights were heavier for the high social dominance strain within the W.L. and for the low social dominance strain within the R.I.R.

W.L. high strain pullets had greater individual variability for age at first egg and rate of lay and lower mean rate of lay and percentage survival than W.L. lows. Such differences were not found in comparing high and low strains within the R.I.R. breed. It was tentatively concluded that greater social stress in flocks of the high social dominance W.L. strain was responsible for the greater individual variability and lower rates of egg production and survival found there.

REFERENCES

Craig, J. V., 1968. Correlated responses in body weight and egg production traits in chickens selected for social dominance. Poultry Sci. 47: 1033–1035.

Craig, J. V., D. K. Biswas and A. M. Guhl, 1969. Agonistic behaviour influenced by strangeness, crowding and heredity in female domestic fowl (Gallus domesticus). Anim. Behav. (in press).

Craig, J. V., L. L. Ortman and A. M. Guhl, 1965. Genetic selection for social dominance ability in chickens. Anim. Behav. 13: 114–131.

Guhl, A. M., 1953. Social behavior of the domestic fowl. Kansas Agric. Exper. Sta. Tech. Bul. 73.

Guhl, A. M., 1968. Social inertia and social stability in chickens. Anim. Behav. 16: 219–232.

Guhl, A. M., and W. C. Allee, 1944. Some measurable effects of social organization in flocks of hens. Physiol. Zool. 17: 320–347.

McBride, G., 1958. The relationship between aggressiveness, peck order and some characters of selective significance in the domestic hen. Proc. Roy. Phys. Soc. Edin. 27: 56–60.

McBride, G., 1960. Poultry husbandry and the peck order. Brit. Poultry Sci. 1: 65–68.

McBride, G., 1962. Behaviour and a theory of poultry husbandry. Proc. XIIth World's Poultry Congress (Symposia), pp. 102–105.

McBride, G., 1968. Behavioral measurement of social stress. In Adaptation of Domestic Animals, ed. E.S.E. Hafez pp. 360–366. Lea and Febiger, Philadelphia.

Morgan, W. C., and B. J. Bonzer, 1959. Stresses associated with moving cage layers to floor pens. Poultry Sci. 38: 603–606.

Sanctuary, W. C., 1932. A study of avian behavior to determine the nature and persistency of the order of dominance in the domesticated fowl and to relate these to certain physiological reactions. M.S. Thesis, Mass. State College, Amherst.

Schjelderup-Ebbe, T., 1935. Social behavior of birds. In A Handbook of Social Psychology, ed. C. Murchison, Clark Univ. Press, Worcester, Mass., pp. 947–972.

Tindell, D., and J. V. Craig, 1959. Effects of social competition on laying house performance in the chicken. Poultry Sci. 38: 95–105.

References

Allee, W. C. 1938. *The Social Life of Animals.* Revised ed., 1958, Beacon Press, Boston, 233 pp.

Allee, W. C., and Collias, N. E. 1938. Effects of injections of epinephrine on the social order in small flocks of hens. *Anatomical Record Supplement,* 72:119.

Allee, W. C., and Collias, N. E. 1940. The influence of estradiol on the social organization of flocks of hens. *Endocrinology,* 27:87–94.

Allee, W. C., and Douglis, M. B. 1945. A dominance order in the hermit crab, *Pagurus longicarpus,* Say. *Ecology,* 26:411–412.

Allee, W. C., Collias, N. E., and Beeman, E. 1940. The effect of thyroxin on the social order in flocks of hens. *Endocrinology,* 27:827–835.

Allee, W. C., Emerson, A. E., Park, O., Park, T., and Schmidt, K. P. 1949. *Principles of Animal Ecology.* Saunders, Philadelphia, pp. 413–415.

Allee, W. C., Foreman, D., Banks, E. M., and Holaberd, C. 1955. Effects of an androgen on dominance and subordination in six common breeds of *Gallus gallus. Physiological Zoology,* 28:89–115.

Altmann, M. 1952. Social behavior of elk, *Cervus canadensis nelsoni,* in the Jackson Hole area of Wyoming. *Behaviour,* 4(2):116–143.

Alverdes, F. 1927. *Social Life in the Animal World.* Harcourt Brace Jovanovich, New York, 216 pp.

Banks, E. M. 1956. Social organization in red jungle fowl hens *(Gallus gallus* subsp.*). Ecology,* 37(2):239–248.

Bennett, M. A. 1939. The social hierarchy in ring doves. *Ecology,* 20:337–357.

Birch, H. G., and Clark, G. 1950. Hormonal modification of social behavior: IV. The mechanism of estrogen-induced dominance in chimpanzees. *Journal of Comparative and Physiological Psychology,* 43:181–193.

Biswas, D. K., and Craig, J. V. 1971. Social tension indexes and egg production traits in chickens. *Poultry Science,* 50(4):1063–1065.

Bovbjerg, R. V. 1956. Some factors affecting aggressive behavior in crayfish. *Physiological Zoology,* 29(2):127–136.

Carpenter, C. R. 1942a. Characteristics of social behavior in nonhuman primates. *Transactions of the New York Academy of Science,* 4(8):248–258.

Carpenter, C. R. 1942b. Societies of monkeys and apes. *Biological Symposia,* 8:177–204.

Collias, N. E. 1944. Aggressive behavior among vertebrate animals. *Physiological Zoology,* 17(1):83–123.

Craig, J. V. 1968. Correlated responses in body weight and egg production traits in chickens selected for social dominance. *Poultry Science,* 47:1033–1035.

Craig, J. V., and Baruth, R. A. 1965. Inbreeding and social dominance ability in chickens. *Animal Behaviour,* 13(1):109–113.

Craig, J. V., and Guhl, A. M. 1969. Territorial behavior and social interactions of pullets kept in large flocks. *Poultry Science,* 48(5):1622–1628.

Craig, J. V., Biswas, D. K., and Guhl, A. M. 1969. Agonistic behavior influenced by strangeness, crowding and heredity in female domestic fowl *(Gallus gallus). Animal Behaviour,* 17:498–506.

Craig, J. V., Ortman, L. L., and Guhl, A. M. 1965. Genetic selection for social dominance ability in chickens. *Animal Behaviour,* 13(1):114–131.

Darling, F. F. 1937. *A Herd of Red Deer.* Oxford University Press, New York, 215 pp.

389

References

Davis, D. E. 1957. Aggressive behavior in castrated starlings. *Science,* 126:253.

Douglis, M. B. 1946. Some evidence of a dominance subordinance relationship among lobsters, *Homarus americanus. Anatomical Record,* 94:57.

Douglis, M. B. 1948. Social factors influencing the hierarchies of small flocks of the domestic hen: interactions between resident and part-time members of organized flocks. *Physiological Zoology,* 21:147–182.

Drews, D. R. 1973. Group formation in captive *Galago crassicaudatus:* notes on the dominance concept. *Zeitschrift für Tierpsychologie,* 32:425–435.

Espinas, A. V. 1878. *Des Sociétés animales.* 2nd ed., Bailliere, Paris 588 pp. (Reprinted by G. E. Stechert, New York, 1924. 389 pp.)

Fennell, R. A. 1945. The relation between heredity, sexual activity and training to dominance subordination in game cocks. *American Naturalist,* 79:142–151.

Fischel, W. 1927. Beitrage zur Sozialogie des Haushuhns. *Biologische Zentralblatte,* 47:678–695.

Grzimek, G. 1949. Rangordnungsversuche mit Pferden. *Zeitschrift für Tierpsychologie,* 6:455–464.

Guhl, A. M. 1942. Social discrimination in small flocks of the common domestic fowl. *Journal of Comparative Psychology,* 34:127–148.

Guhl, A. M. 1949. Heterosexual dominance and mating behavior in chickens. *Behaviour,* 2:106–120.

Guhl, A. M. 1950. Social dominance and receptivity in the domestic fowl. *Physiological Zoology,* 23(4):361–366.

Guhl, A. M. 1958. The development of social organization in the domestic chick. *Animal Behaviour,* 6:92–111.

Guhl, A. M. 1961. Gonadal hormones and social behavior in infrahuman vertebrates. In *Sex and Internal Secretions,* 3rd ed., ed. by W. C. Young. Vol. 2, pp. 1240–1267. Williams & Wilkins, Baltimore.

Guhl, A. M. 1964. Psychophysiological interrelations in the social behavior of chickens. *Psychological Bulletin,* 61(4):277–285.

Guhl, A. M. 1968. Social inertia and social stability in chickens. *Animal Behaviour,* 16:219–232.

Guhl, A. M., and Allee, W. C. 1944. Some measurable effects of social organization in flocks of hens. *Physiological Zoology,* 17:320–347.

Guhl, A. M., and Eaton, R. C. 1948. Inheritance of aggressiveness in the fowl. *Poultry Science,* 27:665.

Guhl, A. M., and Warren, D. C. 1946. Number of offspring sired by cockerels related to social dominance in chickens. *Poultry Science,* 25:460–472.

James, W. T. 1939. Further experiments in social behavior among dogs. *Journal of Genetic Psychology,* 54:151–164.

James, W. T., and Foenander, F. 1961. Social behaviour studies on domestic animals: Hens in laying cages. *Australian Journal of Agricultural Research,* 12(6):1239–1252.

Maroney, R. J., Warren, J. M., and Sinha, M. M. 1959. Stability of social dominance hierarchies in monkeys *(Macaca mulatta). Journal of Social Psychology,* 50:285–293.

Masure, R. H., and Allee, W. C. 1934b. Flock organizations of the shell parakeet, *Melopsittacus undulatus* Shaw. *Ecology,* 15:388–398.

McBride, G. 1958. The relationship between aggressiveness, peck order and some characters of selective significance in the domestic hen. *Proceedings of the Royal Physiological Society of Edinburgh,* 27:56–60.

McBride, G. 1960. Poultry husbandry and the peck order. *British Poultry Science,* 1:65–68.

McBride, G. 1962. Behaviour and a theory of poultry husbandry. *Proceedings of the 12th World's Poultry Congress* (Symposium), pp. 102–105.

Muller, H. D. 1970. Influences of genotype and environment on the relationship between social rank and egg production in the chicken. *Poultry Science,* 49(4):934–941.

Murchison, C. 1935a. The experimental measurement of a social hierarchy in *Gallus domesticus:* II. The identification and inferential measurement of Social Reflex No. 1, and Social Reflex No. 2 by means of social discrimination. *Journal of Social Psychology,* 6:3–30.

390

Murchison, C. 1935b. The experimental measurement of a social hierarchy in *Gallus domesticus:* III. The direct and inferential measurement of Social Reflex No. 3. *Journal of Genetic Psychology,* 46(1):76–102.

Murchison, C. 1935c. The experimental measurement of a social hierarchy in *Gallus domesticus:* IV. Loss of body weight under conditions of mild starvation as a function of social dominance. *Journal of General Psychology,* 12(1):296–311.

Murchison, C. 1935d. The experimental measurement of a social hierarchy in *Gallus domesticus:* VI. Preliminary identification of social law. *Journal of General Psychology,* 12(1):227–248.

Murchison, C. 1936. The time function in the experimental formation of social hierarchies of different sizes in *Gallus domesticus. Journal of Social Psychology,* 7(1):3–18.

Noble, G. K. 1939. The role of dominance in the social life of birds. *Auk,* 56(3):263–273.

Noble, G. K., Wurm, M., and Schmidt, A. 1938. Social behavior of the black-crowned night heron. *Auk,* 55:7–40.

Ortman, L. L., and Craig, J. V. 1968. Social dominance in chickens modified by genetic selection-physiological mechanisms. *Animal Behaviour,* 16(1):33–37.

Pardi, L. 1948. Dominance order in Polistes wasps. *Physiological Zoology,* 21:1–13.

Potter, J. H. 1949. Dominance relations between different breeds of domestic hens. *Physiological Zoology,* 22:261–280.

Radlow, R., Hale, E. B., and Smith., W. I. 1958. Note on the role of conditioning in the modification of social dominance. *Psychological Reports,* 4:579–581.

Rowell, T. 1966. Hierarchy in the organization of a captive baboon group. *Animal Behaviour,* 14:430–443.

Schein, M. W., Hyde, C. E., and Fohrman, M. H. 1954. The effect of psychological disturbances on milk production of dairy cattle. *Proceedings of the Association of Southern Agricultural Workers,* 52nd Convention (Louisville, Ky.), pp. 79–80.

Schjelderup-Ebbe, T. 1923. Weitere Beitrage zur social und individual Psychologie des Haushuhns. *Zeitschrift für Psychologie,* 92:60–87.

Schloeth, R. 1958. Cycle annuel et comportement social du taureau de Camargue. *Mammalia,* 22(1):121–139.

Scott, J. P. 1945. Social behavior, organization and leadership in a small flock of domestic sheep. *Comparative Psychological Monographs, Serial 96,* 18(4):1–29.

Scott, J. P. 1946. Dominance reactions in a small flock of goats. *Anatomical Record,* 94:380–390.

Shoemaker, H. H. 1939. Social hierarchy in flocks of the canary. *Auk,* 56(4):381–406.

Thomas, J. W., Robinson, R. M., and Marburger, G. R. 1965. Social behavior in a white-tailed deer herd containing hypogonadal males. *Journal of Mammalogy,* 46(2):314–327.

Tindall, D., and Craig, J. V. 1959. Effects of social competition on laying house performance in the chicken. *Poultry Science,* 38:95–105.

Uhrich, J. 1938. The social hierarchy in albino mice. *Journal of Comparative Psychology,* 25(2):373–414.

Author Citation Index

394

Subject Index